ADVANCES IN MOLECULAR AND CELL BIOLOGY

Volume 17 • 1996

PROTEIN TARGETING TO MITOCHONDRIA

ADVANCES IN MOLECULAR AND CELL BIOLOGY

PROTEIN TARGETING TO MITOCHONDRIA

Series Editor: E. EDWARD BITTAR
Department of Physiology
University of Wisconsin

Guest Editor: F. ULRICH HARTL
Howard Hughes Medical Institute and
Memorial Sloan-Kettering Cancer Center

VOLUME 17 • 1996

JAI PRESS INC.

Greenwich, Connecticut *London, England*

55 Old Post Road No. 2
Greenwich, Connecticut 06836

JAI PRESS LTD.
38 Tavistock Street
Covent Garden
London, WC2E 7PB
England

ISBN: 0-7623-0144-9

Transferred to digital printing 2006

CONTENTS

LIST OF CONTRIBUTORS vii

PREFACE
F. Ulrich Hartl ix

TARGETING SIGNALS FOR PROTEIN IMPORT INTO MITOCHONDRIA AND OTHER SUBCELLULAR ORGANELLES
Gunnar von Heijne 1

PROTEIN TRANSPORT INTO MITOCHONDRIA: CYTOSOLIC FACTORS WHICH OPERATE DURING AND AFTER TRANSLATION IN PROTEIN TRAFFICKING
Lisa Estey and Michael G. Douglas 13

PRESEQUENCE BINDING PROTEINS AS CYTOSOLIC IMPORT-STIMULATION FACTORS IN MITOCHONDRIAL PROTEIN IMPORT
Katsuyoshi Mihara and Tsuneo Omura 33

MOLECULAR MECHANISMS OF PROTEIN TRANSLOCATION INTO AND ACROSS THE MITOCHONDRIAL OUTER MEMBRANE
Roland Lill, Andreas Mayer, Harald Steiner, Gyula Kispal, and Walter Neupert 51

TARGETING AND INSERTION OF PROTEINS INTO THE MITOCHONDRIAL OUTER MEMBRANE
Gordon C. Shore, Heidi M. McBride, Douglas G. Millar, Nancy A.E. Steenaart, and Mai Nguyen 71

TARGETING AND TRANSLOCATION OF PREPROTEINS BY THE TOMs OF THE MITOCHONDRIAL RECEPTOR COMPLEX
Petra Keil, Angelika Hönlinger, and Nikolaus Pfanner 81

MITOCHONDRIAL IMPORT OF CYTOCHROME *C*
Mark E. Dumont 103

TRANSLOCATION OF PREPROTEINS ACROSS THE MITOCHONDRIAL INNER MEMBRANE: TIMs AND HSP70
Michiel Meijer, Ammy Maarse, Michael Kübrich, and Nikolaus Pfanner 127

UNRAVELING THE PROTEIN TRANSLOCATION MACHINERY IN THE MITOCHONDRIAL INNER MEMBRANE
Nafsika G. Kronidou and Martin Horst 149

PROTEOLYTIC PROCESSING OF MITOCHONDRIAL PRECURSOR PROTEINS
Wayne A. Fenton and Frantisek Kalousek 163

SORTING OF PROTEINS TO THE MITOCHONDRIAL INTERMEMBRANE SPACE
Rosemary A. Stuart, Heike Fölsch, Albrecht Gruhler, and Walter Neupert 193

ENERGETICS OF MITOCHONDRIAL PROTEIN IMPORT AND INTRAMITOCHONDRIAL PROTEIN SORTING
Sabine Rospert 221

EXPORT OF PROTEINS FROM MITOCHONDRIA
Robert O. Poyton, Kevin A. Sevarino, Edward E. McKee, David J. M. Duhl, Vicki Cameron, and Bradley Goehring 247

PROTEIN FOLDING IN MITOCHONDRIA
Jörg Höhfeld 281

ASSEMBLY OF MULTISUBUNIT COMPLEXES IN MITOCHONDRIA
Mark Prescott, Rodney J. Devenish, and Phillip Nagley 299

THE DIVISION AND INHERITANCE OF MITOCHONDRIA
Michael P. Yaffe 341

INDEX 351

LIST OF CONTRIBUTORS

Vicki Cameron	Department of Biology Ithaca College
Rodney J. Devenish	Department of Biochemistry and Molecular Biology Monash University Victoria, Australia
Michael G. Douglas	Sigma Corporation St. Louis, Missouri
David J.M. Duhl	Howard Hughes Medical Institute Beckman Center Stanford University
Mark Dumont	Department of Biochemistry University of Rochester
Lisa Estey	Department of Molecular Endocrinology Glaxco Wellcome Research Triangle Park, North Carolina
Wayne A. Fenton	Department of Genetics Yale University School of Medicine
Heike Fölsch	Institut für Physiologische Chemie der Universität München, Germany
Bradley Goehring	Department of Molecular Cellular and Developmental Biology University of Colorado, Boulder
Albrecht Gruhler	Institut für Physiologische Chemie der Universität München München, Germany

F. Ulrich Hartl	Howard Hughes Medical Institute Memorial Sloan-Kettering Cancer Center
Jörg Höhfeld	ZMBH Universität Heidelberg Heidelberg, Germany
Angelika Hönlinger	Institut für Biochemie und Molekularbiologie Universität Freiburg Freiburg, Germany
Martin Horst	Biozentrum University of Basel Basel, Switzerland
Frantisek Kalousek	Department of Genetics Yale University School of Medicine
Petra Keil	Institut für Biochemie und Molekularbiologie Universität Freiburg Freiburg, Germany
Gyula Kispal	Institut für Physiologische Chemie der Universität München München, Germany
Nafsika G. Kronidou	Biozentrum University of Basel Basel, Switzerland
Michael Kübrich	Institut für Biochemie und Molekularbiologie Universität Freiburg Freiburg, Germany
Roland Lill	Institut für Physiologische Chemie der Universität München München, Germany
Ammy Maarse	Institute for Molecular Cell Biology BioCentrum Amsterdam The Netherlands

Andreas Mayer — Institut für Physiologische Chemie der Universität München, München, Germany

Heidi M. McBride — Department of Biochemistry, McGill University

Edward E. McKee — Indiana University School of Medicine and University of Notre Dame

Michiel Meijer — Institute for Molecular Cell Biology, BioCentrum Amsterdam, The Netherlands

Katsuyoshi Mihara — Department of Molecular Biology, Graduate School of Medical Sciences, Kyushi University, Fukuoka, Japan

Douglas G. Millar — Department of Biochemistry, McGill University

Phillip Nagley — Department of Biochemistry and Molecular Biology, Monash University, Victoria, Australia

Walter Neupert — Institut für Physiologische Chemie der Universität München, München, Germany

Mai Nguyen — Department of Biochemistry, McGill University

Tsuneo Omura — Department of Molecular Biology, Graduate School of Medical Sciences, Kyushi University, Fukuoka, Japan

Nikolaus Pfanner — Institut für Biochemie und Molekularbiologie, Universität Freiburg, Freiburg, Germany

Robert O. Poyton — Department of Molecular Cellular and Developmental Biology
University of Colorado, Boulder

Mark Prescott — Department of Biochemistry and Molecular Biology
Monash University
Victoria, Australia

Sabine Rospert — Biozentrum
University of Basel
Basel, Switzerland

Kevin A. Sevarino — Division of Molecular Psychiatry
Department of Psychiatry
Yale University School of Medicine

Gordon C. Shore — Department of Biochemistry
McGill University

Harald Steiner — Institut für Physiologische Chemie der Universität München
München, Germany

Nancy A. E. Steenaart — Department of Biochemistry
McGill University

Rosemary A. Stuart — Institut für Physiologische Chemie der Universität München
München, Germany

Gunnar von Heijne — Department of Biochemistry
Stockholm University
Stockholm, Sweden

Michael P. Yaffe — Department of Biology
University of California, San Diego

PREFACE

A large number of newly-synthesized polypeptides must cross one or several intracellular membranes to reach their functional locations in the eukaryotic cell. The mechanisms of protein trafficking, in particular the post-translational targeting and membrane translocation of proteins, are of fundamental biological importance and are the focus of intensive research world-wide. For more than 15 years, mitochondria have served as the paradigm organelle system to study these processes. Although key questions, such as how precisely proteins cross a membrane, still remain to be answered, exciting progress has been made in understanding the basic pathways of protein import into mitochondria and the components involved. In addition to a fascinating richness and complexity in detail, the analysis of mitochondrial protein import has revealed mechanistic principles of general significance: Major discoveries include the demonstration of the requirement of an unfolded state for translocation and of the essential role of molecular chaperones on both sides of the membranes in maintaining a translocation-competent conformation and in protein folding after import. It is becoming clear how a polypeptide chain is "reeled" across the membrane in an ATP-dependent process by the functional cooperation of membrane proteins, presumably constituting part of a transmembrane channel, with peripheral components at the trans-side of the membrane.

In this volume, eminent experts in the field take the time to review the central aspects of mitochondrial biogenesis. The logical order of the 16 chapters is determined by the sequence of steps during protein import, starting with the events taking place in the cytosol, followed by the recognition of targeting signals, the translocation of precursor proteins across the outer and inner membranes, their proteolytic processing and intramitochondrial sorting, and finally their folding and oligomeric assembly. In addition, the mechanisms involved in the export of mitochondrially encoded proteins as well as recent advances in understanding the division and inheritance of mitochondria will be discussed.

F. Ulrich Hartl
Guest Editor

TARGETING SIGNALS FOR PROTEIN IMPORT INTO MITOCHONDRIA AND OTHER SUBCELLULAR ORGANELLES

Gunnar von Heijne

I. Introduction 2
II. Background: Two Beasts in the Sorting Signal Zoo 2
A. Signal Peptides 2
B. Transit Peptides 3
III. Mitochondrial Targeting Peptides 4
A. Signals for Matrix Targeting 4
B. Signals for Intramitochondrial Sorting 5
C. Sorting Signals in Inner and Outer Membrane Proteins 7
IV. Conclusions 7
Acknowledgment 8
References 8

Advances in Molecular and Cell Biology,
Volume 17, pages 1–12.

ISBN: 0-7623-0144-9

I. INTRODUCTION

Intracellular protein sorting is effected by a set of organelle-specific import machineries that each recognizes a specific kind of sorting signal in proteins destined for that particular compartment. Our understanding of the different sorting signals has in general progressed from an early demonstration that a particular segment of a nascent polypeptide contains the sorting information, via statistical analysis, biophysical characterization, and mutagenesis studies of larger numbers of signals that have led to a definition of their basic design, to more recent biochemical investigations on their interactions with receptor proteins or lipids in the target membrane.

In the case of mitochondrial targeting peptides, their existence was first suggested by Maccecchini and colleagues (1979); the concept of the amphiphilic helix as an important design principle was suggested by von Heijne (1986) based on statistical sequence analysis, and by Roise and colleagues (1986) based on biophysical characterization of a synthetic model peptide and was later confirmed by mutagenesis, CD, and NMR studies; and their interactions with lipids and receptors are still under scrutiny.

In this chapter, our current picture of the mitochondrial targeting peptide will be displayed against a background of what is known about two other kinds of related sorting signals, namely secretory signal peptides and chloroplast transit peptides.

II. BACKGROUND: TWO BEASTS IN THE SORTING SIGNAL ZOO

A. Signal Peptides

The secretory signal peptide (SP) was the first kind of sorting signal to be identified (Milstein et al., 1972) and sequenced (Brownlee et al., 1973; Milstein et al., 1974; Schechter et al., 1974). SPs target proteins for secretion in both prokaryotic and eukaryotic cells, and can be described by a common design that varies little between different organisms (von Heijne, 1985; von Heijne and Abrahmsén, 1989). SPs are located at the N-terminus of secretory proteins, and they are generally between 15 and 25 residues long, have a short, positively charged N-terminal tail (n-region), a 7-15 residues long central hydrophobic stretch (h-region), and a C-terminal, more polar segment (c-region) that contains the cleavage-site for the signal peptidase enzyme. Beyond these general characteristics there is little sequence homology between different SPs, and simplified designs with homopolymeric h-regions and c-regions built according to known positional amino acid preferences have been found to work well in both *E. coli* and mammalian cells (Kendall et al., 1986; Chou and Kendall, 1990; Laforet and Kendall, 1991; Nilsson and von Heijne, 1991; Doud et al., 1993). If the h-region becomes too long or if the signal peptidase recognition site is destroyed, the SP cannot be removed and instead permanently anchors the protein to the membrane (Chou and Kendall, 1990; Sakaguchi et al., 1992).

Structural characterization of SPs by CD and NMR spectroscopy has revealed that they can insert perpendicularly to a lipid monolayer, and that the h-region has a helical conformation in detergent micelles (Bruch et al., 1989; Gierasch, 1989; Keller et al., 1992; Rizo et al., 1993). The interaction with lipid monolayers or liposomes is promoted by anionic phospholipids (de Vrije et al., 1988; Phoenix et al., 1993).

Recent biochemical studies suggest that the SP may first be recognized by the signal recognition particle (SRP) in eukaryotic cells, and by the SRP-like Ffh-4.5S RNA complex in bacteria (Zopf et al., 1990; High and Dobberstein, 1991; Luirink et al., 1992; Lutcke et al., 1992). It may then interact with both the SecA and SecY proteins in the *E. coli* inner membrane or the eukaryotic SecY-homologue SEC61p (Trun et al., 1988; High et al., 1991; Osborne and Silhavy, 1993), and is finally cleaved by the signal peptidase enzyme (Lep in *E. coli* or SEC11p in yeast) (Dalbey and von Heijne, 1992).

B. Transit Peptides

Chloroplast transit peptides (cTPs) target proteins for import into the stromal compartment. The first published amino acid sequence of a cTP was from the small subunit of Rubisco from *Chlamydomonas reinhardtii* (Schmidt et al., 1979). cTPs are also N-terminally located, tend to be enriched in hydroxylated amino acids (Ser in particular), and contain few acidic amino acids (von Heijne et al., 1989). They thus often have a net positive charge, though examples of completely uncharged cTPs are known (von Heijne et al., 1991). Some weak patterns of conserved residues can be found in the cleavage-site region (Gavel and von Heijne, 1990b), but they have limited predictive value.

cTPs have not been much studied by mutagenesis, and no clear domain structure has been discovered so far. Likewise, it is not clear if cTPs have a conserved secondary structure (von Heijne and Nishikawa, 1991), and results from biophysical studies are conflicting (Pilon et al., 1992; Theg and Geske, 1992). It does appear, however, that cTPs have a significant affinity for galactolipids, a class of lipid that is specific for the chloroplast envelope membranes (van't Hof et al., 1993).

Proteins destined for the lumen of the thylakoids have a bi-partite targeting signal composed of an N-terminal, stroma-targeting cTP hooked to a thylakoid transfer domain (TTD) that has many features in common with a classical signal peptide (von Heijne, et al., 1989). In fact, both SecA and SecY homologues have been found in chloroplasts (Douglas, 1992; Scaramuzzi et al., 1992a, 1992b), and it has also been shown that the cleavage specificity of the thylakoid processing peptidase that removes the TTD from the mature protein is very similar to that of the signal peptidases (Halpin et al., 1989; Shackleton and Robinson, 1991). It is, thus, likely that transport into the thylakoid lumen is based on many of the same principles as those operating in the secretory machinery, although very recent data on the energy requirements for thylakoid import have suggested that there may be more than one mechanism in this case (Cline et al., 1993).

III. MITOCHONDRIAL TARGETING PEPTIDES

A. Signals for Matrix Targeting

Mitochondrial matrix-targeting signals (mTPs) are enriched for positively charged residues (Arg in particular), and are generally devoid of negatively charged amino acids. Their amphiphilic, membrane-active nature was first suggested by experiments where a synthetic analog of the yeast COX IV mTP was shown to induce leakage in liposomes and to fold into a helical conformation in the presence of detergent micelles (Roise, et al., 1986), and by theoretical hydrophobic-moment analysis of a set of known mTP sequences (von Heijne, 1986).

Many mutational studies have corroborated the amphiphilic helix model. The most complete set of data is available for a truncated "minimal" version of the yeast ATPase $F_1\beta$ mTP, where ~100 mutants were isolated using a saturation mutagenesis protocol, and where the import activity of the different mutants was found to correlate surprisingly well with their position on either the charged or the hydrophobic face in a helical model of the peptide (Bedwell et al., 1989). In a complementary approach, peptides with measurable import activity were selected from a library of random DNA sequences; again, net positive charge and a tendency for charged and hydrophobic amino acids to segregate to opposite sides of a helical model were found in the active peptides (Baker and Schatz, 1987; Lemire et al., 1989).

Direct evidence for a helical conformation in the presence of lipids or detergents have been obtained by circular dichroism (CD) and NMR spectroscopy (Epand et al., 1986; Roise et al., 1988; Endo et al., 1989; Goormaghtigh et al., 1989; Karslake et al., 1990; Tamm and Bartoldus, 1990; Bruch and Hoyt, 1992; Thornton et al., 1993). The mTP from rat liver mitochondrial aldehyde dehydrogenase was recently shown by NMR to be composed of two separate 6–8 residues long helical segments, neither of which was sufficient to promote import (Karslake, et al., 1990; Wang and Weiner, 1993). When the short intervening linker was deleted, it was found that a continuous helix of 12–14 residues length was formed and that the linker-deleted mTP was active in import (Thornton, et al., 1993). These studies suggest that one function of the mTP may be to provide a sufficiently hydrophobic surface, and that this may be accomplished either by two short amphiphilic helices or by one long.

Synthetic mTPs have also been shown to bind to both artificial liposomes and to intact mitochondria (Swanson and Roise, 1992), and even to be translocated far enough across a liposome to become accessible to internally trapped proteases when a potential of the appropriate sign (negative inside) is imposed across the bilayer (Maduke and Roise, 1993).

mTPs are removed from the imported protein by at least two distinct matrix proteases: the matrix-processing peptidase/processing enhancing protein (MPP+PEP) complex (corresponding to the MAS1 and MAS2 genes in yeast), and the intermediate processing peptidase (MIP; Isaya et al., 1992a; Kalousek et al., 1992). In some organisms, MPP is part of the inner membrane cytochrome c reductase complex, and may serve a dual function (Eriksson and Glaser, 1992; Emmermann et al., 1993). Three distinct amino acid patterns have been found around the cleavage sites in mTPs: Arg-Xxx↓Xxx, Arg-Xxx-Tyr↓Xxx, and Arg-Xxx-Phe-Xxx_7↓Xxx (Hendrick et al., 1989; Gavel and von Heijne, 1990a). As far as is known, MPP+PEP cleaves the first two types. Whether the second pattern results from a one-step cleavage after Tyr, or whether Tyr is removed in a second step is unknown. The third type results from two consecutive cleavage events, first by MPP+PEP after the Arg-Xxx pattern and then by MIP, that removes the final eight (or sometimes seven) residues. The piece removed by MIP is not well conserved, but an amino-terminal hydrophobic residue seems to be required for efficient cleavage (Isaya et al., 1992b). It is clear both from the poor conservation of the cleavage sites and from a range of experimental data that other parts of the mTP contribute to the cleavage specificity; possibly, the amphiphilic helix is one element that is recognized by the processing enzyme (Thornton et al., 1993). For this reason, it is often difficult to predict the cleavage site, except in cases where there is an Arg-Xxx-Tyr pattern present some 12–65 residues downstream of the initiator Met (Gavel and von Heijne, 1990a).

B. Signals for Intramitochondrial Sorting

Most intermembrane space (IMS) proteins have a bipartite targeting sequence composed of an N-terminal positively charged part followed by a mildly hydrophobic region (Figure 1). Compared to matrix-targeting mTPs, the initial part is surprisingly long in most cases. The apolar part has been compared both to the h-region in signal peptides (von Heijne et al., 1989) and to the apolar part of a transmembrane segment. In the examples listed in Figure 1, the apolar stretch is some 16–20 residues long; this is longer that the typical h-region and more akin to a transmembrane segment. On the other hand, the apolar stretch is distinctly less hydrophobic than both signal peptides and transmembrane segments (von Heijne et al., 1989). Bipartite signals are cleaved by the IMP1-IMP2 protease located in the inner membrane (Schneider et al., 1991; Nunnari et al., 1993); IMP1 and IMP2 have weak but significant similarity to bacterial and eukaryotic signal peptidases (Behrens et al., 1991), suggesting a common evolutionary origin.

The targeting function of bipartite targeting sequences has been analyzed genetically in a couple of cases. Thus, the N-terminal positively charged part of cytochrome c_1 can target a passenger protein into the matrix (Hurt et al., 1985; van Loon et al., 1987), suggesting that the apolar segment serves either as a translocation signal for back-transfer from the matrix towards the IMS ("conservative sorting";

cytochrome c_1

```
        MSLGKKIRIGFDGFGRINRFITRGAAQRNDSKLPSRNDALKHGLDGLGSAGSKSFRALAAIGAGVSGLLSFATIAYS*DEAEHGLE
MAAAAASLRGVVLGPRGAGLPGARARGLLCSARPGQLPLRTPQAVALSSKSGLSRGRKVMLSALGMLAAGGAGLAVALHSAVSA*SDLELHPP
                MLARTCLRSTRTFASAKNGAFKFAKRSASTQSSGAAAESPLRLNIAAAAATAVAAGSIAWYYHLYGFASA*MTPAEEGL
                          MFSNLSKRWAQRTLSKSFYSTATGAASKSGKLTQKLVTAGVAAAGITASTLLYADSLTAEA*MTAAEHGL
```

cytochrome b_2

```
        MFKSQLRTATARSSFRSLASKLNPQRFNSSKTPLLNATRGSNRSKNSLIALAISLSAVSSSYYLYQKDKFISA*DVPHWKDI
MLKYKPLLKISKNCEAAILRASKTRLNTIRAYGSTVPKSKSFEQDSRKRTQSWTALRVGAILAATSSVAYLNWHNGQIDN*EPKLDMNK
```

cytochrome c peroxidase (cleavage site not known)

```
MTTAVRLLPSLGRTAHKRSLYLFSAAAAAAAATFAYSQSQKRSSSSPGGGSNHG
```

creatine kinase

```
MASIFSKLLTGRNASLLFATMGTSVLTTGYLLNRQKVCA*EVREQPRLFPP
MAGPFSRLLSARPGLRLLALAGAGSLAAGFLLRPEPVRA*ASERRRLYPPS
```

Figure 1. IMS-targeting signals. Stretches of uncharged residues are underlined. The final cleavage sites are marked by "*".

Hartl and Neupert, 1990), or as a stop-transfer sequence that interrupts the import process and allows the protein to diffuse laterally out of the import machinery into the inner membrane ("stop-transfer" model; van Loon and Schatz, 1987). Mutants that are imported into the matrix rather than into the IMS have been isolated for the cytochrome b_2 and c_1 bipartite signals: positively charged residues immediately upstream of the apolar segment were found to be important and changes that reduced the hydrophobicity of the apolar segment were also detrimental (Jensen et al., 1992; Beasley et al., 1993; Schwarz et al., 1993). Interestingly, mutations to prolines were among the most commonly found ones in the apolar region, suggesting that this region may be required to fold into a specific conformation such as a helix to function properly.

C. Sorting Signals in Inner and Outer Membrane Proteins

Targeting signals in inner and outer membrane proteins have only been defined in a couple of cases. Thus, in the outer membrane protein MAS70, the targeting information is encoded within the 29 N-terminal residues: a short, positively charged segment followed by a stretch of 20 apolar residues (Nakai et al., 1989). Deletion of the apolar stretch leads to import into the matrix (Hase et al., 1984). The orientation of the protein in the outer membrane depends on the flanking charged residues (Li and Shore, 1992): when the N-terminus is extended or when extra positively charged residues are added to it, the protein inserts in an inverted N_{cyto}-C_{in} orientation.

Among the inner membrane proteins, the only detailed analysis has been carried out for the ADP/ATP carrier (AAC). The AAC is thought to span the inner membrane six times, and has an internal three-fold repeat. It lacks a cleavable mTP in yeast and *Neurospora*, but not in *Homo*. Targeting information seems to be present in all three repeats (Pfanner et al., 1987). The mTP in human AAC is not necessary for import, but affects the efficiency and specificity of targeting (Zara et al., 1992).

Little is known about targeting signals in other inner membrane proteins, which as a rule seem to lack cleavable prepeptides.

IV. CONCLUSIONS

Our current picture of the mitochondrial matrix-targeting signal is based on the concept of the positively charged amphiphilic α-helix. This model is supported by a substantial body of evidence: statistical, biophysical, and genetical. The amphililic helix is required not only for correct targeting, but apparently also for correct cleavage, where it might serve to roughly position the cleavage enzyme in relation to the rather loosely defined cleavage site pattern.

It is still unclear how the targeting peptide is recognized. Biophysical studies have demonstrated that targeting peptides have strong lipid-interacting properties, suggesting that direct protein-lipid interaction may be important at some stage. Both cytoplasmic and mitochondrial receptors that bind targeting peptides have been identified, and it was recently suggested that mitochondrial hsp70 also binds to targeting peptides as they emerge in to the matrix space (Gambill et al., 1993; Voos et al., 1993).

The positive charge of the targeting peptide is thought to play a role during its transfer across the inner membrane, and an electrophoretic model where the electrical component of the membrane electrochemical potential is one possibility.

The targeting signals responsible for routing proteins to the intermembrane space are composed of an N-terminal matrix-targeting part and a C-terminal apolar part. Whether targeting proceeds via a conservative sorting or a stop-transfer mechanism clearly cannot be decided only the basis of sequence analysis; it is true that the length of the apolar stretch is more like a typical ER stop-transfer sequence than a signal peptide, but this may be a compensation for its low average hydrophobicity rather than an indication of stop-transfer function.

ACKNOWLEDGMENT

This work was supported by grants from the Swedish Natural Sciences Research Council (NFR).

REFERENCES

Baker, A., & Schatz, G. (1987). Sequences from a prokaryotic genome or the mouse dihydrofolate reductase gene can restore the import of a truncated precursor protein into yeast mitochondria. Proc. Natl. Acad. Sci. USA 84, 3117–3121.

Beasley, E. M., Muller, S., & Schatz, G. (1993). The signal that sorts yeast cytochrome-b2 to the mitochondrial intermembrane space contains 3 distinct functional regions. EMBO J. 12, 2303–2311.

Bedwell, D. M., Strobel, S. A., Yun, K., Jongeward, G. D., & Emr, S. D. (1989). Sequence and structural requirements of a mitochondrial protein import signal defined by saturation cassette mutagenesis. Mol. Cell. Biol. 9, 1014–1025.

Behrens, M., Michaelis, G., & Pratje, E. (1991). Mitochondrial inner membrane protease-1 of saccharomyces-cerevisiae shows sequence similarity to the escherichia-coli leader peptidase. Mol. Gen. Genet. 228, 167–176.

Brownlee, G. G., Cartwright, E. M., Cowan, N. J., Jarvis, J. M., & Milstein, C. (1973). Purification and sequence of messenger RNA for immunoglobulin light chains. Nature New Biol. 244, 236–240.

Bruch, M. D., & Hoyt, D. W. (1992). Conformational analysis of a mitochondrial presequence derived from the F1-ATPase beta-subunit by CD and NMR spectroscopy. Biochim. Biophys. Acta 1159, 81–93.

Bruch, M. D., Mcknight, C. J., & Gierasch, L. M. (1989). Helix formation and stability in a signal sequence. Biochem. 28, 8554–8561.

Chou, M. M., & Kendall, D. A. (1990). Polymeric sequences reveal a functional interrelationship between hydrophobicity and length of signal peptides. J. Biol. Chem. 265, 2873–2880.

Cline, K., Henry, R., Li, C., & Yuan, J. (1993). Multiple pathways for protein transport into or across the thylakoid membrane. EMBO J. 12, 4105–4114.

Dalbey, R. E., & von Heijne, G. (1992). Signal peptidases in prokaryotes and eukaryotes—a new protease family. Trends Biochem. Sci. 17, 474–478.

de Vrije, T., de Swart, R., Dowhan, W., Tommassen, J., & de Kruijff, B. (1988). Phosphatidylglycerol is involved in protein translocation across *Escherichia coli* inner membranes. Nature 334, 173–175.

Doud, S. K., Chou, M. M., & Kendall, D. A. (1993). Titration of protein transport activity by incremental changes in signal peptide hydrophobicity. Biochemistry 32, 1251–1256.

Douglas, S. E. (1992). A secY homologue is found in the plastid genome of Cryptomonas Φ. FEBS Lett. 298, 93–96.

Emmermann, M., Braun, H. P., Arretz, M., & Schmitz, U. K. (1993). Characterization of the bifunctional cytochrome-c reductase-processing peptidase complex from potato mitochondria. J. Biol. Chem. 268, 18936–18942.

Endo, T., Shimada, I., Roise, D., & Inagaki, F. (1989). N-terminal half of a mitochondrial presequence peptide takes a helical conformation when bound to dodecylphosphocholine micelles—a proton nuclear magnetic resonance study. J. Biochem. 106, 396–400.

Epand, R. M., Hui, S. W., Argan, C., Gillespie, L. L., & Shore, G. C. (1986). Structural analysis and amphiphilic properties of a chemically synthesized mitochondrial signal peptide. J. Biol. Chem. 261, 10017–10020.

Eriksson, A. C., & Glaser, E. (1992). Mitochondrial processing proteinase—a general processing proteinase of spinach leaf mitochondria is a membrane-bound enzyme. Biochim. Biophys. Acta 1140, 208–214.

Gambill, B. D., Voos, W., Kang, P. J., Miao, B. J., Langer, T., Craig, E. A., & Pfanner, N. (1993). A dual role for mitochondrial heat shock protein-70 in membrane translocation of preproteins. J. Cell Biol. 123, 109–117.

Gavel, Y., & von Heijne, G. (1990a). Cleavage-site motifs in mitochondrial targeting peptides. Protein Eng. 4, 33–37.

Gavel, Y., & von Heijne, G. (1990b). A conserved cleavage-site motif in chloroplast transit peptides. FEBS Lett. 261, 455–458.

Gierasch, L. M. (1989). Signal sequences. Biochemistry 28, 923–930.

Goormaghtigh, E., Martin, I., Vandenbranden, M., Brasseur, R., & Ruysschaert, J.-M. (1989). Secondary structure and orientation of a chemically synthesized mitochondrial signal sequence in phospholipid bilayers. Biochem. Biophys. Res. Comm. 158, 610–616.

Halpin, C., Elderfield, P. D., James, H. E., Zimmermann, R., Dunbar, B., & Robinson, C. (1989). The reaction specificities of the thylakoidal processing peptidase and *escherichia coli* leader peptidase are identical. EMBO J. 8, 3917–3921.

Hartl, F. U., & Neupert, W. (1990). Protein sorting to mitochondria—evolutionary conservations of folding and assembly. Science 247, 930–938.

Hase, T., Müller, U., Riezman, H., & Schatz, G. (1984). A 70-kd protein of the yeast mitochondrial outer membrane is targeted and anchored via its extreme amino terminus. EMBO J. 3, 3157–3164.

Hendrick, J. P., Hodges, P. E., & Rosenberg, L. E. (1989). Survey of amino-terminal proteolytic cleavage sites in mitochondrial precursor proteins: Leader peptides cleaved by two matrix proteases share a three-amino acid motif. Proc. Natl. Acad. Sci. USA 86, 4056–4060.

High, S., & Dobberstein, B. (1991). The signal sequence interacts with the methionine-rich domain of the 54-kD protein of signal recognition particle. J. Cell Biol. 113, 229–233.

High, S., Gorlich, D., Wiedmann, M., Rapoport, T. A., & Dobberstein, B. (1991). The identification of proteins in the proximity of signal-anchor sequences during their targeting to and insertion into the membrane of the ER. J. Cell Biol. 113, 35–44.

Hurt, E., Müller, U., & Schatz, G. (1985). The first twelve amino acids of a yeast mitochondrial outer membrane protein can direct a nuclear-encoded cytochrome oxidase subunit to the mitochondrial inner membrane. EMBO J. 4, 3509–3518.

Isaya, G., Kalousek, F., & Rosenberg, L. (1992a). Sequence analysis of rat mitochondrial imtermediate peptidase: Similarity to zinc metallopeptidase and to a putative yeast homologue. Proc. Natl. Acad. Sci. USA 89, 8317–8321.

Isaya, G., Kalousek, F., & Rosenberg, L. E. (1992b). Amino-terminal octapeptides function as recognition signals for the mitochondrial intermediate peptidase. J. Biol. Chem. 267, 7904–7910.

Jensen, R. E., Schmidt, S., & Mark, R. J. (1992). Mutations in a 19-amino-acid hydrophobic region of the yeast cytochrome c1 presequence prevent sorting to the mitochondrial intermembrane space. Mol. Cell Biol. 12, 4677–4686.

Kalousek, F., Isaya, G., & Rosenberg, L. (1992). Rat liver mitochondrial intermediate peptidase (MIP): Purification and initial characterization. EMBO J. 11, 2803–2809.

Karslake, C., Piotto, M. E., Pak, Y. K., Weiner, H., & Gorenstein, D. G. (1990). 2D NMR and structural model for a mitochondrial signal peptide bound to a micelle. Biochemistry 29, 9872–9878.

Keller, R. C. A., Killian, J. A., & de Kruijff, B. (1992). Anionic phospholipids are essential for α-helix formation of the signal peptide of prePhoE upon interaction with phospholipid vesicles. Biochemistry 31, 1672–1677.

Kendall, D. A., Bock, S. C., & Kaiser, E. T. (1986). Idealization of the hydrophobic segment of the alkaline phosphatase signal peptide. Nature 321, 706–708.

Laforet, G. A., & Kendall, D. A. (1991). Functional limits of conformation, hydrophobicity, and steric constraints in prokaryotic signal peptide cleavage regions—wild type transport by a simple polymeric signal sequence. J. Biol. Chem. 266, 1326–1334.

Lemire, B. D., Fankhauser, C., Baker, A., & Schatz, G. (1989). The mitochondrial targeting function of randomly generated peptide sequences correlates with predicted helical amphiphilicity. J. Biol. Chem. 264, 20206–20215.

Li, J. M., & Shore, G. C. (1992). Reversal of the orientation of an integral protein of the mitochondrial outer membrane. Science 256, 1815–1817.

Luirink, J., High, S., Wood, H., Giner, A., Tollervey, D., & Dobberstein, B. (1992). Signal-sequence recognition by an escherichia coli ribonucleoprotein complex. Nature 359, 741–743.

Lutcke, H., High, S., Romisch, K., Ashford, A. J., & Dobberstein, B. (1992). The methionine-rich domain of the 54 kDa subunit of signal recognition particle is sufficient for the interaction with signal sequences. EMBO J. 11, 1543–1551.

Maccecchini, M. L., Rudin, Y., Blobel, G., & Schatz, G. (1979). Import of proteins into mitochondria: Precursor forms of the extramitochondrially made F1-ATPase subunits in yeast. Proc. Natl. Acad. Sci. USA 76, 343–347.

Maduke, M., & Roise, D. (1993). Import of a mitochondrial presequence into protein-free phospholipid vesicles. Science 260, 364–367.

Milstein, C., Brownlee, G. G., Cartwright, E. M., Jarvis, J. M., & Proudfoot, N. J. (1974). Sequence analysis of immunoglobulin light chain messenger RNA. Nature 252, 354–359.

Milstein, C., Brownlee, G. G., Harrison, T. M., & Mathews, M. B. (1972). A possible precursor of immunoglobulin light chains. Nature New Biol. 239, 117–120.

Nakai, M., Hase, T., & Matsubara, H. (1989). Precise determination of the mitochondrial import signal contained in a 70 kDa protein of yeast mitochondrial outer membrane. J. Biochem. (Tokyo) 105, 513–519.

Nilsson, I. M., & von Heijne, G. (1991). A *de novo* designed signal peptide cleavage cassette functions *in vivo*. J. Biol. Chem. 266, 3408–3410.

Nunnari J., Fox T. D., & Walter P. (1993). A mitochondrial protease with two catalytic subunits of nonoverlapping specificities. Science 262, 1997–2004.

Osborne, R. S., & Silhavy, T. J. (1993). PrlA suppressor mutations cluster in regions corresponding to three distinct topological domains. EMBO J. 12, 3391–3398.

Pfanner, N., Hoeben, P., Tropschug, M., & Neupert, W. (1987). The carboxyl-terminal two-thirds of the ADP/ATP carrier polypeptide contains sufficient information to direct translocation into mitochondria. J. Biol. Chem. 262, 14851–14854.

Phoenix, D. A., de Wolf, F. A., Staffhorst, R. W. H. M., Hikita, C., Mizushima, S., & de Kruijff, B. (1993). Phosphatidylglycerol dependent protein translocation across the *Escherichia Coli* inner membrane is inhibited by the anti-cancer drug doxorubicin—evidence for an electrostatic interaction between the signal sequence and phosphatidylglycerol. FEBS Lett. 324, 113–116.

Pilon, M., Rietveld, A. G., Weisbeek, P. J., & de Kruijff, B. (1992). Secondary structure and folding of a functional chloroplast precursor protein. J. Biol. Chem. 267, 19907–19913.

Rizo, J., Blanco, F. J., Kobe, B., Bruch, M. D., & Gierasch, L. M. (1993). Conformational behavior of *Escherichia Coli* OmpA signal peptides in membrane mimetic environments. Biochemistry 32, 4881–4894.

Roise, D., Horvath, S. J., Tomich, J. M., Richards, J. H., & Schatz, G. (1986). A chemically synthesized pre-sequence of an imported mitochondrial protein can form an amphiphilic helix and perturb natural and artificial phospholipid bilayers. EMBO J. 5, 1327–1334.

Roise, D., Theiler, F., Horvath, S. J., Tomich, J. M., Richards, J. H., Allison, D. S., & Schatz, G. (1988). Amphiphilicity is essential for mitochondrial presequence function. EMBO J. 7, 649–653.

Sakaguchi, M., Tomiyoshi, R., Kuroiwa, T., Mihara, K., & Omura, T. (1992). Functions of signal and signal-anchor sequences are determined by the balance between the hydrophobic segment and the N-terminal charge. Proc. Natl. Acad. Sci. USA 89, 16–19.

Scaramuzzi, C. D., Hiller, R. G., & Stokes, H. W. (1992a). Identification of a chloroplast-encoded secA gene homologue in a chromophytic alga—Possible role in chloroplast protein translocation. Curr. Genetics 22, 421–427.

Scaramuzzi, C. D., Stokes, H. W., & Hiller, R. G. (1992b). Characterization of a chloroplast-encoded sec-Y homologue and atpH from a chromophytic alga-evidence for a novel chloroplast genome organisation. FEBS Lett. 304, 119–123.

Schechter, I., McKean, D. J., Guyer, R., & Terry, W. (1974). Partial amino acid sequence of the precursor of immunoglobulin light chain programmed by messenger RNA *in vitro*. Science 188, 160–162.

Schmidt, G. W., Devillers-Thiery, A., Desruisseaux, H., Blobel, G., & Chua, N. H. (1979). NH_2-terminal amino acid sequences of precursor and mature forms of the ribulose-1,5-bisphosphate carboxylase small subunit from *Chlamydomonas reinhardtii*. J. Cell Biol. 83, 615–622.

Schneider, A., Behrens, M., Scherer, P., Pratje, E., Michaelis, G., & Schatz, G. (1991). Inner membrane protease-I, an enzyme mediating intramitochondrial protein sorting in yeast. EMBO J. 10, 247–254.

Schwarz, E., Seytter, T., Guiard, B., & Neupert, W. (1993). Targeting of cytochrome-b2 into the mitochondrial intermembrane space—Specific recognition of the sorting signal. EMBO J. 12, 2295–2302.

Shackleton, J. B., & Robinson, C. (1991). Transport of proteins into chloroplasts—the thylakoidal processing peptidase is a signal-type peptidase with stringent substrate requirements at the -3-position and -1-position. J. Biol. Chem. 266, 12152–12156.

Swanson, S. T., & Roise, D. (1992). Binding of a mitochondrial presequence to natural and artificial membranes—role of surface potential. Biochemistry 31, 5746–5751.

Tamm, L. K., & Bartoldus, I. (1990). Secondary structure of a mitochondrial signal peptide in lipid bilayer membranes. FEBS Lett. 272, 29–33.

Theg, S. M., & Geske, F. J. (1992). Biophysical characterization of a transit peptide directing chloroplast protein import. Biochemistry 31, 5053–5060.

Thornton, K., Wang, Y., Weiner, H., & Gorenstein, D. G. (1993). Import, processing, and 2-dimensional NMR structure of a linker-deleted signal peptide of rat liver mitochondrial aldehyde dehydrogenase. J. Biol. Chem. 268, 19906–19914.

Trun, N., Stader, J., Lupas, A., kumamoto, C., & Silhavy, T. (1988). Two cellular components, PrlA and SecB, that recognize different sequence determinants are required for efficient protein export. J. Bacteriol. 170, 5928–2930.

van Loon, A., Brändli, A., Pesold-Hurt, B., Blank, D., & Schatz, G. (1987). Transport of proteins to the mitochondrial intermembrane space: The "matrix-targeting" and the "sorting" domains in the cytochrome c1 presequence. EMBO J. 6, 2433–2439.

van Loon, A., & Schatz, G. (1987). Transport of proteins to the mitochondrial intermembrane space: The "sorting" domain of the cytochrome c1 presequence is a stop-transfer sequence specific for the mitochondrial inner membrane. EMBO J. 6, 2441–2448.

van't Hof, R., van Klompenburg, W., Pilon, M., Kozubek, A., de Kortekool, G., Demel, R. A., Weisbeek, P. J., & de Kruijff, B. (1993). The transit sequence mediates the specific interaction of the precursor of ferredoxin with chloroplast envelope membrane lipids. J. Biol. Chem. 268, 4037–4042.

von Heijne, G. (1985). Signal sequences. The limits of variation. J. Mol. Biol. 184, 99–105.

von Heijne, G. (1986). Mitochondrial targeting sequences may form amphiphilic helices. EMBO J. 5, 1335–1342.

von Heijne, G., & Abrahmsén, L. (1989). Species-specific variation in signal peptide design: Implications for protein secretion in foreign hosts. FEBS Lett. 244, 439–446.

von Heijne, G., Hirai, T., Klösgen, R.-B., Steppuhn, J., Bruce, B., Keegstra, K., & Herrmann, R. (1991). CHLPEP—A database of chloroplast transit peptides. Plant Mol. Biol. Reporter 9, 104–126.

von Heijne, G., & Nishikawa, K. (1991). Chloroplast transit peptides—The perfect random coil? FEBS Lett. 278, 1–3.

von Heijne, G., Steppuhn, J., & Herrmann, R. G. (1989). Domain structure of mitochondrial and chloroplast targeting peptides. Eur. J. Biochem. 180, 535–545.

Voos, W., Gambill, B. D., Guiard, B., Pfanner, N., & Craig, E. A. (1993). Presequence and mature part of preproteins strongly influence the dependence of mitochondrial protein import on heat shock protein-70 in the matrix. J. Cell Biol. 123, 119–126.

Wang, Y., & Weiner, H. (1993). The presequence of rat liver aldehyde dehydrogenase requires the presence of an α-helix at its N-terminal region which is stabilized by the helix at its C-termini. J. Biol. Chem. 268, 4759–4765.

Zara, V., Palmieri, F., Mahlke, K., & Pfanner, N. (1992). The cleavable presequence is not essential for import and assembly of the phosphate carrier of mammalian mitochondria but enhances the specificity and efficiency of import. J. Biol. Chem. 267, 12077–12081.

Zopf, D., Bernstein, H. D., Johnson, A. E., & Walter, P. (1990). The methionine-rich domain of the 54 kd protein subunit of the signal recognition particle contains an RNA binding site and can be crosslinked to a signal sequence. EMBO J. 9, 4511–4517.

PROTEIN TRANSPORT INTO MITOCHONDRIA:
CYTOSOLIC FACTORS WHICH OPERATE DURING AND AFTER TRANSLATION IN PROTEIN TRAFFICKING

Lisa Estey and Michael G. Douglas

I. Introduction . . . 14
II. Mechanisms of Protein Delivery Vary with Different Precursors . . . 15
A. Role of Cytosolic Components in Post-translational Protein Import . . . 16
III. Other Cytosolic Events which Affect the Efficiency of Protein Import into Mitochondria . . . 21
A. Post-translational Models of Mitochondrial Protein Import . . . 22
B. Coupled Protein Synthesis-Mitochondrial Import . . . 22
C. Genetic Screens Based on the Mistargeting of Mitochondrial Precursors . . 24
D. Is the Management of mRNA Important in the Trafficking Process? . . . 26
Acknowledgments . . . 28
References . . . 28

Advances in Molecular and Cell Biology,
Volume 17, pages 13–32.

ISBN: 0-7623-0144-9

I. INTRODUCTION

The translocation of nascent polypeptides from the cytoplasm to subcellular locations is orchestrated such that correct targeting and delivery can occur before folding or aggregation of the protein occurs. In addition, the proteins in question must be efficiently delivered to their correct location defined by signals encoded in their primary sequence. It is now clear that the operation of soluble components in the cell cytoplasm is necessary to insure that this occurs (Hartl et al., 1994). In the case of proteins that must be delivered to the endoplasmic reticulum, the folding process is prevented by arresting synthesis of the nascent protein until the proper interaction with the translocation complex occurs (Hann and Walter, 1991). However, the transport of proteins into mitochondria exhibits a flexibility that uncouples elongation from localization (Hannavy et al., 1993). To different degrees, the proteins which are destined for mitochondria require that other cytosolic proteins be available to assist in maintaining them in a deliverable and nonaggregated state.

Mitochondrial precursors represent the spectrum of proteins that have to be dealt with by cytosolic machines. Some mitochondrial precursors that are very inefficiently imported, or fail to import post-translationally *in vitro* require a close coupling of synthesis and transport. On the other hand, some precursors can be efficiently imported as purified precursor proteins in the complete absence of additional components. For the group of mitochondrial precursors that need close coupling of translation, protein factors that intimately participate in the initiation process could indirectly play a significant role in the efficiency of protein import. Recent studies in yeast that define components that improve import efficiency have identified activities or proteins that make the process of RNA utilization more efficient. These components are factors with a role in translation or RNA metabolism (see below). What emerges from the vast majority of these studies is the notion that post translational protein import in the cell will require the assistance of preexisting components that are able to regulate protein folding or enhance specific interactions in some way to better couple the process of delivery and/or transport. It is generally accepted that the process is rendered more efficient because the participating factors function to insure this. At the present time, molecular chaperones of the HSP70 family, a general class of chaperones, appear to provide this function for transport into mitochondria of lower eukaryotes (Craig et al., 1993; Hohfeld and Hartl, 1994). In higher eukaryotes, there is mounting data in support for additional components with chaperone-like properties that may be restricted perhaps to specific mitochondrial precursors (Murakami and Mori, 1990; Hachiya et al., 1993).

Finally, the field of protein import has not to date been able to provide a completely satisfactory model to explain why basic, amphiphilic sequences at the amino terminal ends of mitochondrial precursors provide the specific targeting observed *in vivo*. For example, the amino terminal ends of a substantial number of nonmitochondrial proteins should be capable of targeting to mitochondria. The specificity of targeting observed, however, indicates that mechanisms are operating in addition to presequence recognition by mitochondrial surface receptors to insure efficient localization. The presequences of mitochondrial precursors exhibit high affinity for acidic membrane lipids and must be assisted by the action of cytosolic components to insure productive interaction with their correct target. Also, there is a small but significant population of mitochondrial proteins that localize to the organelle surface via their carboxy-terminal end (Kutay et al., 1993). How this occurs and the extent to which these carboxy terminal signals exploit the same factors required of some amino-terminal targeted proteins has not been determined.

Another way to insure efficient targeting is to regulate translation initiation of mitochondrial precursor mRNAs. This would prevent competition for the presequence signals and provide the specificity desired (Verner, 1993). Thus, we have included in this review some discussion of the cytosolic factors that have been identified or implicated in protein import that appear to be involved with mRNA processing or translation machinery. First, some of the molecular chaperones that have been shown to assist in the maintenance of proteins for entry into mitochondria appear to associate with ribosomes in an elongation-dependent fashion (Nelson et al., 1992; Zhong and Arndt, 1993; Hendrick et al., 1993). Second, some genetic screens that demand reduced efficiency of protein import begin to uncover proteins, some of which may function in the regulation of protein synthesis and the management of mRNA in the cell (see following).

II. MECHANISMS OF PROTEIN DELIVERY VARY WITH DIFFERENT PRECURSORS

Studies on protein import over the past 15 years have revealed that different proteins that are capable of post-translational entry into different organelles are unique in their requirements for cytosolic factors and nucleotide. For some precursors, it could be clearly demonstrated by *in vitro* import studies in mitochondria that ATP was required outside the organelle (Pfanner and Neupert, 1986; Eilers et al., 1987; Chen and Douglas, 1987b). Although the action of nucleotide hydrolysis in the import mechanism was proposed to keep precursors in the import competent state, the mechanism and components participating in this remained unresolved (Weinhues et al., 1991). The importance of the regulation of nascent protein folding by an ATP-dependent process on post-translational protein import was supported by independent observations. These studies demonstrated that conditions that reduce the potential for a precursor to adopt a tightly folded structure or aggregate would reduce the dependence on nucleotide hydrolysis outside mitochondria for protein entry. By introducing deletions (Chen and Douglas, 1988) or destabilizing mutations (Vestweber and Schatz, 1988) within internal domains of an ATP-dependent precursor or by arresting elongation early to prevent structure formation (Verner and Schatz, 1987), precursors exhibit little or no dependence on nucleotide hydrolysis for import. Further, a number of studies using precursors apparently free of soluble components need only be partially unfolded by dilution from denaturants (Iwahashi et al., 1992; Becker et al., 1992) or pH extremes (Pfanner et al., 1988) for import.

In its simplest form the import of proteins into mitochondria is a kinetic process in which the timing of translation and membrane insertion of a nascent precursor will define the time for a protein to aggregate or fold into an import incompetent structure. The folding profile of some proteins during translation probably does not necessitate the intervention of additional factors. *In vivo*, the kinetics of post-translational protein import are quite different and much easier to demonstrate for some precursors than for others (Reid and Schatz, 1982a; Reid et al., 1982). The differences observed *in vivo* are likely linked to interaction of molecular chaperone proteins and their ATP-dependent reactions that have been identified (see Gething and Sambrook, 1992; Craig et al., 1993; Caplan et al., 1993; Hartl et al., 1994). These chaperones that operate outside the organelle share a common cytosolic compartment and also participate in the translocation of nascent proteins through other cellular membranes (Gething and Sambrook, 1992). Currently, it is not known whether different isoforms of chaperones and their interacting factors that occupy the same subcellular space participate in specific processes *in situ*.

A. Role of Cytosolic Components in Post-translational Protein Import

Essentially all of post-translational localization described thus far appear to involve soluble factors and ATP. This post-translational delivery is not restricted to mitochondria. Selected proteins like yeast mating pheromone prepro-α factor (ppαf), which is targeted to the endoplasmic reticulum, may be transported following completion of translation (Hansen et al., 1986; Rothblatt and Meyer, 1986; Waters and Blobel, 1986; Waters et al., 1986). The ATP- and factor-dependent events were shown through both genetic (Deshaies et al., 1988) and biochemical (Chirico et al., 1988) studies to involve cytosolic Hsp70 homologues (Ssa1p/Ssa2p). The most surprising finding in the *in vivo* studies was the observation that depletion of Hsp70 homologues in the cytosol caused the accumulation of precursor forms of mitochondrial precursors as well as ppαf, presumably in the cytoplasm. What was not certain, however, was the role of these chaperones in either the maintenance of protein unfolding or the competence of the membrane associated translocation complexes.

Yeast *S. cerevisiae* contain four cytosolic Hsp70 homologues, SSA1-SSA4. The functional redundancy exhibited by these genes is consistent with their high degree of conservation. Studies performed in the Craig lab (reviewed in Craig et al., 1993) show that single deletions of the major constitutive Hsp70s, SSA1 or SSA2, yield no observable effect on the growth of yeast, whereas deletions of both of these proteins together yield thermosensitive strains unable to grow at elevated temperatures (Craig and Jacobsen, 1984). The loss of additional SSA genes beyond SSA1 and SSA2 causes a lethal phenotype. The triple deletion strain ssa1ssa2 ssa4 fails to grow at any temperature but can be rescued by constitutive expression of the normally inducible SSA3 gene product (Werner-Washburne et al., 1987).

The effect of depletion of cytosolic Hsp70 on mitochondrial protein entry was first demonstrated using an inviable ssa1ssa2ssa4 deletion mutant maintained for viability by a plasmid encoded SSA1 protein behind the regulatable GAL1 promoter. This host-vehicle combination, which grew normally in galactose, was used to examine the effect on protein delivery during loss of the cytoplasmic homologue. A shift of cells to glucose to block induction on Ssa1p caused the loss of the minimal levels of chaperone necessary to support efficient protein translocation. Under these conditions, precursor forms of ppαf destined for ER and the precursor to the $F_1\beta$-subunit of the mitochondrial ATPase also accumulated in cells (Deshaies et al., 1988). Also, the cytosol- and ATP-dependent translocation of precursors into isolated microsomes confirmed the role for Ssa1p in restoring protein translocation in depleted cytosolic lysates (Deshaies et al., 1988; Waters et al., 1988). In more recent work, the direct interaction of purified Hsp70 homologues with ppαf has been demonstrated by co-immunoprecipitation using anti-Hsp70 antibody (Chirico, 1992).

These and other *in vitro* translocation systems helped define the Hsp70 chaperones required for protein entry. They also suggested that other components present in the cytosol were also necessary. In earlier *in vitro* studies, small amounts of cytosol were necessary in addition to purified Hsp70 to restore post-translational transport competency to the microsomal systems (see also Brodsky et al., 1993). At present, one of the most attractive candidates for the identity of the additional components are the eukaryotic homologues of the *E. coli* proteins, DnaJ and GrpE. These two proteins are known to regulate the activity of the *E. coli* Hsp70 homologue, DnaK (Zylicz et al., 1989; Liberek et al., 1991; Langer et al., 1992; Hartl et al., 1994), and its interaction with unfolded substrate (Langer et al., 1992, Hartl et al., 1994). The identification of homologues to DnaJ in yeast and their involvement with protein import was the first confirmation that a homologous chaperone-dependent pathway was operating in the yeast cytosol.

There are at least eight different DnaJ-like proteins in yeast and at least two of these share the same cytosolic compartment (Caplan et al., 1993; Cyr et al., 1994). Two of these homologues, when defective in yeast, exhibit defects in mitochondrial protein import. First, YDJ1/MAS5 in yeast encodes an abundant cytosolic protein that is required for efficient protein import. This protein was initially identified in genetic screens that demanded the accumulation of the mitochondrial precursor to the F_1-ATPase β-subunit (Attencio and Yaffe, 1992). *In vivo*, the Ydj1p is necessary for post-translational protein translocation of cytosolic precursors across both ER and mitochondrial membranes (Caplan et al., 1992). These data in combination with those showing the role of the SSA1 and SSA2 gene products in post-translational protein entry further support a mechanism similar to that defined in *E. coli* for the maintenance of loosely folded structures and transport. Studies *in vitro* demonstrate a direct role for the Ydj1p and Ssa1p in protein folding (see below).

The second DnaJ homologue demonstrated to exhibit an effect on mitochondrial import is Sis1p. This gene product, which was uncovered for reasons entirely different from those related to protein translocation, has now been shown to be required for the initiation of protein translation (Zhong and Arndt, 1993). The Sis1p exhibits some overlap with the functions of the YDJ1 protein. When it is overexpressed in a YDJ1 deletion mutant, Sis1p is able to restore growth and partially rescue translocation defects (Caplan and Douglas, 1992). The involvement of Sis1p in the initiation of translation could be significant as discussed previously, since this would be one potential mechanism to regulate the extent to which proteins are allowed to adopt structure before initiating translocation. Thus, there appear to be at least two DnaJ homologues that operate at early stages in the import process to ensure the regulation of Hsp70 chaperones which operate in the maintenance of protein folding.

Biochemical studies of the action of the major eukaryotic DnaJ homologues support a role for this co-chaperone that is the same as that established for the prokaryotic protein (Liberek et al., 1991; Langer et al., 1992; Cyr et al., 1992). Like DnaJ, the YDJ1 protein from yeast directly interacts with its Hsp70 counterpart and activates the latent ATPase activity of the Hsp70 homologue. In the yeast system, purified Ydj1p will stimulate the ATPase activities of the major cytoplasmic Hsp70 homologues, Ssa1p and Ssa2p, to approximately the same extent, 8 to 10-fold. In addition, Ydj1p can directly regulate the interaction of different Hsp70 homologues with unfolded protein substrate (Cyr et al., 1992; Cyr and Douglas, 1994). The substrates bound by the different Hsp70 homologues include the presequence containing ends of mitochondrial precursors. Each of the cytoplasmic forms of Hsp70 can bind unfolded protein substrate. These different Hsp70s include the two soluble Ssa1p and Ssa2p homologues as well as the ribosomally associated Ssb1p and Ssb2p. *In vitro*, Ydj1p exhibits a specificity for the SSA family over the SSB family in stimulating ATPase activity and promoting the release of bound substrate (Cyr and Douglas, 1994). Ydj1p maximally stimulates the ATPase activity and the release of bound substrate at a ratio of approximately one mole of Ydj1p to one mole of Hsp70 (Cyr et al., 1992). Protein or peptide substrate release by Hsp70 is an ATP-dependent process that responds differently to Ydj1p. The SSA family members do not release bound substrate in the presence of ATP or Ydj1p alone. In the presence of both, however, bound substrate is released in an ATP-dependent manner. This release of bound substrate requires the hydrolysis of ATP since nonhydrolyzable ATP analogs fail to promote this release reaction. The SSB family members on the other hand do not appear to be as dependent on the action of Ydj1p for ATP dependent release. Ssb1/2p-bound substrate is released by ATP alone and not significantly stimulated by the addition of Ydj1p (Cyr and Douglas, 1994).

Studies with purified *E. coli* chaperone proteins confirm that the interaction of the different chaperones with unfolded substrate is dependent upon the ability of the chaperones to alter their binding and release properties, depending on the extent of protein folding (Hartl et al., 1994; Cyr et al., 1994). Although DnaJ can promote the ATP-dependent release of DnaK from an extended polypeptide undergoing a folding reaction, it can independently associate with proteins that have adopted some structure and stabilize the binding of DnaK. This switch from release to stabilization of DnaK binding by DnaJ occurs as domains of the partially folded protein reach a critical concentration relative to the extended chain regions so that bound DnaJ and DnaK are juxtaposed in a stable complex. This stable arrangement of the DnaK, DnaJ chaperones bound to substrate is discharged upon the interaction of GrpE (Langer et al., 1992).

To date, biochemical analysis of the interaction of purified chaperones from eukaryotes with unfolded substrate indicate that the same paradigm of Hsp70 cycles and a YDJ1-dependent switch from release to stabilization of Hsp70 is occurring. Further, studies in eukaryotes indicate that different chaperone family members within the cytosolic compartment exhibit different behavior in the binding and release cycles of nascent proteins. Thus, the management of a growing polypeptide en route to receptors of the organelle surface may involve the concerted action of different chaperones. The nascent mitochondrial precursor on the ribosome will initially encounter the ribosomally associated SSB1/2p gene products to assist the efficient transfer of the growing protein from the ribosome to the membrane (Figure 1). It is likely that SSA family members that also interact with extended chains and in conjunction with Ydj1p undergo cycles of binding and release until sufficient secondary structure is attained to stabilize the binding of Ydj1p (Langer et al., 1992; Hartl et al., 1994; Cyr et al., 1994). DnaJ homologs have recently been shown to associate with polysomes in a puromycin sensitive manner (Hendrick et al., 1993; Kudlicki et al., 1994).

The SSA/YDJ-stabilized precursor generated in this manner is competent for interaction with receptors on the organelle surface or the completion of folding. This is likely to be controlled by additional cytosolic factors. The fate of the chaperone-stabilized precursor and the factors that regulate it is the focus of new work in the area of protein import. In *E. coli*, the interaction of a third factor, GrpE, promotes the release of the Hsp70 and DnaK chaperones from unfolded substrate (Langer et al., 1992). In the cytosol, most likely a eukaryotic GrpE homologue will be identified that functions in a manner similar to the *E. coli* protein, namely in promoting the release of chaperone-bound protein. At present, although there are reports of GrpE homologues in mitochondria (Bolliger et al., 1994; Voos et al., 1994), no such cytoplasmic homologue has been identified. One intriguing possibility to provide an important vectorial component to the cytoplasmic movement of precursor to mitochondria would be to localize a GrpE-like function at the mitochondrial membrane surface in order to catalyze the release of chaperones associated with either completed or nascent proteins at membrane receptor complexes. The association of these cytosolic factors with precursors has been shown to enhance protein import by reducing the extent of direct precursor binding to the membrane phospholipid (Hajek and Bedwell, 1994).

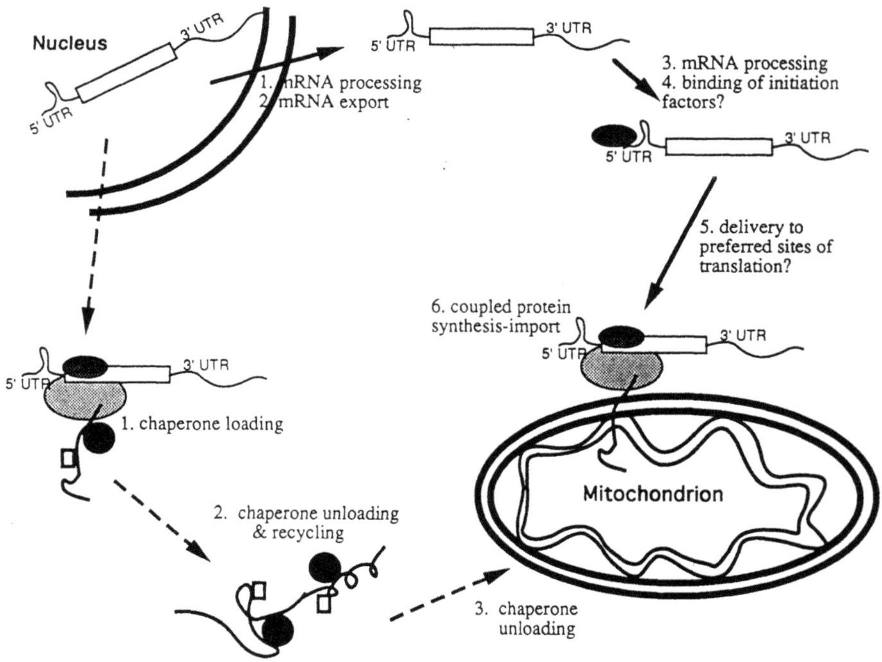

Figure 1. Model for coupled protein synthesis-mitochondrial import. Post-translational import pathways for a precursor are shown with the dashed arrows; cotranslational import with the solid arrows. *In vivo,* the cotranslational pathway may be more efficient; however, formation of a productive initiation complex may be reduced if the mRNA is not localized to the mitochondrial periphery prior to the initiation of translation.

The cytoplasmic chaperones identified in recent studies and shown to be important for the maintenance of proteins for mitochondrial import have well-characterized counterparts in the prokaryotic secretion pathway (Walter and Johnson, 1994). In mammalian cells, two recently described proteins, which to date do not have prokaryotic counterparts, appear to be important for the efficient import of presequence containing precursors. The description will be brief here because each of these protein complexes will be described in adjoining chapters in the volume. In rat liver, a factor termed the presequence binding factor (PBF) has been shown to bind to presequences of mitochondrial precursors and to be required for the import of these proteins into mitochondria (Murakami and Mori, 1990). This cytosolic protein of 50kDa is required for the post-translational import of a subset of mitochondrial presequence containing proteins from rabbit reticulocyte lysate. Using depletion-readdition studies, PBF appears to be part of a complex that is necessary for post-translational protein entry (Murakami et al., 1992). It is proposed that this factor may either assist in or protect the presequences of some precursors for efficient binding to mitochondria or interaction with other cytosolic components. In independent studies, Omura and colleagues have characterized an independent cytosolic factor that interacts with mitochondrial precursors to stimulate protein import into isolated mitochondria. This factor, termed the mitochondrial import stimulation factor (MSF) has been characterized and shown to consist of two nonidentical but highly homologous proteins of 30 and 32kDa. MSF exhibits an ATPase activity that is stimulated by the presence of an import-incompetent mitochondrial protein. It is likely that MSF is a chaperone-like protein with a fairly defined role in protein import (Hachiya et al., 1993).

III. OTHER CYTOSOLIC EVENTS WHICH AFFECT THE EFFICIENCY OF PROTEIN IMPORT INTO MITOCHONDRIA

In vivo, the presence of mitochondrial precursors can only be demonstrated under conditions in which the cells have been treated to block the maintenance of a membrane potential across the mitochondrial inner membrane (see following). These precursors are normally very short-lived and can only be trapped by rapid quenching methods that destroy cellular integrity. Although these precursors are presumably extra-mitochondrial, the loss of cellular integrity makes this assignment difficult. The question of whether mitochondrial import occurs by a cotranslational mechanism *in vivo* is still one of the unresolved aspects in the early steps of the mitochondrial import pathway (Verner, 1992, 1993). As discussed previously, the data available thus far show that the import of mitochondrial precursors can occur post-translationally and that chaperones are required for the proper delivery of some mitochondrial precursors. However, the mechanism for coupling translation to the translocation process for mitochondria has not been defined. At this time, there is no evidence that would support a cotranslational model analogous to that seen for the import of ER proteins. That is, cotranslational import of mitochondrial precursors does not require an SRP-dependent arrest of elongation, nor have receptors for ribosome binding been identified on the outer surface of mitochondria. To date, the consensus model of the mitochondrial import community suggests that protein synthesis is loosely coupled to mitochondrial import. As suggested by Neupert and colleagues (Hartl et al., 1989), the observations of cotranslational import may be interpreted as the initiation of import prior to the completion of protein synthesis. The association of cytoplasmic ribosomes with the outer mitochondrial membrane, therefore, may reflect a difference in the kinetics of protein synthesis and mitochondrial import: in the case of some mitochondrial precursors, rates of protein synthesis may be much slower than the rates at which these precursors associate with the import apparatus (Schatz and Butow, 1983). For the sake of simplicity, we shall use the term "coupled protein synthesis/import" to describe both kinetic and coupling arguments for "cotranslational" import. This discussion will serve as the basis to explain studies from several laboratories that indicate that the components that participate or control the initiation of protein translation may be critical factors in protein import.

A. Post-translational Models of Mitochondrial Protein Import

Although *in vitro* studies show that nascent precursors can be imported by post-translational mechanisms, they do not reflect the conditions present in the cell. *In vivo*, the concentration of protein in the cytoplasm is much higher than that encountered in the most competent *in vitro* translation/import systems. As mentioned above, pools of mitochondrial precursors have been detected in yeast spheroplasts (Maccecchini et al., 1979; Reid and Schatz, 1982a, 1982b) as well as *Neurospora crassa* (Hallermayer et al., 1977), however, under steady-state conditions it cannot be absolutely demonstrated where these pools reside. Precursor pools do form when cells are perturbed with carbonyl cyanide m-chlorophenylhydrazone (CCCP), a reagent that dissipates the membrane potential of mitochondria. Under such conditions, stable precursor pools form that can enter the mitochondrial matrix post-translationally when CCCP is inactivated (Reid and Schatz, 1982b). *In vivo* studies with $F_1\beta$-copper metallothionein fusion proteins also illustrate that synthesis of the fusion protein is tightly coupled to its import (Chen and Douglas, 1987a). These studies serve to illustrate that detectable cytosolic pools of mitochondrial precursors result under conditions in which some perturbation has occurred to disrupt the coupling of the translation and translocation processes. Thus, the coupled translation of mitochondrially targeted proteins may serve as the primary mode of protein import *in vivo*; however, this mechanism is effectively supported by a chaperone-mediated process to ensure correct targeting of precursors uncoupled from this process.

B. Coupled Protein Synthesis-Mitochondrial Import

Evidence for the coupling of protein synthesis to mitochondrial import was initially described by Butow and colleagues (Kellems and Butow, 1974). Electron microscopic analysis of yeast spheroplasts demonstrated that a fraction of cytosolic ribosomal particles was associated with the outer mitochondrial membrane and enriched at sites where the inner and outer membrane were in close proximity (i.e., contact sites). Contact sites appear to cofractionate with mitochondrially associated cytosolic ribosomes (Pon et al., 1989) and thus may represent the sites required for coupling of translation and import. A portion of these ribosomes was released from isolated mitochondria by high salt washes, while the remaining ribosomes were released by the addition of puromycin. These results suggested that the cytosolic ribosomes were bound to mitochondria via nascent polypeptide chains, presumably in a cotranslational manner (Kellems et al., 1974). Microscopic evidence for the association of cytosolic ribosomes has also been reported in higher eukaryotes (Loncar, 1990). Using immunoelectron microscopy and *in situ* RNA localization techniques, Loncar examined the localization of uncoupler protein (UCP) in brown adipose tissue. Under cold stress, the increase in UCP expression lead to the accumulation of the protein on the mitochondrial outer membrane as well as its normal localization with the cristae. The protein was shown to colocalize with ribosomal clusters and specific RNA on the outer membrane. The general pattern of ribosome-UCP-RNA localization differed from the observations of Kellems and Butow in that these particles were not clustered at contact sites (Loncar, 1990).

The coupling of protein synthesis and mitochondrial import is also supported by studies showing that several precursor-specific mRNAs are associated with mitochondrial membranes. Approximately 59% of translatable $F_1\beta$-subunit mRNAs were shown to be associated with a sub-population of ribosomes bound to yeast mitochondria; these ribosomes also showed enrichment in other mitochondrially directed protein mRNAs including the $F_1\gamma$-subunit and cytochrome *c* peroxidase (Suissa and Schatz, 1982). Not all mitochondrial precursor mRNAs exhibited such a distribution, since 99% of COX V and VI mRNAs were associated with free cytosolic polysomes.

Data in support of *in vivo* models for coupled protein synthesis and mitochondrial import come from studies that examine these processes in an *in vitro* homologous yeast system (Verner and Weber, 1989; Fujiki and Verner, 1991, 1993; Price and Verner, 1993). In this system, translation of COXIV-DHFR fusion protein was performed in yeast lysates before addition to an import reaction. Addition of methotrexate to the reaction inhibited import of the COXIV-DHFR fusion when the precursor was translated in reticulocyte lysates, however, methotrexate sensitivity was not observed with COXIV-DHFR translated in yeast lysates (Verner and Weber, 1989). Since this drug stabilizes the tertiary structure of DHFR, it was concluded that translation of the precursor in yeast lysates protected the nascent chain from interacting with methotrexate and forming mature protein structures that would prevent post-translational import of the mitochondrial precursor. Similar observations were made *in vivo*: methotrexate treated yeast cultures were shown to efficiently import COXIV-DHFR fusions, yet when CCCP was used to create post-translationally imported pools of precursor, methotrexate treatment resulted in cytosolic accumulation of COXIV-DHFR (Fujiki and Verner, 1993). Interestingly, the post-translational import of COXIV-DHFR fusions was not as efficient in the homologous system when cycloheximide was added to translation lysates prior to the addition of mitochondria (Fujiki and Verner, 1991). In these experiments, the synthesis of the fusion protein was allowed to proceed for 20 min before the addition of cycloheximide. Although 80% of the total fusion protein was available for import, very little of the precursor was imported into the added yeast mitochondria post-translationally. Mitochondrial import was more efficient when mitochondria were present during protein synthesis. The addition of cycloheximide to the coupled import system blocked both protein synthesis and mitochondrial import when added after the start of the reaction. This inhibition was specific for the coupled translation/import system, since the addition of cycloheximide had no effect on post-translational import reactions consisting of reticulocyte lysate translated fusion protein and isolated yeast mitochondria. The import of porin, an outer membrane protein, showed similar increased import efficiency in the homologous yeast system (Fujiki and Verner, 1991). The import of authentic $F_1\beta$ precursors also appeared to be tightly coupled to protein synthesis *in vivo*; yet unlike the COXIV-DHFR precursors studied in the homologous yeast system, $F_1\beta$ could be imported by post-translational mechanisms (Fujiki and Verner, 1993). These data suggest that under physiological conditions, at least some precursors may enter mitochondria via a mechanism that is tightly coupled to protein synthesis.

C. Genetic Screens Based on the Mistargeting of Mitochondrial Precursors

Several genes that are necessary for protein export in *E. coli* have been successfully identified using a genetic screen that exploited mutations that confer conditional lethal defects in protein translocation (Bieker et al., 1990). Modifications of this screen have been used in *Saccharomyces* to identify potential components of the mitochondrial import apparatus (Garrett et al., 1991; Kenna et al., 1993; Zoladek et al., 1995; Estey et al., submitted), yet have failed to reveal mutations that directly affect the import process. What these genetic screens do show, however, is that in yeast, the processes of protein synthesis and trafficking may be more tightly controlled than the prevailing view in the field of mitochondrial import. These studies reveal a general class of cytosolic factors that indicate that the process of "RNA management," including translation initiation and its regulation, is a considerable factor in the efficiency of the trafficking process. The different screens described ahead have revealed what at first may appear to be a diverse collection of cellular players. However, these components may have a common role at different stages along the trafficking pathway.

To uncover gene products associated with early events in mitochondrial targeting, several laboratories have exploited the observation that import of an $F_1\beta$-β-galactosidase hybrid protein into mitochondria disrupts respiratory competence in *Saccharomyces* (Douglas et al., 1984). A selection scheme was designed for the identification of yeast mutants that exhibit reduced efficiency of mitochondrial import yet are capable of growth on non-fermentable carbon sources. This genetic screen is similar to that used for the isolation of the sec mutations in the protein export pathway of *E. coli* (Emr et al., 1980). MFT1 was the first gene characterized from such an $F_1\beta$-β-galactosidase fusion screen (Garrett et al., 1991). Mutants in MFT1 continued to synthesize the $F_1\beta$-β-galactosidase protein, but the fusion was no longer associated with mitochondria. Interestingly, the Mft1 mutants failed to show defective import of $F_1\beta$ precursors *in vivo*. An open reading frame of 392 codons for MFT1 was confirmed by Ito and colleagues (1993). Disruption of the correct open reading frame results in temperature-sensitive growth defects on rich media. The predicted amino acid sequence for Mft1p suggests soluble protein rich in acidic residues with no significant homology to known proteins.

Using a similar missorting screen, Kenna and colleagues (1993) identified a gene, HKE1 (homology to KEM1), which allowed growth of yeast on nonfermentable carbon sources in the presence of the toxic $F_1\beta$-β-galactosidase fusion protein. Hke1p is encoded by an essential gene and contains a $5'\rightarrow 3'$ exoribonuclease activity presumably involved in mRNA processing. Unlike MFT1, no defect in mitochondrial import could be identified in HKE1 mutants. Conditional hke1 mutants did exhibit a defect in protein synthesis at the nonpermissive temperature that could not be ascribed to changes in poly (A) RNA levels. It has been proposed that this protein is important in the presentation of mature mRNA, by either transport or processing, to the translation apparatus. In independent studies, Amberg and colleagues (1992) isolated the same gene, termed RAT1 (RNA transport), as a mutant that exhibited defects in mRNA transport from the nucleus at the restrictive temperature. One rationale for the identification of this gene in the mistargeting screen is that some alleles could result in a modest reduction of translation efficiency that might be exaggerated in the gene fusion, $F_1\beta$-β-galactosidase, employed for the genetic screen.

Another temperature-sensitive mutant (mpt1, mitochondrial protein targeting) uncovered in the $F_1\beta$-β-galactosidase missorting screen also identifies a protein that appears to have a role in the management and expression of RNA. The levels of mRNA for mitochondrial precursors in the mutant at the nonpermissive temperature are the same as in wild type; however, this RNA is not translated. Further analysis revealed that the block in expression occurred at the level of translation and was not restricted to just mitochondrial precursors (Estey and Douglas, submitted). The polysome profiles of the temperature-sensitive mutant grown at the nonpermissive temperature also showed a dramatic decrease in the levels of 80S monosomes, suggesting that the temperature-sensitive mpt1 allele may facilitate the dissociation of monosomes, or prevent reassociation of 40S and 60S subunits in the initiation process. This observation, in addition to the observed enrichment of Mpt1p in isolated 40S ribosomal complexes, suggests that MPT1 is involved in the initiation of protein synthesis. MPT1 encodes an essential protein with a predicted molecular weight of 42.3 kDa that contains several potential kinase modification sites (Estey and Douglas, submitted).

An independent genetic screen to identify mitochondrial import mutants has also uncovered genes with a role in the management of cytoplasmic RNA (Zoladek et al., 1995). In this screen, a tRNA modification enzyme coded by the MOD5 gene was employed. Mod5p is a soluble protein that localizes to both mitochondrial and cytoplasmic compartments in *Saccharomyces* (Gillman et al., 1991; Slusher et al., 1991). Using an allele of MOD5 that localizes to only mitochondria, temperature-sensitive mutants that retain Mod5p activity in the cytoplasm were selected and characterized. Three genetic complementation groups were identified; these mutants were all temperature-sensitive for growth on nonfermentable carbon sources and exhibited defects in the import of MOD5-COXIV fusion proteins into mitochondria. Two of the mutations (MIP1 and MIP3) have been cloned; MIP1 was found to be ORF41 on *Saccharomyces cerevisiae* chromosome V. MIP1 shows homology to mouse nedd4, human hape6, and rat rn100, three proteins whose functions have yet to be determined. MIP3 was identical to PAN1, a yeast poly(A) binding protein-dependent ribonuclease (Sachs and Deardorff, 1992). Conditional mutations in PAN1 lead to altered patterns of poly (A) tail lengths and affect protein synthesis at initiation steps in the translation process (Sachs and Deardorff, 1992).

The mislocalization screen employing MOD5, like those employing toxic fusion proteins, has identified a set of cytosolic components with a common theme. Over the years, some molecular chaperones have been characterized for their role as cytosolic factors that are necessary to assist in the formation and maintenance of native protein structure in the cell. These chaperones serve as components to assist import beyond the requirements of coupled translation and translocation. As we now start to probe the more subtle control of mitochondrial trafficking, we are beginning to uncover the importance of various aspects of RNA management for effective organelle biogenesis. In the yeast systems that were employed to initially identify these factors, RNA management may not be as critical, however, in neuronal projections where large distances exist between sites of synthesis and utilization of specific mRNAs, mechanisms for coupling translation and membrane translocation would be essential.

D. Is the Management of mRNA Important in the Trafficking Process?

As one begins to access the vast amount of experimental and genetic data that defines what we understand about the process of protein import, one is immediately struck by the ambiguity of the information that is available to describe the events outside mitochondria. This, we understand now, probably results from the fact that the protein import process in the cell cannot be defined by a single post-translational mechanism. As described earlier, the differences in each mitochondrial precursor will dictate the extent to which other factors in the cytoplasm will come into play to insure efficient trafficking. For the most part, the cytoplasmic factors that have been defined and characterized are the molecular chaperones that deal with the different detractions and aggregations that may be suffered by an immature protein en route to the mitochondrial membrane. These post-translational events are probably minimal under normal growth conditions. When required, chaperones or cytoplasmic factors are available to insure the completion of the targeting process that the precursor, for whatever reason, is unable to complete on its own. The requirement for chaperones and other factors that provide similar function can be circumvented if conditions such as temperature and translation rates are appropriately reduced. Indeed, the compendium of data indicate that in some manner it is the window between the initiation of translation and transport that defines what factors must participate. Recent work is beginning to reveal what relative contributions to the import process are made by the intervention of cytosolic factors at earlier points in the localization process (Figure 1). The genes identified in genetic screens based on the mistargeting of mitochondrial precursors suggest that in some manner the regulated utilization of specific messenger RNA molecules is important. In the genetic screen described, the mitochondrially targeted toxic fusion product continues to be synthesized at the permissive temperature, yet in some of the mutants, the fusion products do not effectively compete with authentic precursor proteins for import into mitochondria. What are the features responsible for the recognition of authentic protein or its transcript over the gene fusion protein that give it a competitive advantage for mitochondrial entry? Given that both fusion and authentic transcripts contain the same information in their 5′ untranslated regions, the steps involved in the formation of translation initiation complexes on the two types of transcripts should be identical. Maybe the information provided by the 3′ untranslated region of the authentic transcripts may allow preferential synthesis and subsequent targeting of authentic precursors to the mitochondria. The identification of PAN1, a poly A-dependent ribonuclease as a defective component in the mistargeting of a protein to mitochondria is an indication that this mechanism is operating. Until recently, the role of the 3′ untranslated region in mRNA structure had been underestimated, however, current evidence suggests that in higher eukaryotes, this region contains signals for not only polyadenylation and mRNA stability, but also for mRNA localization (Wilhelm and Vale, 1993) and control of translational initiation (Jackson, 1993). In the case of α- and β-isoforms of actin, the localization of these proteins to specific subcellular regions occurs via specific signals contained within the 3′ untranslated regions of their respective mRNAs (Kislauskis et al., 1993). Also, the localization of specific mRNAs in oocytes, including those for the mitochondrial $F_1\alpha$-subunit precursor, is dependent on an active process in which a specific region at the 3′ end of the message is recognized (Weeks and Melton, 1987; Gottlieb, 1992). Whether lower eukaryotes such as *Saccharomyces* employ such signals in the sorting of mRNAs is unknown. This could provide an explanation for the enrichment of precursor-specific mRNAs found with mitochondrial associated ribosomes (Suissa and Schatz, 1982), however, other explanations are possible. A mechanism of this kind would also support the proposal that protein synthesis and import are coupled events for at least some mitochondrial protein precursors. Thus, some of the mutants identified in the mislocalization screens could reveal factors and activities that are necessary for the post-transcriptional deposition or processing of some precursor mRNAs.

Another possible mechanism for how the management of mRNA might provide an overall more efficient import process is through control of the translation initiation of some imported mitochondrial precursors. Several of the mutants identified in the fusion screen phenotypically present themselves as defects in the initiation of protein synthesis. Several of the gene products that are imported more efficiently in a cotranslational manner are regulated in some way at the level of translation. The expression of the mitochondrial $F_1\beta$-subunit precursor is regulated at the translational level in neonatal rat liver, probably through a modification of initiation factor 2, eIF-2 (Luis et al., 1993). Unlike cytochrome c_1 or cytochrome oxidase VI transcripts, the levels of $F_1\beta$ transcript in yeast strains carrying defective mitochondrial genomes remain constant, suggesting that another mechanism is responsible for controlling the utilization of $F_1\beta$ mRNA (Parikh et al., 1987). Thus, it is feasible that post-transcriptional mechanisms may be involved in regulating the expression of $F_1\beta$ and some other nuclear-encoded mitochondrial proteins. In some cases, steps in the translation initiation process are responsive to trans-acting factors required for regulation of gene expression (for review see Altmann and Trachsel, 1993). It is possible that some of the mutations uncovered in the fusion mistargeting screen represent defects in RNA processing or trafficking (i.e., HKE1, PAN1) that ultimately affect the translational efficiency of tightly regulated mRNAs. Mutations in MPT1 and MFT1 may also result in compromised rates of translation initiation, however, the mechanisms for these effects is unclear.

The fusion mistargeting screen has identified a number of genes that can indirectly affect the process of mitochondrial import, thus lending support to the idea that protein synthesis and mitochondrial import are closely linked processes. Although the data do not directly support a cotranslational model for mitochondrial import, it is quite possible that the efficiency of mitochondrial import is dependent on the rates of protein synthesis for genes whose expression is controlled by post-transcriptional mechanisms. In such cases, it may be necessary for the transcripts to be delivered to the mitochondrial periphery prior to initiation of protein synthesis, therefore, mitochondrially associated ribosomes merely represent a preferred subcellular location for translation of some nuclear-encoded mitochondrial proteins. There are numerous examples in development where location and time-dependent translation occur under the control of cytoplasmic factors. Future investigations that examine intracellular mRNA sorting as well as the coordinate control of mitochondrial and nuclear-encoded mitochondrial proteins in yeast should aid in the elucidation of mechanisms responsible for the coupling of protein synthesis to mitochondrial import.

ACKNOWLEDGMENTS

The authors would like to acknowledge the National Institutes of Health and the American Heart Association for their support of the research from this laboratory.

REFERENCES

Amberg, D. C., Goldstein, A. L., & Cole, C. N. (1992). Isolation and characterization of RAT1: An essential gene of *Saccharomyces cerevisiae* required for the efficient nucleocytoplasmic trafficking of mRNA. Genes Dev. 6,1173–1189.

Altmann, M., & Trachsel, H. (1993). Regulation of translation initiation and modulation of cellular physiology. Trends Biochem. Sci. 18, 429–432.

Attencio, D. P., & Yaffe, M. P. (1992). *MAS5*, a yeast homologue of DnaJ involved in mitochondrial protein import. Mol. Cell. Biol. 12, 283–291.

Becker, K., Guiard, B. , Rassow, J., Söllner, T., & Pfanner, N. (1992). Targeting of a chemically pure preprotein to mitochondria does not require the addition of a cytosolic signal recognition factor. J. Biol. Chem. 267, 5637–43.

Bieker, K. L., Phillips, G. L., & Silhavy, T. J. (1990). The *sec* and *prl* genes of *Escherichia coli*. J. Bioenerg. Biomembr. 22, 291–310.

Bolliger, L., Deloche, O., Glick, B. S., Georgopoulos, C., Jeno, P., Kronidou, N., Horst, M., Morishima, N., & Schatz, G. (1994). A mitochondrial homolog of bacterial GrpE interacts with mitochondrial hsp 70 and is essential for viability. EMBO J. 13, 1998–2006.

Brodsky, J. L., Hamamoto, S., Feldheim, D., & Schekman, R. (1993). Reconstitution of protein translocation from solubilized yeast membranes reveals topologically distinct roles for BiP and cytosolic hsc70. J. Cell. Biol. 120, 95–102.

Caplan, A. J., & Douglas, M. D. (1991). Characterization of YDJ1: A yeast homolog of the bacterial dnaJ protein. J. Cell Biol. 114, 609–621.

Caplan, A. J., Cyr, D. M., & Douglas, M. G. (1992). YDJ1 facilitates polypeptide translocation across different intracellular membranes by a conserved mechanism. Cell 71, 1143–1155.

Caplan, A. J., Cyr, D. M., & Douglas, M. G. (1993). Eukaryotic Homologues of *E. coli* dnaJ: A diverse protein family that functions with hsp70 stress proteins. Mol. Cell Biol. 4, 555–563.

Chen, W-J., & Douglas, M. G. (1987a). The role of protein structure in the mitochondrial import pathway. J. Biol. Chem. 262, 15605–15609.

Chen, W-J., & Douglas, M. G. (1987b). Phosphodiester bond cleavage outside mitochondria is required for the completion of protein import into the mitochondrial matrix. Cell 49, 651–658.

Chen, W-J.; & Douglas, M. G. (1988). An F1-ATPase β-subunit precursor lacking an internal tetramer-forming domain is imported into mitochondria in the absence of ATP. J. Biol. Chem. 263 4997–5000.

Chirico, W. J., Waters, M. G., & Blobel (1988). 70K heat shock related proteins stimulate protein translocation into microsomes. Nature 322, 805–810.

Chirico, W. (1992). Dissociation of complexes between 70kd stress proteins and pre-secretory proteins is facilitated by a cytosolic factor. Biochem. Biophys. Res. Comm. 189, 1150–1156.

Craig, E. A., & Jacobsen, K. (1984). Mutations of the heat inducible 70 kilodalton genes of yeast confer temperature sensitive growth. Cell 38, 841–849.

Craig, E. A., Gambill, D. B., & Nelson, J. R. (1993). Heat shock proteins: Molecular chaperoned of protein biogenesis. Microbiological Reviews 57, 402–414.

Cyr, D. M., Lu, X., & Douglas, M. G. (1992). Regulation of eukaryotic Hsp70 function by a DnaJ Homologue. J. Biol. Chem. 267, 20927–20931.

Cyr, D. M., Langer, T., & Douglas, M. G. (1994). DnaJ-like proteins represent a new class of molecules that function as molecular chaperones and specific regulators of Hsp70. Trends Biochem. Sci. 19, 176–181.

Cyr, D. M., & Douglas, M. D. (1994). Differential regulation of Hsp70 subfamilies by the eukaryotic DnaJ homolog YDJ1. J. Biol. Chem. 269, 9798–9804.

Deshaies, R. J., Koch, B. D., Werner-Washburne, M., Craig, E. A., & Schekman, R. (1988). A subfamily of stress proteins facilitates translocation of secretory and mitochondrial presursor polypeptides. Nature 322, 800–805.

Douglas, M. G., Geller, B. L., & Emr, S. D. (1984). Intracellular targeting and import of an F_1-ATPase β-subunit-β-galactosidase hybrid protein into yeast mitochondria. Proc. Natl. Acad. Sci. USA 81, 3983–3987.

Eilers, M., Opplinger, W., & Schatz, G. (1987). Both ATP and an energized inner membrane are required to import a purified precursor protein into mitochondria. EMBO J. 7, 1139–1145.

Emr, S. D., Hall, M. N., & Silhavy, T. J. (1980). A mechanism of protein localization: The signal hypothesis and bacteria. J. Cell Biol. 86, 701–711.

Fujiki, M., & Verner, K. (1991). Coupling of protein synthesis and mitochondrial import in a homologous yeast *in vitro* system. J. Biol Chem. 266, 6841–47.

Fujiki, M., & Verner, K. (1993). Coupling of cytosolic protein synthesis and mitochondrial protein import in yeast: Evidence for cotranslational import *in vivo*. J. Biol. Chem. 268, 1914–20.

Garrett, J. M., Singh, K. K., von der Haar, R. A., & Emr, S. D. (1991). Mitochondrial protein import: Isolation and characterization of the *Saccharomyces cerevisiae* MFT1 gene. Mol. Gen. Genet. 225, 483–491.

Gething, M-J., & Sambrook, J. (1992). Protein folding in the cell. Nature 355, 33–45.

Gillman, E. C., Slusher, L. B., Martin, N. C., & Hopper, A. K. (1991). MOD5 translation initiation sites determine N_6-isopentyladenosine modification of mitochondrial and cytoplasmic tRNA. Mol. Cell. Biol. 11, 2382–2390.

Gottlieb, E. (1992). The 3′-untranslated region of localized messages contains a conserved motif involved in mRNA localization. Proc. Natl. Acad. Sci. 89, 7164–7168.

Hachiya, M., Alam, R., Sakasegawa, M., Mihara, K., & Omura, T. (1993). A mitochondrial import factor purified from rat liver cytosol is an ATP-dependent conformational modulator for precursor proteins. EMBO J. 12, 1579–1586.

Hajek, P., & Bedwell, D. M. (1994). Characterization of the mitochondrial binding and import properties of purified yeast F_1 ATPase β subunit precursor: Import requires external ATP. J. Biol. Chem. 269, 7192–7200.

Hallermayer, G., Zimmermann, R., & Neupert, W. (1977). Kinetic studies on the transport of cytoplasmically synthesized proteins into the mitochondria in intact cells of *Neuorspora crassa*. Eur. J. Biochem. 81, 523–532.

Hann, B. C., & Walter, P. (1991). The signal recognition particle in *S. cerevisiae*. Cell 67, 131–144.

Hannavy, K., Rospert, S., & Schatz (1993). Protein import into mitochondria: A paradigm for the translocation of polypeptides across membranes. Curr. Opin. Cell Biol. 5, 694–700.

Hansen, W., Garcia, P., & Walter, P. (1986). *In vitro* protein translocation across the yeast endoplasmic reticulum: ATP-dependent post-translational translocation of the prepro-α-factor. Cell 45, 397–406.

Hartl, F-U., Pfanner, N., Nicholson, D. W., & Neupert, W. (1989). Mitochondrial protein import. Biochim. Biophys. Acta 988, 1–45.

Hartl. F-U., Hlodan, R., & Langer, T. (1994). Molecular chaperones in protein folding: The art of avoiding sticky situations. Trends Biochem. Sci.19, 20–25.

Hendrick, J. P., Langer, T., Davis, T. A., Hartl, F-U., & Wiedmann, M. (1993). Control of folding and membrane translocation by binding of the chaperone DnaJ to nascent polypeptides. Proc. Natl. Acad. Sci. USA 90, 10216–10220.

Hohfeld, J., & Hartl, F-U. (1994). Post-translational protein import and folding. Curr. Opin. Cell Biol. 6, 499–509.

Ito, M., Yasui, A., & Komamine, A. (1993). Precise mapping and molecular characterization of the MFT1 gene involved in import of a fusion protein into mitochondria in *Saccharomyces cerevisiae*. FEBS Lett. 320, 125–129.

Iwahashi, J., Furuya, S., Mihara, K., & Omura, T. (1992). Characterization of adrenodoxin precursor expressed in *Escherichia coli*. J. Biochem. 111, 451–455.

Jackson, R. J. (1993). Cytoplasmic regulation of mRNA function: The importance of the 3′ untranslated region. Cell 74, 9–14.

Kellems, R. E., Allison, V. F., & Butow, R. A. (1974). Cytoplasmic type 80S ribosomes associated with yeast mitochondria, evidence for the association of cytoplasmic ribosomes with the outer mitochondrial membrane *in situ*. J. Biol. Chem. 249, 3297–3303.

Kellems, R. E., & Butow, R. A. (1974). Cytoplasmic type 80S ribosomes associated with yeast mitochondria: Changes in the amount of bound ribosomes in response to changes in metabolic state. J. Biol. Chem. 249, 3304–3310.

Kenna, M., Stevens, A., McCammon, M. M., & Douglas, M. G. (1993). An essential yeast gene with homology to the exonuclease-encoding XRN1/KEM1 gene also encodes a protein with exoribonuclease activity. Mol. Cell. Biol. 13, 341–350.

Kislauskis, E. H., Li, Z., Singer, R. H., & Taneja, K. L. (1993). Isoform-specific 3′-untranslated sequences sort α-cardiac and β-cytoplasmic actin messenger RNAs to different cytoplasmic compartments. J. Cell Biol. 123, 165–172.

Kudlicki, W., Odom, O. W., Kramer, G., & Hardesty, B. (1994). Chaperone-dependent folding and activation of ribosome-bound nascent rhodanese. J. Mol. Biol. 244, 319–331.

Kutay, U., Hartmann, E., & Rapoport, T. (1993). A class of membrane proteins with a C-terminal anchor. Trends in Cell Biol. 3, 72–74.

Langer, T., Lu, C. Echols, H., Flanagan, J., Hayer, M., & Hartl, F-U. (1992). Successive action of dnaK, dnaJ and GroEL along the pathway of chaperone-mediated folding. Nature 356, 683–689.

Liberek, K., Marszalek, J., Ang, D., Georgopoulos, C., & Zylicz, M. (1991). *Escherichia coli* dnaJ & grpE heat shock proteins jointly stimulate ATPase activity of dnaK. Proc. Natl. Acad. Sci. USA 88, 2874–2878.

Loncar, D. (1990). Immunoelectron microscopical studies on synthesis and localization of uncoupling protein in brown adipocytes: Evidence for cotranslational transport of uncoupling protein into mitochondria. J. Struct. Biol. 105, 133–145.

Luis, A. M., Izquierdo, J. M., Ostronoff, L. K., Salinas, M., Santaren, J. F., & Cuezva, J. M. (1993). Translational regulation of mitochondrial differentiation in neonatal rat liver. Specific increase in the translational efficiency of the nuclear-encoded mitochondrial β-F_1-ATPase mRNA. J. Biol. Chem. 268, 1868–1875.

Maccecchini, M-L., Rudin, Y., Blobel, G., & Schatz, G. (1979). Import of proteins into mitochondria: Precursor forms of the extramitochondrially made F_1-ATPase subunits in yeast. Proc. Natl. Acad. Sci. USA 76, 343–347.

Murikami, K., & Mori, M. (1990). Purified presequence binding factor (PBF) forms an import-competent complex with a purified mitochondrial precursor protein. EMBO J. 9, 3201–3208.

Murikami, K., Tanase, S., Morino, Y., & Mori, M. (1992). Presequence binding factor-dependent and-independent import of proteins. J. Biol. Chem. 267, 13119–13122.

Nelson, R. J., Ziegelhoffer, T., Nicolet, C. Werner-Washburne, M., & Craig, E. A. (1992). The translation machinery and 70 kd heat shock protein cooperate in protein synthesis. Cell 71, 97–105.

Parikh, V. S., Morgan, M. M., Scott, R., Clements, L. S., & Butow, R. A. (1987). The mitochondrial genotype can influence nuclear gene expression in yeast. Science 235, 576–580.

Pfanner, N., & Neupert, W. (1986). Transport of the F_1-ATPase subunit beta into mitochondria depends on both a membrane potential and nucleoside triphosphates. FEBS Lett. 209, 152–156.

Pfanner, N., Pfaller, R., Kleene, R., Ito, M., Tropschug, M., & Neupert, W. (1988). Role of ATP in mitochondrial protein import. J. Biol. Chem. 263, 4049–4051.

Pon, L., Moll, T., Vestweber, D., Marshallsay, B., & Schatz, G. (1989). Protein import into mitochondria: ATP-dependent protein translocation activity in a submitochondrial fraction enriched in membrane contact sites and specific proteins. J. Cell Biol. 109, 2603–16.

Price, J., & Verner, K. (1993). Puromycin inhibits protein import into mitochondria by interfering with an intramitochondrial ATP-dependent reaction. Biol. Biophys. Acta 1150, 89–97.

Reid, G. A., & Schatz, G. (1982a). Import of proteins into mitochondria: Yeast cells grown in the presence of carbonyl cyanide m-chlorophenylhydrazone accumulate massive amounts of some mitochondrial precursor polypeptides. J. Biol. Chem. 257, 13056–13061.

Reid, G. A., & Schatz, G. (1982b). Import of proteins into mitochondria: Extramitochondrial pools and post-translational import of mitochondrial protein precursors *in vivo*. J. Biol. Chem. 257, 13062–67.

Reid, G. A., Yonetani, T., & Schatz, G. (1982). Import of protein into mitochondria: Import and maturation of the mitochondrial imtermembrane space enzymes cytochrome b_2 and cytochrome c peroxidase in intact yeast cells. J. Biol. Chem. 257, 13068–13074.

Rothblatt, J. A., & Meyer, D. (1986). Secretion in yeast: Reconstitution of the translocation and glycosylation of α-factor and invertase in the homologous cell-free system. Cell 44, 619–628.

Sachs A. B., & Deardorff, J. A. (1992). Translation initiation requires the PAB-dependent poly (A) ribonuclease in yeast. Cell 70, 961–973.

Schatz, G., & Butow, R. (1983). How are proteins imported into mitochondria. Cell 32, 316–318.

Slusher, L. B., Gillman, E. C., Martin, N. C., & Hopper, A. K. (1991). mRNA leader length and initiation codon context determine alternative AUG selection for the yeast gene MOD5. Proc. Natl. Acad. Sci. USA 88, 9789–9793.

Suissa, M., & Schatz, G. (1982). Import of proteins into mitochondria. J. Biol. Chem. 257, 13048–13055.

Verner, K. (1992). Early events in yeast mitochondrial protein targeting. Mol. Microbiol. 6, 1723–1728.

Verner, K. (1993). Co-translational protein import into mitochondria: An alternative view. Trends Biochem. Sci. 18, 366–371.

Verner, K., & Schatz, G. (1987). Import of an incompletely folded precursor protein into isolated mitochondria requires an energized inner membrane, but no added ATP. EMBO J. 6, 2449–2456.

Verner, K., & Weber, M. (1989). Protein import into mitochondria in a homologous yeast *in vitro* system. J. Biol. Chem. 264, 3877–79.

Vestweber, D., & Schatz, G. (1988). Point mutations destabilizing a procursor protein enhance it post-translational import into mitochondria. EMBO J. 7, 1147–1151.

Voos, W., Gambill, B. D., Laloraya, S., Ang, D., Craig, E. A., & Pfanner, N. (1994). Mitochondrial GrpE is present in a complex with hsp70 and preproteins in transit across membranes. Mol. Cell. Biol. 14, 6627–6634.

Walter, P., & Johnson, A. E. (1994). Signal sequence recognition and protein targeting to the endoplasmic reticulum membrane. Ann. Rev. Cell Biol. 10, 87–119.

Waters, M. G., & Blobel, G. (1986). Secretory protein translocation in a yeast cell-free system can occur post-translationally and requires ATP hydrolysis. J. Cell Biol. 102, 1543–1550.

Waters, M. G., Chirico, W. J., & Blobel, G. (1986). Protein translocation across the yeast microsomal membrane is stimulated by a soluble factor. J. Cell Biol. 103, 2629–2636.

Waters, M. G., Evans, E. A., & Blobel, G. (1988). Prepro-α-factor has a cleavable signal sequence J. Biol. Chem. 263, 6209–6214.

Weeks, D. L., & Melton, D. A. (1987). A maternal mRNA localized to the animal pole of *Xenopus* eggs encodes a subunit of the mitochondrial ATPase. Proc. Natl. Acad. Sci. 89, 2798–2802.

Weinhues, U., Becker, K., Schleyer, M., Guiard, B., Tropschug, M., Horwich, A. L., Pfanner, N., & Neupert, W. (1991). Protein folding causes an arrest of preprotein translocation into mitochondria *in vivo*. J. Cell Biol. 115, 1601–1609.

Werner-Washburne, M., Stone, D. E., & Craig, E. A. (1987). Complex interactions among members of an essential subfamily of hsp 70 genes in *Saccharomyces cerevisiae*. Mol. and Cell Biol. 7, 2568–2577.

Wilhelm, J. E., & Vale, R. D. (1993). RNA on the move: The mRNA localization pathway. J. Cell Biol. 123, 269–74.

Zhong, T., & Arndt, K.T. (1993). The yeast S1S1 protein, a DnaJ homolog, is required for the initiation of translation. Cell 73, 1175–1186.

Zoladek, T., Vaduva, G., Hunter, L. A., Boguta, M., Go, B. D., Martin, N. C., & Hopper, A. K. (1995). Mutations altering the mitochondrial-cytoplasmic distribution of Mod5p implicate the actin cytoskeleton and mRNA 3′ ends and/or protein synthesis in mitochondrial delivery. Mol. Cell. Biol. 15, 6884–6894.

Zylicz, M., Ang, D., Liberek, K., & Georgopoulos, C. (1989). Initiation of λ DNA replication with purified host and bacteriophage encoded proteins: The role of the dnaK, dnaJ, and grpE heat shock proteins. EMBO J. 8,1601–1608.

PRESEQUENCE BINDING PROTEINS AS CYTOSOLIC IMPORT-STIMULATION FACTORS IN MITOCHONDRIAL PROTEIN IMPORT

Katsuyoshi Mihara and Tsuneo Omura

I. Introduction 34
II. "Targeting Factor" of Rabbit Reticulocyte Lysate 35
III. "Presequence Binding Factor (PBF)" Purified from Rabbit Reticulocyte Lysate 35
IV. "Mitochondrial Import Stimulation Factor (MSF)" Purified from Rat Liver Cytosol 36
A. Purification of MSF from Rat Liver Cytosol 36
B. Specificity of MSF 37
C. MSF Catalyzes Depolymerization of Precursors 37
D. NEM Inhibits a Function of MSF 38
E. Dual Functions of MSF 39
F. Precursor Protein-induced ATPase Activity of MSF 41
G. Presequence-induced ATPase Activity of MSF 43

Advances in Molecular and Cell Biology,
Volume 17, pages 33–49.

ISBN: 0-7623-0144-9

H. Binding of the MSF-Precursor Complex to the Outer Mitochondrial Membrane 44
I. Molecular Properties of MSF Subunits 44
V. Presequence-binding Cytosolic Factors in Mitochondrial Protein Import 46
References 47

I. INTRODUCTION

A requirement of cytosolic factors for the *in vitro* import of precursor proteins into mitochondria was already described in papers published in 1983–1984 (Miura et al., 1983; Argan et al., 1983; Ohta and Schatz, 1984). Later studies confirmed the participation of a cytosolic molecular chaperone, hsp70, in mitochondrial protein import (Deshaies et al., 1988; Murakami et al., 1988a). Hsp70 seems to maintain the newly-synthesized mitochondrial precursor proteins in import-competent conformations in an ATP-dependent manner, but the chaperone function of hsp70 is evidently not specific for mitochondrial precursor proteins. Hsp70 also facilitates the post-translational translocation of secretory proteins across the membrane of the endoplasmic reticulum in yeast (Deshaies et al., 1988; Chirico et al., 1988). An N-ethylmaleimide (NEM)-sensitive cytosolic factor, distinct from hsp70, was also reported to be involved in protein import into yeast mitochondria (Murakami et al., 1988a). This NEM-sensitive factor apparently acts synergistically with hsp70. It is not known, however, whether the factor is specific for mitochondrial protein import.

Mitochondrial proteins encoded by nuclear genes are synthesized by free ribosomes in the cytosol and usually possess an amino-terminal peptide extension, the presequence, which acts as the mitochondrial targeting signal. Cytosolic factors specifically required for mitochondrial protein import would be expected to recognize the mitochondrial targeting signal in the presequence. Ono and Tuboi purified a cytosolic factor from rabbit reticulocyte lysate, and named it "targeting factor" because the factor recognized the presequence of mitochondrial precursor proteins and stimulated both the binding of the precursor proteins to mitochondria and their import (Ono and Tuboi, 1988, 1990a, 1990b). Murakami and Mori purified another cytosolic factor from rabbit reticulocyte lysate, which they named "presequence binding factor (PBF)" (Murakami and Mori, 1990). PBF seemed to bind to the presequence portion of mitochondrial precursor, maintaining them in an import-competent form in cooperation with hsp70. However, the molecular characterization of these cytosolic factors isolated from reticulocyte lysate has yet to be reported.

We also purified a mitochondrial import stimulation factor, termed MSF, from rat liver cytosol and analyzed its function in the *in vitro* import of precursor proteins into mitochondria (Hachiya et al., 1993). MSF recognizes the targeting signal in the presequences of mitochondrial precursor proteins, catalyzes their unfolding in an ATP-dependent manner, and exhibits a significant ATPase activity that is dependent on the presence of import-incompetent precursor proteins. The primary structures of the subunits of MSF revealed that they belong to the 14-3-3 protein family, whose physiological functions are not yet clearly understood (Alam et al., 1994).

In this paper, we first describe briefly the properties of "targeting factor" and "presequence binding factor" reported by the two other research groups, and then describe in more detail our studies on MSF.

II. "TARGETING FACTOR" OF RABBIT RETICULOCYTE LYSATE

Ono and Tuboi reported the requirement for a cytosolic factor in reticulocyte lysate for the *in vitro* import of a 34 residue synthetic peptide corresponding to the presequence portion of ornithine aminotransferase (1988). The cytosolic factor was purified to homogeneity from rabbit reticulocyte lysate by affinity chromatography using the synthetic peptide, indicating its affinity for the mitochondrial targeting sequence. The purified factor was significantly inactivated during the purification, but its identity with the cytosolic factor in the original reticulocyte lysate was confirmed by the use of an antibody raised against the purified preparation. Since the cytosolic factor not only stimulated the import of mitochondrial precursor proteins but also increased their binding to the mitochondrial surface, Ono and Tuboi proposed to call it "targeting factor" (1990a, 1990b).

The activity of "targeting factor" was detected in the cytosol fraction of rat liver, and also in the high salt wash of isolated liver mitochondria, suggesting that the targeting factor was bound to the mitochondrial surface. Addition of large amounts of "targeting factor" to the import assay mixture resulted in a significant inhibition of the import of precursor proteins. This seemed to indicate competition between free and precursor-bound "targeting factor" for a receptor(s) on the mitochondrial outer membrane.

Purified "targeting factor" is different from hsp70, having a relative molecular mass of 28 kDa on SDS-PAGE (Ono and Tuboi, 1990a). It is also clearly separated from hsp70 by ATP-agarose column chromatography. The molecular nature of "targeting factor" has not yet been elucidated.

III. "PRESEQUENCE BINDING FACTOR (PBF)" PURIFIED FROM RABBIT RETICULOCYTE LYSATE

Murakami and colleagues studied the *in vitro* import into mitochondria of the purified precursor of ornithine carbamoyltransferase (OTC) expressed in *E. coli* and found the import to be dependent on the addition of rabbit reticulocyte lysate (Murakami et al., 1988b). The import stimulating activity in the reticulocyte lysate was purified and named "presequence binding factor" (PBF) because the factor binds to the purified precursor protein (pOTC), but not the mature OTC from which the presequence had been cleaved. Association of PBF with the precursor protein to form a soluble complex can be inhibited by synthetic peptides corresponding to the presequences of OTC and two other mitochondrial matrix proteins, malate dehydrogenase and aspartate aminotransferase (Murakami and Mori, 1990). Purified PBF migrates as a single polypeptide of Mr 50,000 on SDS-PAGE. PBF-dependent import of pOTC was found to be further stimulated by the addition of hsp70.

PBF was also isolated from the cytosol fractions of rat liver and heart by affinity chromatography on pOTC-Sepharose, suggesting that it occurs ubiquitously in various animal tissues. The precursors of OTC, aspartate aminotransferase and malate dehydrogenase, were not imported into mitochondria when they were synthesized in reticulocyte lysate that had been depleted of PBF by treatment with pOTC-Sepharose. Addition of purified PBF to the PBF-depleted reticulocyte lysate resulted in full restoration of the import. However, rat 3-oxoacyl CoA thiolase, which has no cleavable presequence at its amino-terminus (Mori et al., 1985), did not require PBF for import into mitochondria *in vitro*. It was also observed that PBF bound to the precursor proteins with cleavable presequences, but did not associate with 3-oxoacyl CoA thiolase synthesized *in vitro*, suggesting the existence of PBF-dependent and -independent pathways for mitochondrial protein import (Murakami et al., 1992).

IV. "MITOCHONDRIAL IMPORT STIMULATION FACTOR (MSF)" PURIFIED FROM RAT LIVER CYTOSOL

A. Purification of MSF from Rat Liver Cytosol

It has been demonstrated that mitochondrial precursor proteins synthesized in the reticulocyte lysate system are efficiently imported into isolated mitochondria *in vitro*, whereas those synthesized in the wheat germ lysate system have a tendency to aggregate and are imported into mitochondria with low efficiency (Murakami et al., 1988a). In fact, we found that adrenodoxin precursor (pAd), synthesized in the wheat germ system, was not imported at all into isolated rat liver mitochondria. However, the addition of rat liver cytosol to the translation products stimulated their import depending on the amount of cytosol added. The binding of pAd to mitochondria was also increased significantly under these conditions. Thus, rat liver cytosol seemed to contain a factor that stimulated post-translational mitochondrial import of pAd synthesized in the wheat germ lysate system by increasing the amount of the import-competent precursor.

This system allowed the isolation of the stimulating factor from rat liver cytosol. The factor was purified 10,000-fold on an affinity matrix containing a COX IV presequence-β-galactosidase fusion protein (COX-β). The purified factor, termed "mitochondrial import stimulation factor" (MSF), was composed of approximately stoichiometric amounts of two polypeptides with relative molecular masses of 30 kDa and 32 kDa on SDS-PAGE. The two polypeptides coeluted from a Sepharose 6B column as a symmetrical peak with an apparent molecular weight of about 60 kDa, suggesting that MSF is a heterodimer (Hachiya et al., 1993).

B. Specificity of MSF

The import-stimulating activity of MSF was not specific to pAd. MSF also stimulated the import of mitochondrial superoxide dismutase, porin, and of a fusion protein between the yeast COXIV presequence and porin, all of which were synthesized in the wheat germ lysate system (Hachiya et al., 1993). These observations suggest that MSF generally recognizes mitochondrial precursor proteins and stimulates their import into mitochondria.

Murakami and Mori (1990) showed that hsp70 stimulated the import of proteins into mitochondria in the presence of the purified presequence binding factor (PBF). It was also shown by Murakami and colleagues (1988) that hsp70 and an NEM-sensitive component contained in yeast post-ribosomal supernatant acted synergistically in stimulating mitochondrial protein import. However, no significant stimulation of the activity of MSF was observed in the presence of hsp70 (Hachiya et al., 1993). We conclude that the activity of MSF does not require the synergistic action of hsp70 in the *in vitro* assay system.

MSF stimulated the import into mitochondria of a pAd synthesized in the rabbit reticulocyte lysate system from which the stimulation factor had been depleted using the COX-β affinity matrix. We also found that antibodies against both subunits of MSF depleted the stimulation activity from the cytosol and also inhibited the import activity. These results indicated that MSF was not only active in the heterologous wheat germ cell-free system, but also in the homologous reticulocyte lysate system.

C. MSF Catalyzes Depolymerization of Precursors

Upon sucrose gradient centrifugation, mitochondrial precursors synthesized in the wheat germ cell-free system generally sediment to the bottom of the gradient as large aggregates. We examined whether this aggregation correlated with the import-incompetence of the precursor proteins. As expected, pAd synthesized in the wheat germ system sedimented in the density gradient as large aggregates with an $S_{20,w}$ value of about 40S. Strikingly, when the precursor was incubated with 200 ng of MSF, the large aggregates of the precursor were depolymerized and sedimented with a $S_{20,w}$ value of 6.2S (Figure 1).

That MSF recognizes the presequence portion of the precursors, was supported by several observations: (1) When an excess amount of purified, cold pAd was added to the wheat germ-synthesized radioactive pAd, the depolymerization of the *in vitro* synthesized pAd was completely inhibited, whereas no inhibition was observed upon addition of cold, mature adrenodoxin; (2) MSF was unable to depolymerize the import incompetent pAd in which the basic amino acids in the presequence portion had been mutated; and (3) a synthetic peptide corresponding to the presequence portion of pre-cytochrome P-450(SCC), which is the potent inhibitor of mitochondrial protein import (Furuya et al., 1987), also inhibited the

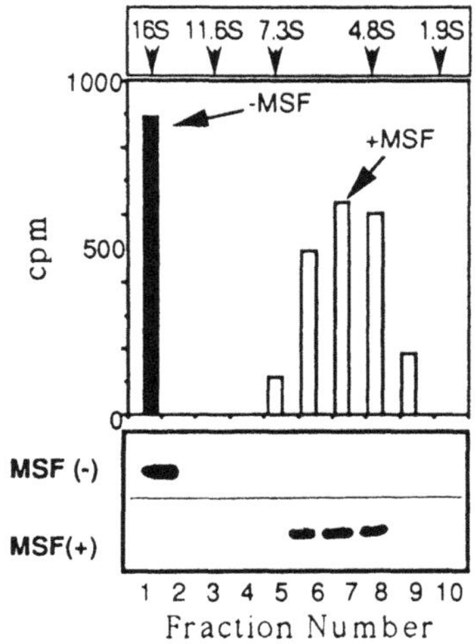

Figure 1. Depolymerization of pAd by MSF as examined by sucrose density gradient centrifugation (Hachiya et al., 1993). pAd synthesized in wheat germ lysate was incubated with (open bars) or without (closed bar) 200 ng of MSF in buffer A at 30°C for 50 min. The reaction mixtures were applied to the top of a sucrose density gradient centrifugation, centrifugated, and fractionated from the bottom. The fractions were TCA precipitated and subjected to SDS-PAGE followed by the counting with Fuji Bioimage Analyzer. The direction of sedimentation is from right to left.

depolymerization activity of MSF; (4) MSF was affinity purified with a fusion protein carrying the presequence of yeast COX IV.

We next examined whether the depolymerization activity of MSF was dependent on ATP-hydrolysis (Hachiya et al., 1993). When pAd translated in the wheat germ system was treated with apyrase prior to the incubation with MSF, no depolymerization of the aggregated pAd was observed. The presence of an excess amount of AMP-PNP also strongly inhibited the depolymerization reaction. These lines of evidence indicated that the depolymerization activity of MSF is dependent on ATP-hydrolysis. Furthermore, in the presence of MSF and ATP, pAd was almost completely degraded by treatment with a low concentration of trypsin. In contrast, only limited proteolysis was observed in the presence of MSF and AMP-PNP, suggesting that the depolymerization was accompanied by a conformational change of the precursor. We thus conclude that MSF modulates the conformational states of mitochondrial precursor proteins via ATP hydrolysis.

D. NEM Inhibits a Function of MSF

As shown in Figure 2, the import stimulating activity of MSF was strongly inhibited by the treatment of MSF with NEM. On the other hand, the unfolding activity of MSF was not affected by the NEM treatment. These results, in conjunc-

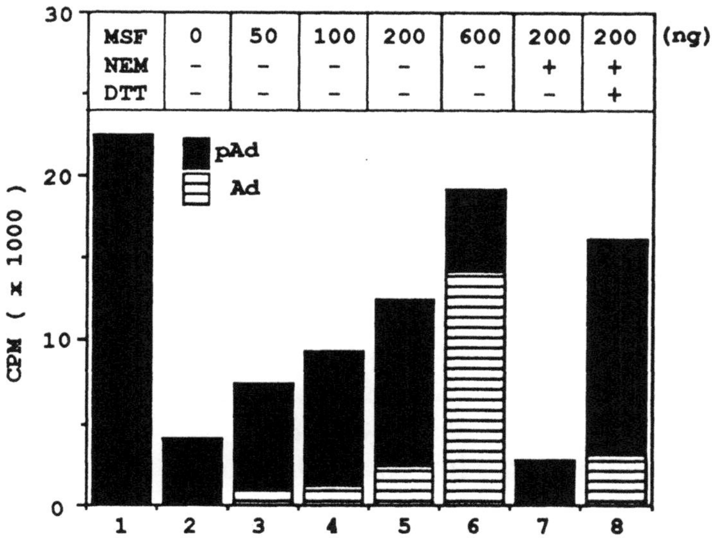

Figure 2. The effect of NEM treatment on the mitochondrial binding- and import-stimulation activities of MSF. pAd synthesized in wheat germ lysate (total amount indicated in column 1) was incubated with mitochondria in the presence of increasing amounts of MSF (Lanes 2–6) or NEM-treated MSF (Lanes 7 and 8). Amounts of pAd bound to the mitochondria (closed bars) and of imported mature Ad (hatched bars) were determined. For NEM treatment, 200 ng of MSF was incubated with 10 mM NEM at 25°C for 20 min followed by the addition of 20 mM of DTT. As a control, MSF was treated with a mixture of 10 mM NEM and 20 mM DTT before incubation with MSF (lane 8).

tion with the data described previously, indicated that NEM inhibited the step(s) following recognition and binding of the precursor proteins by MSF, most likely the dissociation of the precursors from MSF or their targeting to mitochondria.

E. Dual Functions of MSF

To gain further insight into the mechanism of action of MSF, we used *E. coli*-expressed and purified pAd (Iwahashi et al., 1992) as a substrate with which to analyze the function of MSF. When the purified recombinant pAd dissolved in 7 M urea was diluted 50-fold into import buffer and immediately incubated with mitochondria, the precursor was rapidly imported into mitochondria (Figure 3). However, when the diluted solution of the precursor was incubated at 30°C for 30 min or more prior to the addition of mitochondria, no import of the precursor was observed (Figure 3; -MSF). The decrease in the import-competence of pAd seemed to correspond to changes in the conformation from the unfolded state to another, aggregation-prone conformation. The loss of import-competence was prevented

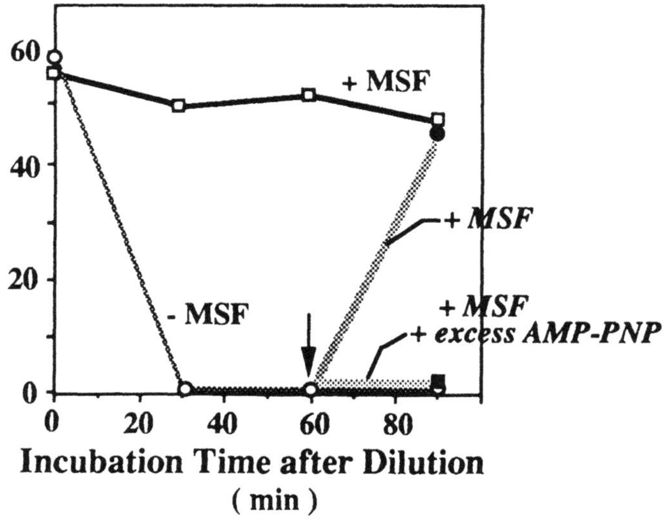

Figure 3. The effect of MSF on the import competence of purified pAd (Hachiya et al., 1994). pAd was expressed and metabolically labeled with [^{35}S] SO4^{--} in *E. coli* cells, solubilized with 7 M urea, and purified to homogeneity. Denatured pAd was diluted 50-fold and incubated at 30°C in the presence (open squares) and absence (open circles) of MSF. The import of pAd was assayed at various time points after the dilution as indicated in the figure. At 60 min after dilution in the absence of MSF, MSF + ATP or MSF + ATP + AMP-PNP were added. The reaction mixtures were incubated for 30 min and then the amounts of imported pAd were measured (expressed in % of total pAD added).

almost completely when MSF was present in the incubation mixture (Figure 3; +MSF).

These observations suggested that MSF interacted with unfolded pAd to form a complex that stabilized the import-competent, unfolded conformation of the precursor. The other characteristic feature of MSF was that, when added to the import-incompetent pAd and incubated in the presence of ATP (arrow in Figure 3), the precursor recovered import-competence to almost the original level. The presence of an excess amount of AMP-PNP prevented the restoration of import-competence, indicating that ATP hydrolysis was required for the reaction.

Complex formation between pAd and MSF was demonstrated by sucrose density gradient centrifugation analysis. pAd was diluted from 7 M urea into buffer in the absence or presence of MSF, incubated at 30°C for 30 min, and analyzed by 5–20% sucrose density gradient centrifugation. As shown in Figure 4, MSF itself was detected in the top fractions, while the urea-denatured pAd diluted in the absence of MSF aggregated and sedimented to the bottom of the tube. In contrast, when pAd was diluted into buffer containing MSF, it was recovered with MSF in the top

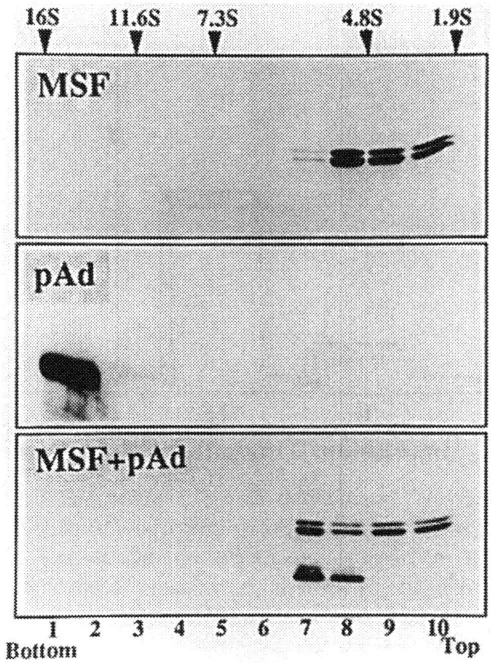

Figure 4. Binding of unfolded pAd to MSF (Hachiya et al., 1994). Reaction mixtures containing either four pmoles of MSF or 20 pmoles of denatured pAd diluted into buffer with or without four pmoles of MSF were incubated at 30°C for 30 min, then placed on a 5–20% sucrose density gradient and centrifuged at 130,000 × g for 12 h at 4°C. After centrifugation, MSF and pAd in the gradient fractions were detected by Western blotting.

fractions, indicating that MSF formed a stable complex with unfolded pAd and prevented the precursor from self-aggregating. Titration experiments showed that MSF was apparently saturated with a 5 molar excess of pAd. We found further that ATP-hydrolysis induced the dissociation of pAd from MSF and all of the pAd present in the mixture sedimented to the bottom of the gradient while the free MSF was recovered in the top fractions.

These results indicated that MSF has dual activities: (1) It recognizes the unfolded mitochondrial precursor and maintains it in an import-competent state, and (2) it recognizes the aggregated import-incompetent precursor and restores its import-competence in an ATP-hydrolysis dependent reaction.

F. Precursor Protein-induced ATPase Activity of MSF

We found that the import-incompetent form of pAd induced the ATPase activity of MSF. The ATPase activity measured was dependent on the folded state of the

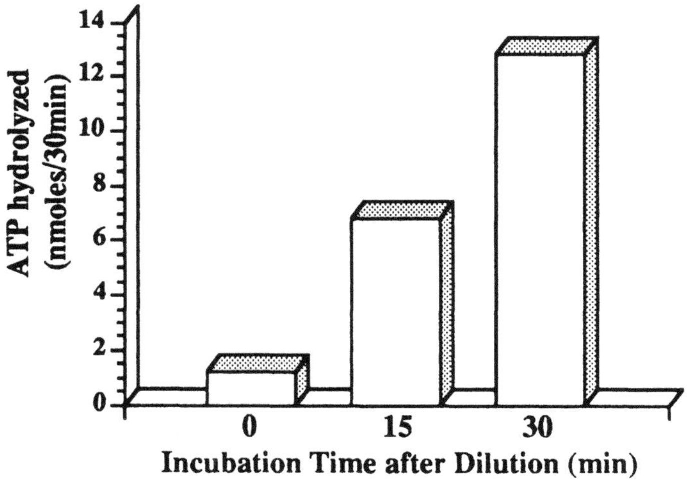

Figure 5. pAd-induced ATPase activity of MSF (Hachiya et al., 1994). One mg of purified pAd in 7 M urea was diluted 50-fold into 100 ml of 100 mM HEPES buffer (pH 7.4) containing 10 mM magnesium acetate and incubated at 30°C. After the times indicated, 1 mg of MSF and 5 mM ATP were added to each of the reactions and incubation at 30°C was continued for 30 min during which the ATPase activity was determined by measuring free phosphate with malachite green reagent.

precursor. The import-competent pAd immediately after dilution from 7 M urea solution had only a low ATPase inducing activity, whereas pAd preparations incubated at 30°C after dilution induced higher ATPase activities in MSF (Figure 5). We then investigated whether the ATPase activity of MSF could also be induced by other proteins that had been unfolded in 7 M urea. Purified proteins with different subcellular localizations were treated with 7 M urea, dialyzed overnight and then analyzed in the MSF ATPase assay (Table 1). It should be noted that porin is the major mitochondrial outer membrane protein. It lacks a cleavable signal sequence (Mihara and Sato, 1985) and is imported via the MAS20/MOM19 pathway (Söllner et al., 1989), whereas apocytochrome *c* is also a protein without a cleavable signal sequence but is imported into the mitochondrial intermembrane space via a unique pathway (Nicholson et al., 1988; Stuart et al., 1990). As shown in the Table 1, mitochondrial precursor proteins, irrespective of the presence or absence of the cleavable signal sequence, induced significant ATPase activity of MSF. However, the ATPase activity of MSF was neither induced by P450(2B1), a type I microsomal membrane protein with an NH_2-terminal signal-anchor sequence (Sato et al., 1990), carboxyesterase E1, a microsomal luminal protein with a modified COOH-terminal KDEL motif (Murakami et al., 1993), peroxisomal catalase, which carries putative internal and COOH-terminal peroxisome-targeting signal (Kragler et al., 1993), or a plant ferredoxin precursor, which carries a 48

Table 1. Mitochondrial Precursor- dependent ATPase Activity of MSF

Proteins	*ATPase Activity (μmole / min / mg)*
pAd	0.70
Porin	0.56
Cytochrome c (apo-form)	0.86
Ad	0.02
Cytochrome c (holo-form)	0.05
P-450(2B1)	<0.01
Carboxyesterase E1	<0.01
Catalase	<0.01
Preferredoxin	<0.01
R-CMLA	<0.01
Casein	<0.01
Rhodanese	<0.01

Note: MSF ATPase activity was assayed in the presence of 2.3 μM effector proteins which had been treated with 7M urea and dialyzed overnight.

residue NH_2-terminal chloroplast-targeting signal peptide (Pilon et al., 1990). Furthermore, no ATPase activity was induced by reduced and carboxymethylated α-lactalbumin (R-CMLA), casein, or rhodanese, proteins which maintain the conformation of stable folding intermediates or refold inefficiently, resulting in aggregate formation (Martin et al., 1991; Langer et al., 1992).

Thus the ATPase activity of MSF seems to be specifically induced by the proteins that can be imported into mitochondria, whereas non-mitochondrial proteins, even when present as folding intermediates or aggregates, do not induce this activity.

G. Presequence-induced ATPase Activity of MSF

We found that chemically synthesized peptides corresponding to mitochondrial targeting sequences induced the ATPase activity of MSF. The NH_2-terminal 19 amino acids of pre-cytochrome P450(SCC), which has a presequence of 39 residues, has been shown to act as the mitochondrial targeting signal (Kumamoto et al., 1989). The synthetic peptide corresponding to this region, SCC1-19, significantly induced the ATPase activity of MSF. On the other hand, the mutated version of the peptide (SCC1-19M) in which three basic amino acid residues were changed to serine or threonine residues, was inactive. It should be noted that SCC1-19, but not SCC1-19M, inhibited protein import into mitochondria *in vitro* (Furuya et al., 1987). Furthermore, gel-filtration HPLC revealed that SCC1-19 bound to MSF and that ATP-hydrolysis induced its release from MSF, whereas SCC1-19M did not bind to MSF.

Taken together, these findings indicate that basic amino acid residues in the mitochondrial targeting signal are critical for recognition by MSF. Synthetic peptides corresponding to the targeting sequences of adrenodoxin and COX IV precursors also significantly activated the ATPase activity. On the other hand, peptides derived from several non-mitochondrial proteins, including a basic peptide having the potential to form an amphophilic α-helix, and peptides containing either nuclear localization or peroxisomal targeting signals were unable to induce the ATPase activity. In conclusion, the primary target of recognition by MSF seems to be the mitochondrial targeting signals of the precursor proteins.

H. Binding of the MSF-Precursor Complex to the Outer Mitochondrial Membrane

The binding of MSF-pAd complex to the isolated mitochondrial outer membrane (OM) was examined by sucrose density gradient centrifugation (Figure 6). Unfolded pAd in 7 M urea was diluted 50-fold into buffer containing MSF at a molar ratio of MSF to pAd of 1:5. After incubation with OM or trypsin-treated OM (tOM), the mixtures were examined by sucrose density gradient centrifugation. OM, MSF, and pAd in fractions of the gradient were visualized by Western blotting. As shown in Figure 6, unfolded pAd and MSF formed a complex that was recovered in the top fractions of the gradient (panel A). In the presence of OM, MSF and pAd cosedimented with OM (GST as a marker) to the bottom fraction, leaving only small amounts of free MSF in the top fractions (panel B). In marked contrast, no binding of the MSF-pAd complex was observed with tOM (panel C) or when the complex was formed with NEM-treated MSF (nMSF; panel D). In a control reaction MSF-pAd complex did not bind to liver microsomes (not shown in Figure 6). Furthermore, in the absence of pAd, only a small amount of MSF was associated with OM (panel E), indicating that the binding of MSF required the presence of unfolded precursor protein. From these results, we concluded that MSF acts to target protein precursors to the mitochondrial outer membrane and NEM blocks this activity (Hachiya et al., 1994).

I. Molecular Properties of MSF Subunits

To characterize the molecular properties of MSF, cDNA clones corresponding to the two subunits of MSF were obtained from a rat liver cDNA library (Alam et al., 1994). The deduced amino acid sequence of the 30 kDa and 32 kDa subunits shows no homology to the known chaperones of the heat-shock protein (hsp) families, but are highly homologous to the 14-3-3 protein family, a group of highly conserved, acidic proteins widely distributed in eukaryotes (Aitken et al., 1992), and whose participation in intracellular protein trafficking had not been described before. The cDNA clones for both subunits of MSF were expressed separately in *E. coli* and

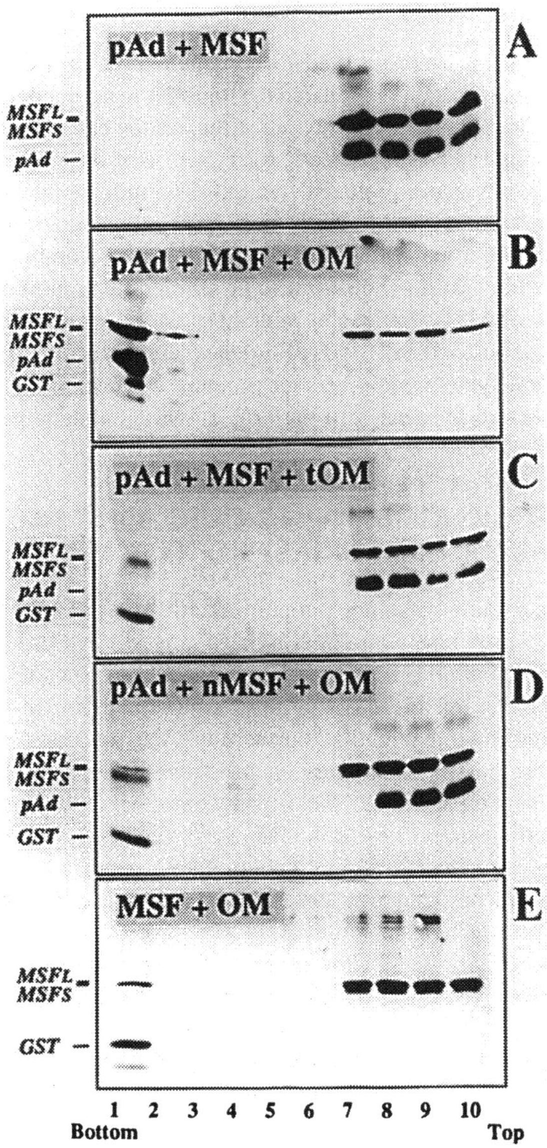

Figure 6. Binding of MSF-pAd complex to mitochondrial outer membranes (OM; Hachiya et al., 1994). Four pmoles of MSF and 20 pmoles of pAd were incubated in a 50 ml reaction at 30°C for 30 min. The mixture was incubated for 30 min at 30°C either without OM (panel A), with 2.5 mg of OM (panel B), or with 2.5 mg of trypsin-treated OM (panel C). MSF-pAd-OM complexes were analyzed by sucrose density gradient centrifugation. Panel D shows the fractionation pattern of the complex between pAd and NEM-treated MSF (10 mM NEM) that had been incubated with 2.5 mg of OM. Panel E represents the fractionation pattern of MSF alone in the presence of OM. All reactions were subjected to SDS-PAGE and Western blotting.

the expressed proteins were purified to homogeneity (Alam et al., 1994). The purified recombinant proteins stimulated the import into mitochondria of precursor proteins synthesized in wheat germ lysate, although the extent of stimulation was lower than that observed with authentic MSF purified from rat liver cytosol. The recombinant subunits also recognized urea-unfolded mitochondrial precursor proteins and formed complexes with them. However, precursor-induced ATPase activity was not observed with either the individual or combined recombinant subunits, suggesting that some post-translational modification(s) may be required for full activity of MSF. The rabbit antibodies raised against the recombinant subunits inhibited both cytosol- and MSF-dependent import into mitochondria of wheat germ lysate-synthesized precursor proteins, confirming the identity of the cloned and expressed proteins with the original MSF subunits purified from rat liver cytosol.

V. PRESEQUENCE-BINDING CYTOSOLIC FACTORS IN MITOCHONDRIAL PROTEIN IMPORT

Three presequence-binding factors, "targeting factor" (Ono and Tuboi, 1990a), PBF (Murakami and Mori, 1990), and MSF (Hachiya et al., 1993), have so far been purified from either rabbit reticulocyte lysate or rat liver cytosol. All of the three factors were detected on the basis of their requirement for *in vitro* import of precursor proteins into mitochondria. Furthermore, they seem to associate with the precursors by recognizing their presequences and maintaining the proteins in import-competent conformations. The majority of mitochondrial precursors seems to require the participation of cytosolic factors for their *in vitro* import. However, certain precursor proteins (Eilers and Schatz, 1986) or relatively short synthetic peptides containing a mitochondrial targeting signal (Pak and Weiner, 1990; Furuya et al., 1991) appear to be exceptions. A 40 kDa protein from yeast cytosol was reported to be needed for the *in vitro* import of some precursor proteins (Ohta and Schatz, 1984). However, it was not clear whether the factor recognized the presequence.

Of the three presequence-binding factors, only MSF has been characterized at the molecular level. The functional properties of "targeting factor" purified from rabbit reticulocyte lysate are similar to those of MSF, whereas PBF is apparently distinct from MSF in many respects. Further characterization of "targeting factor" and PBF is needed to clarify the structural and functional relationships among the three presequence-binding factors.

In addition to their molecular chaperone activity, stabilizing precursor proteins in import-competent conformations, both "targeting factor" and PBF are proposed to have a mitochondrial targeting function, although on the basis of indirect evidence (Ono and Tuboi, 1990b; Murakami and Mori, 1990). More direct evidence for a function in targeting has been presented for MSF. The import-incompetent form of pAd induced the ATPase activity of MSF concomitant with the conversion of the precursor protein to the MSF-bound, import-competent form, and the MSF-pAd complex was found to bind to mitochondrial outer membrane (OM). Either the NEM treatment of MSF or the trypsin treatment of OM abolished the binding of MSF-pAd complex to OM (Hachiya et al., 1994). The latter observation agrees with the known inhibition of the import-stimulating activity of MSF by NEM, and indicates that NEM blocks the targeting activity of MSF without affecting its ATP-dependent molecular chaperone function on mitochondrial precursor proteins.

The apparent mitochondrial targeting activity of MSF is in support of the hypothesis advanced by Lithgow and colleagues (1993). These authors proposed that the role of cytosolic factors in mitochondrial protein import is not restricted to their action as molecular chaperones, but may also include a function in preventing the mistargeting of precursors to other cell organelles by binding to specific receptors on the surface of the target organelle. The proposed multifunctional role of the cytosolic factors in mitochondrial protein import may be compared with the role of the signal recognition particle in targeting nascent secretory proteins to the membrane of the endoplasmic reticulum.

One important question that remains to be addressed is whether the cytosolic presequence-binding factors actually function in mitochondrial protein import *in vivo*. All three factors described so far were detected on the basis of their import-stimulating activity in *in vitro* import assays. In contrast, the role of hsp70 in intracellular protein sorting is supported by genetic studies in yeast as well as by biochemical experiments *in vitro*. The components of MSF have been identified as members of the 14-3-3 protein family. Since yeast also contain 14-3-3 proteins (van Heusden et al., 1992), whose function is at present unknown, a genetic analysis of these proteins in yeast may be helpful in establishing the role of the cytosolic presequence-binding factors *in vivo*.

REFERENCES

Aitken, A., Collinge, D. B., van Heusden, B. P. H., Isobe, T., Roseboom, P. H., Rosenfeld, G., & Soll, J. (1992). 14-3-3 proteins: A highly conserved, widespread family of eukaryotic proteins. Trends Biochem. Sci. 12, 498–501.

Alam, R., Hachiya, N., Sakaguchi, M., Kawabata, S., Iwanaga, S., Kitajima, M., Mihara, K., & Omura, T. (1994). cDNA cloning and characterization of mitochondrial import stimulation factor (MSF) purified from rat liver cytosol. J. Biochem. 116, 416–425.

Argan, C., Lusty, C. J., & Shore, G. C. (1983). Membrane and cytosolic components affecting transport of the precursor for ornithine carbamyltransferase into mitochondria. J. Biol. Chem. 258, 6667–6670.

Chirico, W. J, Waters, M. G., & Blobel, G. (1988). 70K heat shock related proteins stimulate protein translocation into microsomes. Nature 332, 805–810.

Deshaies, R. J., Koch, B. D., Werner-Washburne, M., Craig, A. E., & Schekman, R. (1988). A subfamily of stress proteins facilitates translocation of secretory and mitochondrial precursor polypeptides. Nature 332, 800–805.

Eilers, M., & Schatz, G. (1986). Binding of a specific ligand inhibits import of a purified precursor protein into mitochondria. Nature 322, 228–232.

Furuya, S., Okada, M., Ito, A., Aoyagi, H., Kanmera, T., Kato. T., Sagara, Y., Horiuchi, T., & Omura, T. (1987). Synthetic partial extension peptides of P-450(SCC) and adrenodoxin precursors: Effects on the import of mitochondrial enzyme precursors. J. Biochem. 102, 821–832.

Furuya, S., Mihara, K., Aimoto, S., & Omura, T. (1991). Cytosolic and mitochondrial surface factor-independent import of a synthetic peptide into mitochondria. EMBO J. 10, 1759–1766.

Hachiya, N., Alam, R., Sakasegawa, Y., Sakaguchi, M., Mihara, K., & Omura, T. (1993). A mitochondrial import factor purified from rat liver cytosol is an ATP-dependent conformational modulator for precursor proteins. EMBO J. 12, 1579–1586.

Hachiya, N., Komiya, T., Alam, R., Iwahashi, J., Sakaguchi, M., Omura, T., & Mihara, K. (1994). MSF, a novel cytoplasmic chaperone that functions in precursor targeting to mitochondria. EMBO J. 13, 5146–5154.

Iwahashi, J., Furuya, S., Mihara, K., & Omura, T. (1992). Characterization of adrenodoxin precursor expressed in Escherichia Coli. J. Biochem. 111, 451–455.

Kragler, F., Langeder, A., Raupachova, J., Binder, M., & Hartig, A. (1993). Two independent peroxisomal targeting signals in catalase A of Saccharomyces cerevisiae. J. Cell Biol. 120, 665–673.

Kumamoto, T., Ito, A., & Omura, T. (1989). Critical region in the extension peptide for the import of cytochrome P-450(SCC) precursor into mitochondria. J. Biochem. 105, 72–78.

Langer, T., Lu, C., Echols, H., Flanagan, J., Hayer, M. K., & Hartl, F-U. (1992). Successive action of DnaK, DnaJ, and GroEL along the pathway of chaperone-mediated protein folding. Nature 356, 683–689.

Lithgow, T., Hoj, P. B., & Hoogenraad, N. J. (1993). Hypothesis: Do cytosolic factors prevent promiscuity at the membrane surface? FEBS Lett. 329, 1–4.

Martin, J., Langer, T., Boteva, R., Schremel, A., Horwich, A. L., & Hartl, F-U. (1991). Chaperonin-mediated protein folding at the surface of groEL through a "molten globule"-like intermediate. Nature 352, 36–42.

Mihara, K., & Sato, R. (1985). Molecular cloning and sequencing of cDNA for yeast porin, an outer mitochondrial membrane protein: A search for targeting signal in the primary structure. EMBO J. 4, 769–774.

Miura, S., Mori, M., & Tatibana, M. (1983). Transport of ornithine carbamoyltransferase precursor into mitochondria: Stimulation by potassium ion, magnesium ion, and a reticulocyte cytosolic protein(s). J. Biol. Chem. 258, 6671–6674.

Mori, M., Matsue, H., Miura, S., Tatibana, M., & Hashimoto, T. (1985). Transport of proteins into mitochondrial matrix: Evidence suggesting a common pathway for 3-ketoacyl-CoA thiolase and enzymes having presequences. Eur. J. Biochem. 149, 181–186.

Murakami, H., Pain, D., & Blobel, G. (1988a). 70-kD Heat shock-related protein is one of at least two distinct cytosolic factors stimulating protein import into mitochondria. J. Cell Biol. 107, 2051–2057.

Murakami, K., Amaya, Y., Takiguchi, M., Ebina, Y., & Mori, M. (1988b). Reconstitution of mitochondrial protein transport with purified ornithine carbamoyltransferase precursor expressed in Escherichia coli. J. Biol. Chem. 263, 18437–18442.

Murakami, K., & Mori, M. (1990). Purified presequence binding factor (PBF) forms an import-competent complex with a purified mitochondrial precursor protein. EMBO J. 9, 3201–3208.

Murakami, K., Tanase, S., Morino, Y., & Mori, M. (1992). Presequence binding factor-dependent and -independent import of proteins into mitochondria. J. Biol. Chem., 257, 13119–13122.

Murakami, K., Takagi, Y., Mihara, K., & Omura, T. (1993). An isozyme of microsomal carboxyesterases, carboxyesterase Sec, is secreted from rat liver into the blood. J. Biochem. 113, 61–66.

Nicholson, D. W., Hergersberg, C., & Neupert, W. (1988). Role of cytochrome c heme lyase in the import of cytochrom c into mitochondria. J. Biol. Chem. 263, 19034–19042.

Ohta, S., & Schatz, G. (1984). A purified precursor polypeptide requires a cytosolic protein fraction for import into mitochondria. EMBO J. 3, 651–657.

Ono, H., & Tuboi, S. (1988). The cytosolic factor required for import of precursors of mitochondrial protein into mitochondria. J. Biol. Chem. 263, 3188–3193.

Ono, H., & Tuboi, S. (1990a). Presence of the cytosolic factor stimulating the import of precursor of mitochondrial proteins in rabbit reticulocytes and rat liver cells. Arch. Biochem. Biophys. 277, 368–373.

Ono, H., & Tuboi, S. (1990b). Purification and identification of a cytosolic factor required for import of precursors of mitochondrial proteins into mitochondria. Arch. Biochem. Biophys. 280, 299–304.

Pak, Y. K., & Weiner, H. (1990). Import of chemically synthesized signal peptides into rat liver mitochondria. J. Biol. Chem. 265, 14298–14307.

Pilon, M., Douwe de Boer, A., Luc Knols, S., Koppelman, M. H. G. M., van der Graaf, R. M., de Kruijff, B., & Weisbeek, P. J. (1990). Expression in *Escherichia coli* and purification of a translocation-competent precursor of the chloroplast protein ferredoxin. J. Biol. Chem. 265, 3358–3361.

Sato, T., Sakaguchi, M., Mihara, K., & Omura, T. (1990). The amino-terminal structures that determine topological orientation of cytochrome P-450 in microsomal membrane. EMBO. J. 9, 2391–2397.

Söllner, T., Griffiths, G., Pfanner, N., & Neupert, W. (1989). MOM19, an import receptor for mitochondrial precursor proteins. Cell 59, 1061–1070.

Stuart, R. A., Nicholson, D. W., & Neupert, W. (1990). Early steps in mitochondrial protein import: Receptor functions can be substituted by the membrane insertion activity of apocytochrome c. Cell 60, 31–43.

van Heusden, G. P. H., Wenzel, T. J., Lagendijk, E. L., de Steensma, H. Y., & van den Berg, J. A. (1992). Characterization of the yeast BMH1 gene encoding a putative protein homologous to mammalian protein kinase II activators and protein kinase C inhibitors. FEBS Lett. 302, 145–150.

MOLECULAR MECHANISMS OF PROTEIN TRANSLOCATION INTO AND ACROSS THE MITOCHONDRIAL OUTER MEMBRANE

Roland Lill, Andreas Mayer, Harald Steiner, Gyula Kispal, and Walter Neupert

I.	Introduction	52
II.	The General Mechanism of Outer Membrane Translocation	52
III.	How Inner Membrane and Matrix Proteins Pass the Outer Membrane	55
IV.	Insertion and Assembly of Outer Membrane Proteins	58
V.	The Import Pathway of Heme Lyases	60
VI.	Lessons Learned from Import Reactions with Outer Membrane Vesicles	62
VI.	Perspectives	66
	Acknowledgments	66
	References	66

Advances in Molecular and Cell Biology,
Volume 17, pages 51–70.

ISBN: 0-7623-0144-9

I. INTRODUCTION

Protein targeting and translocation into mitochondria is a complex process that involves numerous components in both the outer and inner mitochondrial membranes and in the mitochondrial matrix space. Detailed investigations from a number of laboratories have identified several distinct pathways by which various precursor proteins are transported into their final sub-mitochondrial locations (see Pfanner and Neupert, 1990; Glick and Schatz, 1991; and Segui-Real et al., 1993b for recent reviews). Most proteins, in particular those destined for the matrix space, traverse the outer and inner membranes simultaneously. In these cases, translocation takes place at the so-called "translocation contact sites," where both membranes are in close opposition (Pfanner et al., 1992; Glick et al., 1991). Translocation into and across the inner membrane strictly depends on an electrochemical potential across the membrane (Schleyer and Neupert, 1982; Gasser et al., 1982), and usually requires ATP as an energy source (Chen and Douglas, 1987; Pfanner et al., 1987; Wachter et al., 1992; Stuart et al., 1994). Since, under conditions of a deenergized inner membrane, preproteins also do not pass across the outer membrane, it was thought that the main function of the outer membrane translocation machinery was to specifically recognize and select cognate mitochondrial precursor proteins and to present them through a passive pore to the translocation apparatus of the inner membrane. This translocation apparatus would then actively accomplish passage across both membranes. Accordingly, preproteins were expected to become translocated across a continuous, sealed channel spanning both membranes. Recently, however, it became clear that both the outer membrane (Mayer et al., 1993) and the inner membrane (Hwang et al., 1989) contain individual protein translocation machineries each of which can transport preproteins. In addition, it could be demonstrated that matrix-targeted proteins are able to traverse both membranes in consecutive steps demonstrating the independence of the two machineries (Segui-Real et al., 1993a). In this review, we will summarize the current knowledge on the molecular mechanisms underlying protein insertion into and translocation across the outer membrane.

II. THE GENERAL MECHANISM OF OUTER MEMBRANE TRANSLOCATION

After their synthesis in the cytosol, mitochondrial preproteins are believed to be maintained in a translocation competent state by cytosolic factors, for example, by members of the Hsp70 protein family (Deshaies et al., 1988) and by presequence binding factors (Murakami and Mori, 1990; Hachiya et al., 1993). Recognition and translocation across the outer membrane is mediated by a number of proteins organized in a large complex, the so-called mitochondrial receptor complex (Kiebler et al., 1990; Söllner et al., 1992). Initial interaction with the mitochondrial outer membrane is mediated by protease-sensitive surface components of this complex.

N. crassa MOM72 and its *S. cerevisiae* homolog Mas70p were identified by virtue of their important function in the import of the ADP/ATP carrier (AAC), an integral protein of the inner membrane (Söllner et al., 1990; Steger et al., 1990; Hines et al., 1990). Recently, phosphate carrier was also found to predominantly use Mas70p during its import into the mitochondrial inner membrane (Dietmeier et al., 1993). Treatment of mitochondria with various proteases yields two fragments of MOM72/Mas70p, a small membrane-bound fragment and a large soluble fragment (Figure 1). The latter domain was recently shown to contain specific preprotein binding capacity for AAC and phosphate carrier, but not for other preproteins (Schlossmann et al., 1994).

In contrast to the highly selective function of MOM72/Mas70p, *N. crassa* MOM19 (Söllner et al., 1989) and its recently identified *S. cerevisiae* homolog Mas20p (Ramage et al., 1993; Moczko et al., 1993) act as general receptors that are involved in the import of the majority of mitochondrial preproteins. Both receptors are anchored through their hydrophobic N-termini to the outer membrane and possess a hydrophilic C-terminal domain that can be proteolytically digested (Figure 1; Schneider et al., 1991). Both biochemical and genetic information demonstrate that *N. crassa* MOM19 is an important component for the biogenesis of mitochondria. Proteolytic removal of MOM19 or binding of specific antibodies inhibits protein import of most preproteins (Söllner et al., 1989, 1990; Lill et al.,

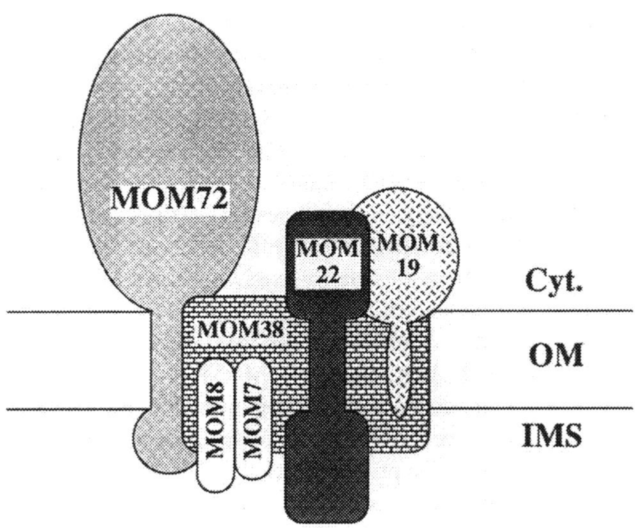

Figure 1. Components of the mitochondrial receptor complex from *Neurospora crassa*. The figure gives the topographical orientation of the mitochondrial outer membrane (MOM) proteins. It should be noted though that most of the interactions between the individual components are not yet known. Cyt., cytosol; OM, outer membrane; IMS, intermembrane space.

1992b; Mayer et al., 1993). A MOM19-deficient mutant displayed a dramatic reduction of growth (Harkness et al., 1994a) yielding a "stop-start" phenotype that is characterized by alternating periods of growth arrest and slow growth (Bertrand and Pittenger, 1972). MOM19-depleted cells contain mitochondria which are largely deficient in cristae membranes, cytochromes, and functional protein synthesis (Harkness et al., 1994b). *In vitro* import of most precursor proteins into mutant mitochondria is drastically reduced, whereas proteins that can use MOM72 as a receptor (like AAC) or that are imported without the participation of protease-sensitive components (like MOM19 itself or apocytochrome *c*; see Lill et al., 1992a for review) are unaffected in their import (Harkness et al., 1994b). Similarly, deletion of the *MAS20* gene leads to a *pet* phenotype (defined as no growth on nonfermentable carbon-sources (Tzagoloff and Dieckmann, 1990)), that is, Mas20p is essential for mitochondrial respiration (Ramage et al., 1993). Comparison to the rather unconspicuous phenotype of a *mas70* deletion suggests that also in yeast this receptor displays a much broader preprotein specificity. These observations demonstrate the important role of MOM19/Mas20p for the well-being of the cell in general and for the biogenesis of mitochondria in particular.

After initial interaction with the receptors, preproteins become inserted into the outer membrane, a step which in the case of AAC is thought to be mediated by MOM22, a component of the receptor complex that so far has only been identified in *N. crassa* (Kiebler et al., 1993). MOM22 has a membrane spanning region in the central part of the protein with the N-terminus facing the cytosol and the C-terminus reaching into the intermembrane space (Figure 1). Genetic and biochemical evidence suggests a functional cooperation of MOM22 with MOM19 during protein import (Harkness et al., 1994b). Upon transfer across a putative translocation channel, preproteins reach a site that seems to be passed by many preproteins, and thus was termed "general insertion pore" (GIP; Pfaller et al., 1988). Transfer of preproteins to the GIP stage is accompanied by acquisition of a moderate resistance to proteolytic attack. Most likely, in this stage parts of the traversing preprotein are already exposed to the intermembrane space. In the case of AAC, a simultaneous interaction with the inner membrane seems possible, since GIP can only be reached in mitochondria, but not in isolated outer membrane vesicles (Mayer et al., 1993). Crosslinking studies suggest that MOM38, MOM8, and MOM7 comprise building blocks of GIP (Söllner et al., 1992; Figure 1). MOM38 was identified as a member of the receptor complex by co-immunoprecipitation with anti-MOM19 antibodies (Kiebler et al., 1990). In *S. cerevisiae* its homolog, Isp42p, represents an essential protein (Baker et al., 1990). The precise function of MOM38/Isp42p not yet clear. However, both proteins can be crosslinked to translocation intermediates spanning the outer membrane (Vestweber et al., 1989; Söllner et al., 1992). This indicates a direct interaction with preproteins in transit, making MOM38/Isp42p a likely candidate for a pore-forming component. *N. crassa* MOM7 and MOM8 were also found by crosslinking to a preprotein arrested in a trans-membrane orientation (Söllner et al., 1992), and may be homologous to Isp6p, which was identified as the gene product of a high copy suppressor of an *isp*42 temperature-sensitive (*isp*42^{ts}) mutant (Kassenbrock et al., 1993). A direct interaction of Isp42p and Isp6p is also suggested by co-immunoprecipitation experiments. How preproteins are then released from the translocation channel and passed on to become either inserted into the outer membrane, released into the intermembrane space, or transferred to the inner membrane translocation machinery, is not yet clear. With the development of a translocation system for isolated outer membranes (Mayer et al., 1993) and after the identification of the first components of the inner mitochondrial membrane (Maarse et al., 1992; Scherer et al., 1992; Dekker et al., 1993; Emtage and Jensen, 1993), such questions can now be addressed directly.

III. HOW INNER MEMBRANE AND MATRIX PROTEINS PASS THE OUTER MEMBRANE

It is well accepted that proteins of the inner membrane and the matrix space, which usually carry an amphipathic signal sequence at their N-terminus (Heijne, 1986), pass the outer and inner membranes simultaneously at the translocation contact sites. At these locations, both membranes are in close opposition leaving almost no empty space between the individual membranes (Schleyer and Neupert, 1985; Rassow et al., 1989; Pfanner et al., 1992; Van der Klei et al., 1994). This topographical arrangement of the two membranes allows the formulation of two extreme models for the translocation process. First, translocation could be achieved by one translocation activity connecting both membranes and forming a continuous, sealed translocation channel reaching from the cytosolic side of the outer membrane to the matrix side of the inner membrane. In a second model, both membranes are predicted to contain their own, independent translocation machineries, which, during import into the matrix space, coordinate their activities in both time and space. Recently, several lines of evidence have accumulated that strongly support the second model. The existence of two individual translocation activities in both the outer and inner membranes was demonstrated in import studies using either purified outer membrane vesicles or mitoplasts, respectively. The functional independence of these two machineries during translocation into the matrix space was suggested by the identification of novel translocation intermediates and by the use of fusion proteins. These studies allowed the dissection of the translocation reaction into a number of steps, and will be discussed in some detail in the text following.

Direct experimental evidence for a translocation activity in the mitochondrial outer membrane was obtained using purified outer membrane vesicles (Mayer et al., 1993; for details see following). Studies with mitoplasts, that is, mitochondria with a disrupted outer membrane, demonstrated that preproteins can directly become targeted to and translocated across the inner membrane (Ohba and Schatz, 1987; Hwang et al., 1989). Since defective translocation into protease-pretreated mitochondria was restored after conversion into mitoplasts, it was concluded that preproteins can directly engage contact with the inner membrane without an obligatory need for passage across the outer membrane. Import into both isolated outer membrane vesicles and mitoplasts resembled the requirements for translocation into intact mitochondria with respect to the energy sources used for import or the presequence characteristics. This suggests the existence of individual preprotein recognition and translocation components in both machineries, and implies that translocation across the two membranes in principle may occur in two consecutive steps.

This expectation was first verified in import experiments employing low matrix ATP concentrations (Hwang et al., 1991; Jascur et al., 1992; Rassow and Pfanner, 1991). Under these conditions, preproteins do not translocate across the inner membrane despite the presence of a membrane potential, but can slowly complete translocation across the outer membrane. This leads to a translocation intermediate termed "ATP depletion intermediate" that is resistant to attack by externally added protease (cf. Figure 2), but becomes accessible after the intermembrane space is selectively opened by hypotonic swelling or digitonin lysis of the outer membrane. First, these observations argue against a sealed and continuous translocation pore spanning both membranes. Second, the fact that the remainder of the preprotein can finish translocation across the inner membrane after restoration of matrix ATP demonstrates that translocation across the two membranes does not have to be coupled and rather can occur in consecutive steps. However, as the ATP depletion intermediate is associated with the inner membrane via a potential-dependent interaction of the signal sequence, these experiments do not address the question of whether in intact mitochondria the two translocation reactions may occur independently of each other. Alternatively, the observed translocation across the outer membrane could require preproteins to be connected to the inner membrane and transit across the outer membrane could be achieved by the relative separation of the two translocation channels by diffusion.

By using a fusion protein between the presequence of the β-subunit of F_1-ATPase and cytochrome *c* heme lyase (CCHL), the translocation reaction into the matrix could be dissected into two both independent and consecutive steps (Segui-Real et al., 1993a). After translocation into the inter-membrane space along the CCHL pathway in the absence of a membrane potential (see following), this fusion protein can be further transported into the matrix space after re-energizing the inner membrane. Thus, translocation does not necessarily have to occur simultaneously across both membranes, and the two translocation machineries may operate independently of each other during the translocation process. In the physiological situation, however, the rates of translocation across the individual membranes may be well balanced so that the process appears to occur in a coupled manner.

The current view of how the translocation machineries of the outer and inner membrane cooperate is illustrated in Figure 2. The initial binding of the precursor and the translocation of the signal sequence across the outer membrane (see below) occurs at the receptor complex, which at this stage is not necessarily in contact with the translocation machinery of the inner membrane. Interaction of the latter with

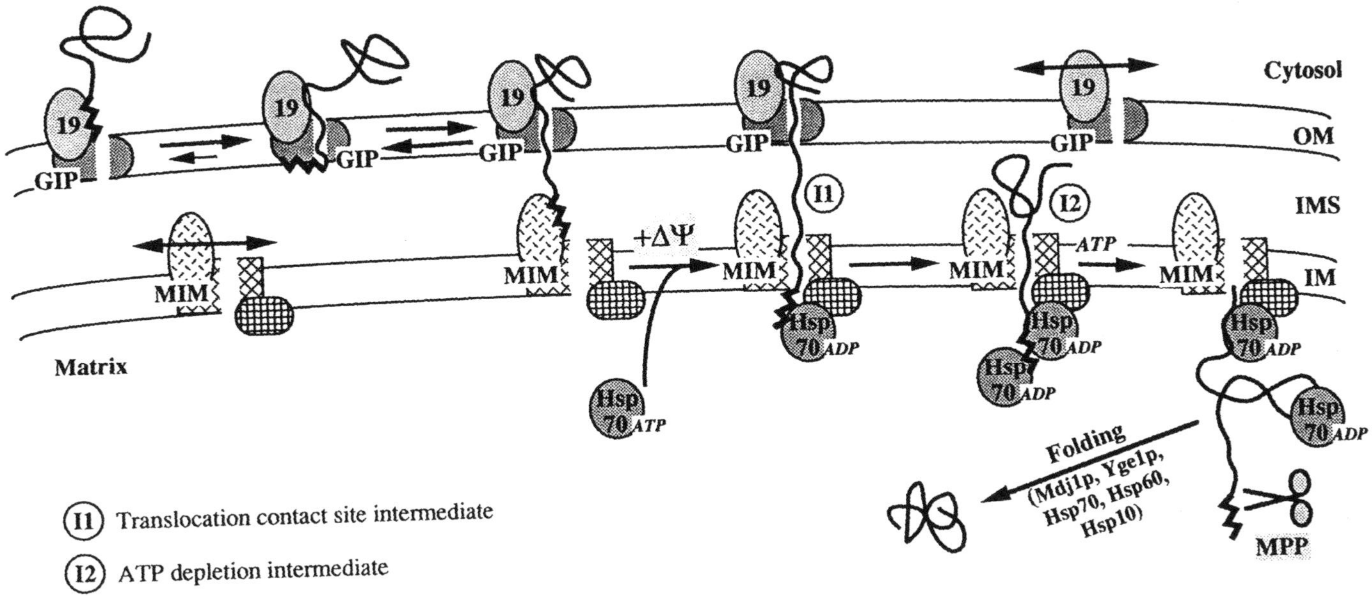

Figure 2. Dynamic cooperation between the translocation machineries of the outer and inner mitochondrial membranes. For details see text. The bold zig-zag portion of the preprotein denotes the N-terminal signal sequence. Abbreviations used: 19, MOM19; GIP, general insertion pore; OM, outer membrane; IM, inner membrane; IMS, intermembrane space; MIM, translocation machinery of the mitochondrial inner membrane; MPP, matrix processing peptidase; Hsp10 (Rospert et al., 1993), Hsp60 (Ostermann et al., 1989), and Hsp70 (Kang et al., 1990), heat shock proteins of 10, 60, 70 kDa; Yge1p (Ikeda et al., 1994), yeast GrpE homolog; Mdj1p (Rowley et al., 1994), mitochondrial DnaJ homolog.

the incoming preprotein, presumably with its signal sequence, becomes possible after the two machineries have moved towards each other by lateral diffusion. Details of this reaction are not yet known. Then, the presequence is passed across the inner membrane in a potential-dependent fashion (Schleyer and Neupert, 1985), thereby allowing Hsp70 to bind to the presequence (Ungermann, unpublished). In the presence of low ATP concentrations, this "translocation contact site intermediate" may be slowly chased across the outer membrane to yield the "ATP depletion intermediate" described previously. When the ATP concentration is sufficiently high, transport into the matrix is resumed, and completed by the cleavage of the signal sequence by the matrix processing peptidase, MPP (Arretz et al., 1991). Finally, the imported protein is folded and assembled into oligomeric complexes with the help of a number of chaperone proteins (Figure 2). The normally observed synchrony of translocation across both membranes raises the interesting, yet unresolved question, of whether the two machineries are reversibly linked either by direct interaction or indirectly through a connecting intermembrane space component, or whether simultaneous translocation is simply explained by laterally diffusing translocation machineries and the kinetic properties of the individual transfer steps.

IV. INSERTION AND ASSEMBLY OF OUTER MEMBRANE PROTEINS

Import into mitochondria is routinely measured by the protection of the translocated, processed protein against externally added protease. In the case of outer membrane proteins, this assay obviously is not suitable. As a method for following the import of most outer membrane proteins, the folding of the polypeptide chain into a protease-resistant conformation has proved to be a reliable test. For instance, the precursor of porin is degraded by low concentrations of protease, whereas membrane-integrated porin is highly resistant to digestion (Kleene et al., 1987). For MOM19, the generation of a characteristic pattern of elastase fragments has been used to track import (Schneider et al., 1991). In other cases, for example for MOM22, import is measured by the generation of a C-terminal proteolytic fragment of 12 kDa corresponding to segments of the protein inside or beyond the outer membrane (Keil and Pfanner, 1993). A further method that may be used to follow membrane insertion is the acquired resistance to alkaline extraction (McBride et al., 1992).

How are outer membrane proteins inserted into the lipid bilayer and assembled into functional complexes? Questions of immediate interest focus on the nature of targeting signals in outer membrane preproteins, the energetics of the process, and the factors assisting in insertion and assembly. The lack of N-terminal, cleavable signal sequences in outer membrane proteins implies that the recognition mechanism by mitochondrial components may be different from that of other preproteins. The outer membrane targeting information has been deciphered and characterized in some detail only for Mas70p (reviewed in Shore et al., 1992). The N-terminus of yeast Mas70p (but not of *N. crassa* MOM72) is similar to a (short) mitochondrial signal sequence and, with low efficiency, directs attached fusion proteins into the matrix space or the inner membrane (Hurt et al., 1985a; 1985b). In case of Mas70p, transfer into the matrix is stopped by a signal-anchor sequence (McBride et al., 1992). For all other outer membrane proteins, however, the nature of the targeting signal remains unknown.

Even less is known about the energy requirements for complete import of outer membrane proteins. ATP is needed for efficient import of several outer membrane proteins (Kleene et al., 1987; Li and Shore, 1992) which most likely is used for the dissociation of chaperones bound to the hydrophobic patches of proteins rather than as an energy source to trigger membrane insertion. The latter reaction by itself may provide enough energy to drive transport into the outer membrane. Even though firm evidence is lacking, it is obvious that the integration of proteins into oligomeric complexes, that is, the specific interaction between two neighboring proteins, may represent a considerable energetic contribution. In that respect, it is important to recall that newly imported proteins can only be placed into empty spaces of partially assembled complexes. The filling of these "holes" by specific complex partners of course is energetically most favorable.

With respect to the components required for import of outer membrane proteins, only the involvement of surface receptors has been thoroughly studied so far. As in the general import pathway, MOM19 participates in the import of most outer membrane proteins studied to date (Söllner et al., 1989, 1990; Mayer et al., 1993). Insertion of MOM38 and MOM22 is additionally affected by blocking MOM72 (Keil et al., 1993; Keil and Pfanner, 1993), a mechanism that has been interpreted to guarantee the high fidelity needed for correct targeting of components that are by themselves involved in the import of many other components. An alternative interpretation might be that for components of the receptor complex the double dependency closely reflects the interactions in the receptor complex.

Nothing is known by which mechanisms preproteins, after initial interaction with surface receptors, are further transported and finally inserted into the membrane. Three possibilities could be envisioned as to how membrane insertion or assembly into a complex might occur. First, preproteins after initial recognition by, for example, MOM19 could insert from the outside into the bilayer thus never really passing through the membrane. Second, insertion could be brought about by lateral release of the preproteins from the translocation channel into the lipid bilayer. This model creates the considerable problem that the membrane spanning regions of outer membrane proteins specifically have to be distinguished from those of the inner membrane. Also, the mechanism would require either a translocation pore that is open to the lipid phase or a tunnel that can dynamically open and close towards the lipid bilayer. Third, translocation could proceed along the general pathway to the inner face of the outer membrane, and insertion would take place from the inside. Detailed biochemical experiments are needed to solve this interesting mechanistic question. Certainly, the topic is the more attractive as it represents a general problem for all membrane translocation systems.

Are components of the outer membrane sufficient to accomplish membrane insertion, or do factors of the intermembrane space or the inner membrane participate in the biogenesis of the outer membrane? Earlier work in intact mitochondria had suggested that after the initial binding to the mitochondrial surface porin precursor associates with the inner membrane before it finally becomes assembled into the outer membrane (Ono and Tuboi, 1987). In this intermediate stage, binding occurs to a membrane fraction containing both outer and inner membranes. Therefore, it was suggested that contact sites may have an important function in outer membrane protein translocation. A recent investigation, however, demonstrated that the outer membrane itself can insert and translocate a number of preproteins (Mayer et al., 1993; see following). The efficiency of import in comparison to intact mitochondria was virtually unchanged, suggesting that direct import into the outer membrane is the default pathway. Peripheral membrane components or soluble proteins of the intermembrane space are unlikely to participate in the translocation reaction as the translocation efficiencies are not decreased even after alkaline extraction of the outer membrane vesicles (Mayer, unpublished).

V. THE IMPORT PATHWAY OF HEME LYASES

Proteins of the intermembrane space reach their functional compartment by two fundamentally different pathways. In one mechanism, proteins like cytochromes c_1 or b_2 use the transport machineries of the inner membrane to get across the outer membrane (Koll et al., 1992; Glick et al., 1992). Another route is characterized by direct translocation across the outer membrane into the intermembrane space without the involvement of the inner membrane. Cytochrome *c* heme lyase (CCHL) from *N. crassa* (Drygas et al., 1989) was the first example following this latter pathway (Lill et al., 1992b; Figure 3). Meanwhile, cytochrome c_1 heme lyase (CC_1HL; Zollner et al., 1992) and CCHL from yeast (Dumont et al., 1987) have been shown to follow the same pathway (H. Steiner, unpublished data). Heme lyases catalyse the covalent attachment of heme to *c*-type cytochromes (Gonzales and Neupert, 1990; Stuart and Neupert, 1990), and are localized at the outer face of the inner membrane (van der Klei and Lill, unpublished). The primary structures of heme lyases are exceptional in that they do not contain N-terminal, cleavable signal sequences. Despite this property, they use components of the general import pathway to reach the intermembrane space. For instance, CCHL from *N. crassa* needs functional MOM19 for its import and passes through GIP (Figure 3; Lill et al., 1992b). Yet, the import pathway is unique, as it differs markedly from the general mechanism with respect to its energy requirement. On the one hand, ATP is not required for import, suggesting that in the cytosol heme lyases might interact with chaperones other than Hsp70. On the other hand, heme lyases reach the intermembrane space even after depletion of the membrane potential across the

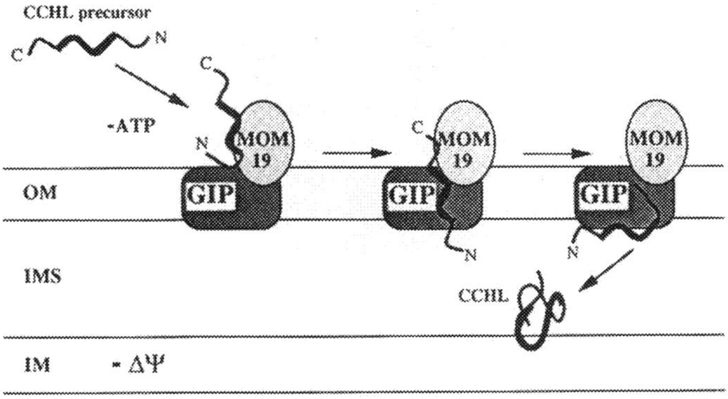

Figure 3. Import pathway of cytochrome *c* heme lyase (CCHL). The bold region in CCHL denotes the internal location of the mitochondrial targeting signal. For details see text. For abbreviations see Figure 2.

inner membrane. The latter observation was taken as evidence that the inner membrane may not be involved in import (Figure 3), a contention that has been confirmed recently by the ability of highly purified outer membrane vesicles to efficiently sequester CCHL (Mayer et al., 1993; see below).

The import of heme lyases raises a number of intriguing questions. For instance, what is the nature of the mitochondrial targeting signal in heme lyases? Deletions of about 70 amino acids at both the N- and C-terminus of *N. crassa* CCHL do not impede import into the intermembrane space (Kispal and Lill, unpublished). Thus, the import signal must be localized in the central part of the protein (Figure 3). The exact nature of the signal, however, has not yet been determined. A targeting signal in the middle of the protein prompts the question in which direction import occurs. Does the N-terminus pass the membrane first as in the case of signal peptide-containing preproteins, or is CCHL translocated starting from the C-terminus? These questions have been addressed using antibodies specific to the N- and C-terminus of *N. crassa* CCHL (Lill, unpublished). Both antibodies inhibit translocation of CCHL as expected. In the case of C-terminal antibodies, a number of N-terminal fragments are observed after protease digestion of the import reactions. Occurrence of these fragments depends on functional MOM19. From the apparent sizes of the fragments, it is evident that more than half of CCHL is inaccessible to external protease, that is, has been translocated (Figure 3). These data show that CCHL with an attached C-terminal antibody gives rise to a membrane-spanning translocation intermediate in which the N-terminus has already been translocated across the outer membrane. Thus, the direction of translocation of CCHL resembles that of signal sequence-bearing preproteins that translocate from the N- to the C-terminus. Since it is not known whether heme lyases are translocated co- or post-translationally, the relevance of the import direction for the mode of translocation remains elusive.

The independence of CCHL import on normally used energy sources like ATP or a membrane potential is a puzzling observation, since the driving force for CCHL membrane passage remains unexplained. As possible models, the folding of CCHL in the intermembrane space or binding to a specific partner, for example, at the outer face of the inner membrane, have been discussed (Lill et al., 1992a). While these suggestions await experimental testing, it is clear from recent data that soluble or peripheral proteins are not involved in transporting CCHL into a protease-protected state in isolated outer membrane vesicles (Mayer, unpublished). The translocation efficiency is virtually unchanged even after extracting the membranes with alkaline buffers to remove peripherally associated proteins. In isolated outer membranes, CCHL might therefore reach a stage similar (or identical) to the GIP stage (Figure 3). In intact mitochondria, release from this stage might then be accompanied by folding of the protein in the intermembrane space and result in association of folded CCHL with the inner membrane. Whether folding of the polypeptide chain occurs spontaneously or is assisted by chaperone proteins of the intermembrane space is not known at present. Likewise, no information is available as to a specific binding partner for heme lyases at the inner membrane.

VI. LESSONS LEARNED FROM IMPORT REACTIONS WITH OUTER MEMBRANE VESICLES

In many complex biochemical problems, considerable advances in our functional knowledge have been obtained by dissecting the process into various subreactions and by using purified components. Recently, a sub-mitochondrial system was developed to allow the study of protein import into isolated outer membranes (Mayer et al., 1993). The purification of outer membranes by density gradient centrifugations resulted in sealed and right-side-out oriented vesicles that were enriched more than 1,000-fold over the inner membrane. A number of important biochemical questions could be answered using this system. Does the isolated outer membrane transport preproteins independently of the inner membrane? Import into the vesicles was found for precursors of various outer membrane proteins and for CCHL, while other preproteins do not become sequestered into the vesicles. Outer membrane proteins like porin, MOM22, and MOM38 are imported in a receptor-dependent fashion (Mayer et al., 1993; A. Mayer, unpublished). After import, precursor proteins behave indistinguishably from endogenous material with respect to their extractability with alkaline buffers, folding (as tested by the intrinsic protease sensitivity), or association with other proteins. The efficiency of the translocation reaction is comparable to that observed in intact mitochondria demonstrating that no components of the translocation machinery have been lost during the purification, and a faithful sub-mitochondrial import system is being investigated.

In contrast to outer membrane proteins and CCHL, preproteins that also use the inner membrane machinery for their import do not become translocated into the lumen of the vesicles, even though the precursors become specifically bound to the vesicles (Mayer et al., 1993). This fits nicely with the observations that in the absence of a membrane potential preproteins are incapable of crossing the outer membrane and reaching the intermembrane space. What distinguishes a protein competent for translocation across the isolated outer membrane (like CCHL) from a protein that cannot enter the vesicles (e.g., cytochrome c_1)? The most likely answer is that for most proteins a driving force to pull them across the membrane is missing. For the latter class of proteins, known driving forces are the membrane potential-dependent translocation of the signal sequence across the inner membrane (Schleyer and Neupert, 1985) and the ATP-dependent action of matrix Hsp70 (Kang et al., 1990). For proteins competent for translocation across the isolated outer membrane the explanation is less clear. One might speculate that for outer membrane proteins the energy of membrane integration or of assembly into oligomeric complexes might be the major contribution. In case of CCHL, specific binding at the inner side of the outer membrane, most likely with components of GIP, may lead to translocation.

It is clear from the preceding paragraph that preproteins that are unable to enter the lumen of isolated outer membrane vesicles, in isolated mitochondria have to pass at least segments across the outer membrane in order to make contact with the inner membrane and to undergo further transport. A likely candidate for such a translocated segment of the preprotein is the N-terminal signal sequence. Experimental evidence for the ability of the outer membrane to present signal sequences to the intermembrane space recently was obtained using isolated outer membranes. To this end, purified matrix processing peptidase (MPP) was enclosed in outer membrane vesicles by a novel freeze-thaw technique (Mayer et al., 1994). Then it was tested whether precursor proteins expose their MPP cleavage site in the lumen of the vesicles and thus become processed by enclosed MPP (Mayer, unpublished). A considerable fraction of the bound preprotein becomes cleaved by lumenal MPP, a reaction that is dependent on functional receptors (Figure 4). A substantial fraction of the processed protein then dissociates from the mitochondrial outer membrane showing that the mature protein has a much weaker affinity for the translocation channel than the signal sequence, and cannot maintain a stable binding interaction. These observations demonstrate the ability of the outer membrane to specifically translocate signal sequences, and imply the existence of a specialized binding site on the inner side of the outer membrane.

A preprotein exposing its signal sequence to the intermembrane space can be considered as a novel translocation intermediate in which the N-terminus of the mature part has already entered the translocation channel, while the rest of the protein still protrudes into the cytoplasm (Figure 4). To continue translocation of the mature part of the protein, the signal sequence apparently has to engage interactions with the inner membrane (Figure 2). It is not yet known whether the

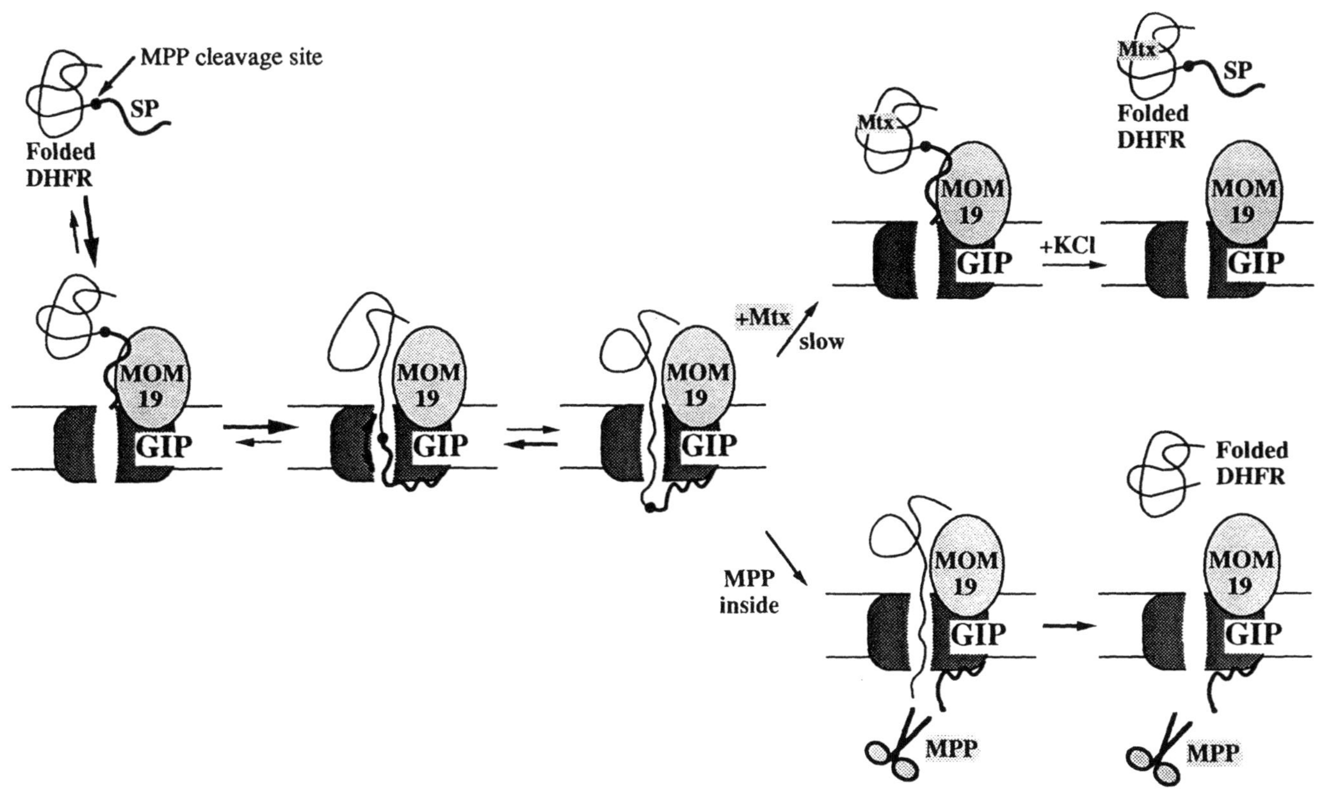

Figure 4. The molecular mechanism of protein translocation across the mitochondrial outer membrane. Note that MPP, a protein of the mitochondrial matrix space, was enclosed inside the outer membrane vesicles for analytical purposes. For further explanations see text. DHFR, dihydrofolate reductase; SP, mitochondrial signal peptide; Mtx, methotrexate. For other abbreviations see Figure 2.

transfer to the inner membrane involves a component specifically recognizing signal sequences, or whether the signal sequence is moved directly into the translocation tunnel of the inner membrane by an electrophoretic effect imposed by the membrane potential (Martin et al., 1991). A prerequisite for the description of these mechanistic details is the identification of components of the outer and inner membranes that recognize mitochondrial signal sequences, for example, by crosslinking studies. An interesting consequence of a putative binding site for the signal sequence inside the outer membrane is that the signal appears to be deciphered at least twice at the outer membrane. A first interaction may occur at the mitochondrial surface with the MOM19 receptor, as suggested by the observation that the signal sequence attached to a cytosolic protein is sufficient for MOM19-dependent import (Becker et al., 1992).

Binding of precursor proteins to isolated outer membranes provides enough energy to drive unfolding of the mature part of preproteins (Mayer, unpublished). In the precursor of Su9(1-69)-DHFR, the DHFR domain immediately follows the signal sequence. After synthesis in reticulocyte lysate, the DHFR moiety of this preprotein is folded. When the preprotein is offered to isolated outer membranes, material bound to the receptor complex is completely unfolded as indicated by a loss of the protease-resistance of the DHFR domain (Figure 4). The energy of specific binding to the inner side of the outer membrane apparently is large enough to drive unfolding of the protein chain. Thus, insertion of the signal sequence into the translocation tunnel and its binding to the inner side of the outer membrane therefore resembles the activity of a protein unfoldase. On the other hand, when the folded structure of DHFR is stabilized in solution by methotrexate and NADPH, specific binding to the outer membrane is diminished (Mayer, unpublished). Obviously, to allow stable binding to mitochondria the signal must be able to insert deeply into the translocation channel. Binding of the signal sequence to the mitochondrial surface receptors (e.g., to MOM19) apparently is labile and does not sustain reisolation of the vesicles. Conversely, addition of methotrexate and NADPH to prebound preSu9-DHFR leads to a slow dissociation of the translocation intermediate from the translocation machinery and is accompanied by refolding of the DHFR domain (Figure 4). This shows the reversibility of the translocation reaction. These data suggest that the individual energetic states of translocation do not differ to a large extent and therefore the equilibrium between these states can easily be shifted by imposing additional driving forces. This mechanism may be of functional advantage to guarantee a high degree of accuracy of protein translocation, since at any of these multiple steps the specificity of the preprotein for the target membrane may be checked. When a noncognate preprotein has entered the translocation channel, it can be released easily by reversing the reactions. To achieve completion of translocation, an irreversible reaction is needed at least during one of the steps (Neupert et al., 1990; Simon et al., 1992). As already discussed, folding, membrane integration, and assembly into oligomeric complexes may render transport of outer membrane proteins irreversible. In case of protein translocation into the mitochondrial matrix such an irreversible step may be mediated by Hsp70-dependent ATP hydrolysis (Kang et al., 1990). Reversibility of the sub-reactions may be a general principle underlying all membrane traffic of proteins. In variations of the theme, reversibility has been observed recently for translocation across the mitochondrial inner membrane (Cyr et al., 1993; Ungermann, unpublished), the endoplasmic reticulum (Ooi and Weiss, 1992), and bacterial plasma membrane (Arkowitz et al., 1993).

VI. PERSPECTIVES

Despite the considerable progress made during the last years in the understanding of how proteins move across the mitochondrial outer membrane, there remain quite a number of open questions. For instance, it is not clear how the various proteins of the receptor complex cooperate to achieve translocation. A prerequisite to answer this question is the description of the three-dimensional arrangement of the various components in the membrane. The reconstitution of the translocation process with purified components will be of importance to understand the minimal requirements for the transport reaction. It is well possible that further components will be discovered in the future. With respect to the unfolding of preproteins before insertion into the translocation channel, it will be interesting to learn whether cytosolic chaperones or components of the receptor complex trigger the unfolding reaction. A mechanistically pressing problem is how the machineries of the outer and inner membranes cooperate during protein transport. Finally, the electrophysiological identification of the translocation channel and the description of its properties will lead to an understanding of the environment a preprotein encounters when passing across the membrane.

ACKNOWLEDGMENTS

We wish to thank Petra Heckmeyer for excellent technical assistance and Dr. Deborah Court for critical reading of the manuscript. Gyula Kispal and Andreas Mayer are generously supported by fellowships from the Alexander-von-Humboldt-Stiftung and the Boehringer Ingelheim Fonds, respectively. The work is supported by grants from the Sonderforschungsbereich 184, Teilprojekt B18 and B19 of the Deutsche Forschungsgemeinschaft, and by the European Science Foundation Network.

REFERENCES

Arkowitz, R. A., Joly, J. C., & Wickner, W. (1993). Translocation can drive the unfolding of a preprotein domain. EMBO J. 12, 243–253.

Arretz, M., Schneider, H., Wienhues, U., & Neupert, W. (1991). Processing of mitochondrial precursor proteins. Biomed. Biochim. Acta 50, 403–412.

Baker, K. P., Schaniel, A., Vestweber, D., & Schatz, G. (1990). A yeast mitochondrial outer membrane protein is essential for protein import and cell viability. Nature 348, 605–609.

Becker, K., Guiard, B., Rassow, J., Söllner, T., & Pfanner, N. (1992). Targeting of a chemically pure preprotein to mitochondria does not require the addition of a cytosolic signal recognition factor. J. Biol. Chem. 267, 5637–5643.

Bertrand, H., & Pittenger, T. H. (1972). Cytoplasmic mutants selected from continuously growing cultures of *Neurospora crassa*. Genetics 71, 521–533.

Chen, W-J., & Douglas, M. G. (1987). Phosphodiester bond cleavage outside mitochondria is required for the completion of protein import into the mitochondrial matrix. Cell 49, 651–658.

Cyr, D. M., Stuart, R. A., & Neupert, W. (1993). A matrix ATP requirement for presequence translocation across the inner membrane of mitochondria. J. Biol. Chem. 268, 23751–23754.

Dekker, P. J. T., Keil, P., Rassow, J., Maarse, A. C., Pfanner, N., & Meijer, M. (1993). Identification of *MIM23*, a putative component of the protein import machinery of the mitochondrial inner membrane. FEBS Lett. 330, 66–70.

Deshaies, R. J., Koch, B. D., & Schekman, R. (1988). The role of stress proteins in membrane biogenesis. TIBS 13, 384–388.

Dietmeier, K., Zara, V., Palmisano, A., Palmieri, F., Voos, W., Schlossmann, J., Moczko, M., Kispal, G., & Pfanner, N. (1993). Targeting and translocation of the phosphate carrier/p32 to the inner membrane of yeast mitochondria. J. Biol. Chem. 268, 25958–25964.

Drygas, M. E., Lambowitz, A. M., & Nargang, F. E. (1989). Cloning and analysis of the *Neurospora crassa* gene for cytochrome c heme lyase. J. Biol. Chem. 264, 17897–17907.

Dumont, M. E., Ernst, J. F., Hampsey, D. M., & Sherman, F. (1987). Identification and sequence of the gene encoding cytochrome c heme lyase in the yeast *S. cerevisiae*. EMBO J. 6, 235–241.

Emtage, J. L. T., & Jensen, R. E. (1993). *MAS6* encodes an essential inner membrane component of the yeast mitochondrial protein import pathway. J. Cell Biol. 122, 1003–1012.

Gasser, S. M., Daum, G., & Schatz, G. (1982). Import of proteins into mitochondria: Energy-dependent uptake of precursors by isolated mitochondria. J. Biol. Chem. 257, 13034–13041.

Glick, B. G., & Schatz, G. (1991). Import of proteins into mitochondria. Annu. Rev. Genet. 25, 21–44.

Glick, B. G., Wachter, C., & Schatz, G. (1991). Protein import into mitochondria: Two systems acting in tandem? Trends Cell Biol. 1, 99–103.

Glick, B. S., Brandt, A., Cunningham, K., Müller, S., Hallberg, R. L., & Schatz, G. (1992). Cytochromes c_1 and b_2 are sorted to the intermembrane space of yeast mitochondria by a stop-transfer mechanism. Cell 69, 809–822.

Gonzales, D. H., & Neupert, W. (1990). Biogenesis of mitochondrial *c*-type cytochromes. J. Bioenerg. Biomembr. 22, 753–768.

Hachiya, N., Alam, R., Sakasegawa, H., Sakaguchi, M., Mihara, N., & Omura, T. (1993). A mitochondrial import factor purified from rat liver cytosol is an ATP-dependent conformational modulator for precursor proteins. EMBO J. 12, 1579–1586.

Harkness, T. A. A., Metzenberg, R. L., Schneider, H., Lill, R., Neupert, W., & Nargang, F. E. (1994a). Inactivation of the *Neurospora crassa* gene encoding the mitochondrial protein import receptor MOM19 by the technique of "sheltered RIP." Genetics 136, 107–118.

Harkness, T. A. A., Nargang, F. E., Van der Klei, I., Neupert, W., & Lill, R. (1994b). A crucial role of the mitochondrial protein import receptor MOM19 for the biogenesis of mitochondria. J. Cell Biol. 124, 637–648.

Heijne, G. V. (1986). Mitochondrial targeting sequences may form amphiphilic helices. EMBO J. 5, 1335–1342.

Hines, V., Brandt, A., Griffith, G., Horstmann, H., Brütsch, H., & Schatz, G. (1990). Protein import into yeast mitochondria is accelerated by the outer membrane protein *MAS70*. EMBO J. 9, 3191–3200.

Hurt, E. C., Pesold-Hurt, B., Suda, K., Oppliger, W., & Schatz, G. (1985a). The first twelve amino acids (less than half of the presequence) can direct mouse cytosolic dihydrofolate reductase into the yeast mitochondrial matrix. EMBO J. 4, 2061–2068.

Hurt, E. C., Müller, U., & Schatz, G. (1985b). The first twelve amino acids of a yeast mitochondrial outer membrane protein can direct a nuclear encoded cytochrome oxidase subunit to the mitochondrial inner membrane. EMBO J. 4, 3509–3518.

Hwang, S. T., Jascur, T., Vestweber, D., Pon, L., & Schatz, G. (1989). Disrupted yeast mitochondria can import precursor proteins directly through their inner membrane. J. Cell Biol. 109, 487–493.

Hwang, S. T., Wachter, C., & Schatz, G. (1991). Protein import into the yeast mitochondrial matrix. J. Biol. Chem. 266, 21083–21089.

Ikeda, E., Yoshida, S., Mitsuzawa, H., Uno, I., & Toh-e, A. (1994). YGE1 is a yeast homologue of *Escherichia coli grpE* and is required for maintenance of mitochondrial functions. FEBS Lett. 339, 265–268.

Jascur, T., Goldenberg, D. P., Vestweber, D., & Schatz, G. (1992). Sequential translocation of an artificial precursor protein across the two mitochondrial membranes. J. Biol. Chem. 267, 13636–13641.

Kang, P., Ostermann, J., Shilling, J., Neupert, W., Craig, E. A., & Pfanner, N. (1990). Requirement for hsp70 in the mitochondrial matrix for translocation and folding of precursor proteins. Nature 348, 137–142.

Kassenbrock, C. K., Cao, W., & Douglas, M. G. (1993). Genetic and biochemical characterization of ISP6, a small mitochondrial outer membrane protein associated with the protein translocation complex. EMBO J. 12, 3023–3034.

Keil, P., & Pfanner, N. (1993). Insertion of MOM22 into the mitochondrial outer membrane strictly depends on surface receptors. FEBS Lett. 321, 197–200.

Keil, P., Weinzierl, A., Kiebler, M., Dietmeier, K., Söllner, T., & Pfanner, N. (1993). Biogenesis of the mitochondrial receptor complex. Two receptors are required for binding of MOM38 to the outer membrane surface. J. Biol. Chem. 268, 19177–19180.

Kiebler, M., Keil, P., Schneider, H., van der Klei, I., Pfanner, N., & Neupert, W. (1993). The mitochondrial receptor complex: A central role of MOM22 in mediating transfer of preproteins from receptors to the general insertion pore. Cell 74, 483–492.

Kiebler, M., Pfaller, R., Söllner, T., Griffith, G., Horstmann, H., Pfanner, N., & Neupert, W. (1990). Identification of a mitochondrial receptor complex required for recognition and membrane insertion or precursor proteins. Nature 348, 610–616.

Kleene, R., Pfanner, N., Pfaller, R., Link, T. A., Sebald, W., Neupert, W., & Tropschug, M. (1987). Mitochondrial porin of *Neurospora crassa*: cDNA cloning, in vitro expression and import into mitochondria. EMBO J. 6, 2627–2633.

Koll, H., Guiard, B., Rassow, J., Ostermann, J., Horwich, A. L., Neupert, W., & Hartl, F. (1992). Antifolding activity of hsp60 couples protein import into the mitochondrial matrix with export to the intermembrane space. Cell 68, 1163–1175.

Li, J., & Shore, G. C. (1992). Protein sorting between mitochondrial outer and inner membranes. Insertion of an outer membrane protein into the inner membrane. Biochim. Biophys. Acta 1106, 233–241.

Lill, R., Hergersberg, C., Schneider, H., Söllner, T., Stuart, R. A., & Neupert, W. (1992a). General and exceptional pathways of protein import into the sub-mitochondrial compartments. In: *Membrane Biogenesis and Protein Targeting* (Neupert, W., & Lill, R., Eds.). Elsevier Science Publishers, Amsterdam, pp. 265–276.

Lill, R., Stuart, R. A., Drygas, M. E., Nargang, F. E., & Neupert, W. (1992b). Import of cytochrome c heme lyase into mitochondria: A novel pathway into the intermembrane space. EMBO J. 11, 449–456.

Lill, R., Kispal, G., Künkele, K. P., Mayer, A., Risse, B., Steiner, H., Heckmeyer, P., Van der Klei, I., & Court, D. A. (1996). Protein transport into and across the mitochondrial outer membrane: Recognition, insertion and translocation of preproteins. In: Proceedings of the NATO/ASI, Cell Biology: Molecular Dynamics of Biomembranes, (Op den Kamp, J. A. F., Ed.), pp. 137–155. Berlin, Springer-Verlag.

Maarse, A. C., Bloom, J., Grivell, L. A., & Meijer, M. (1992). MPI1, an essential gene encoding a mitochondrial membrane protein, is possibly involved in protein import into yeast mitochondria. EMBO J. 11, 3619–3628.

Martin, J., Mahlke, K., & Pfanner, N. (1991). Role of an energized inner membrane in mitochondrial protein import. J. Biol. Chem. 266, 18051–18057.

Mayer, A., Neupert, W., & Lill, R. (1995). Mitochondrial protein import: Reversible binding of the presequence at the *trans* side of the outer membrane drives partial translocation and unfolding. Cell 80, 127–137.

Mayer, A., Driessen, A., Neupert, W., & Lill, R. (1994). Inclusion of proteins into isolated mitochondrial outer membrane vesicles. In: *Cell Biology: A Laboratory Handbook* (Celis, J. E., Ed.). Academic Press, New York.

Mayer, A., Lill, R., & Neupert, W. (1993). Translocation and insertion of precursor proteins into isolated outer membranes of mitochondria. J. Cell Biol. 121, 1233–1243.

McBride, H. M., Millar, D. G., Li, J. M., & Shore, G. C. (1992). A signal-anchor sequence selective for the mitochondrial outer membrane. J. Cell Biol. 119, 1451–1457.

Moczko, M., Gärtner, F., & Pfanner, N. (1993). The protein import receptor MOM19 of yeast mitochondria. FEBS Lett. 326, 251–254.

Murakami, K., & Mori, M. (1990). Purified presequence binding factor (PBF) forms an import-competent complex with a purified mitochondrial precursor protein. EMBO J. 9, 3201–3208.

Neupert, W., Hartl, F., Craig, E. A., & Pfanner, N. (1990). How do polypeptides cross the mitochondrial membranes? Cell 63, 447–450.

Ohba, M., & Schatz, G. (1987). Disruption of the outer membrane restores protein import into trypsin-treated yeast mitochondria. EMBO J. 6, 2116–2122.

Ono, H., & Tuboi, S. (1987). Integration of porin synthesized *in vitro* into outer mitochondrial membranes. Eur. J. Biochem. 168, 509–514.

Ooi, C. E., & Weiss, J. (1992). Bidirectional movement of the nascent polypeptide across microsomal membranes reveals requirements for vectorial translocation of proteins. Cell 71, 87–96.

Ostermann, J., Horwich, A. L., Neupert, W., & Hartl, F. (1989). Protein folding in mitochondria requires complex formation with hsp60 and ATP hydrolysis. Nature 341, 125–130.

Pfaller, R., Steger, H. F., Rassow, J., Pfanner, N., & Neupert, W. (1988). Import pathways of precursor proteins into mitochondria: Multiple receptor sites are followed by a common membrane insertion site. J. Cell Biol. 107, 2483–2490.

Pfanner, N., & Neupert, W. (1990). The mitochondrial protein import apparatus. Annu. Rev. Biochem. 59, 331–353.

Pfanner, N., Rassow, J., van der Klei, I. J., & Neupert, W. (1992). A dynamic model of the mitochondrial protein import machinery. Cell 68, 999–1002.

Pfanner, N., Tropschug, M., & Neupert, W. (1987). Mitochondrial protein import: nucleoside triphosphates are involved in conferring import-competence to precursors. Cell 49, 815–823.

Ramage, L., Junne, T., Hahne, K., Lithgow, T., & Schatz, G. (1993). Functional cooperation of mitochondrial protein import receptors in yeast. EMBO J. 12, 4115–4123.

Rassow, J., Guiard, B., Wienhues, U., Herzog, V., Hartl, F., & Neupert, W. (1989). Translocation arrest by reversible folding of a precursor protein imported into mitochondria: A means to quantitate translocation contact sites. J. Cell Biol. 109, 1421–1428.

Rassow, J., & Pfanner, N. (1991). Mitochondrial preproteins *en route* from the outer membrane to the inner membrane are exposed to the intermembrane space. FEBS Lett. 293, 85–88.

Rospert, S., Glick, B. S., Jenö, P., Schatz, G., Todd, M.J., Lorimer, G.H., & Viitanen, P.V. (1993). Identification and functional analysis of chaperonin 10, the GroES homolog from yeast mitochondria. Proc. Natl. Acad. Sci. USA 90, 10967–10971.

Rowley, N., Prip-Buus, C., Westermann, B., Brown, C., Schwarz, E., Barrell, B., & Neupert, W. (1994). Mdj1p, a novel chaperone of the DnaJ family, is involved in mitochondrial biogenesis and protein folding. Cell.

Scherer, P. E., Manning-Krieg, U. C., Jenö, P., Schatz, G., & Horst, M. (1992). Identification of a 45-kDa protein at the protein import site of the yeast mitochondrial inner membrane. Proc. Natl. Acad. Sci. USA 89, 11930–11934.

Schleyer, M., & Neupert, W. (1982). Requirement of a membrane potential for the posttranslational transfer of proteins into mitochondria. Eur. J. Biochem. 125, 109–116.

Schleyer, M., & Neupert, W. (1985). Transport of proteins into mitochondria: Translocation intermediates spanning contact sites between inner and outer membranes. Cell 43, 330–350.

Schlossmann, J., Dietmeier, K., Pfanner, N., & Neupert, W. (1994). Specific recognition of mitochondrial preproteins by the cytosolic domain of the import receptor MOM72. J. Biol. Chem. 269.

Schneider, H., Söllner, T., Dietmeier, K., Eckerskorn, C., Lottspeich, F., Trülzsch, K., Neupert, W., & Pfanner, N. (1991). Targeting of the master receptor MOM19 to mitochondria. Science 254, 1659–1662.

Segui-Real, B., Kispal, G., Lill, R., & Neupert, W. (1993a). Functional independence of the protein translocation machineries in mitochondrial outer and inner membranes: Passage of preproteins through the intermembrane space. EMBO J. 12, 2211–2218.

Segui-Real, B., Stuart, R. A., & Neupert, W. (1993b). Transport of proteins into the various subcompartments of mitochondria. FEBS Lett. 313, 2–7.

Shore, G. C., Millar, D. G., & Li, J. (1992). Protein insertion into mitochondrial outer and inner membranes via the stop-transfer sorting pathway. In: *Membrane Biogenesis and Protein Targeting* (Neupert, W., & Lill, R., Eds.). Elsevier Science Publishers, Amsterdam, pp. 253–264.

Simon, S. M., Peskin, C. S., & Oster, G. F. (1992). What drives the translocation of proteins? Proc. Natl. Acad. Sci. USA 89, 3770–3774.

Söllner, T., Griffith, G., Pfaller, R., Pfanner, N., & Neupert, W. (1989). MOM19, an import receptor for mitochondrial precursor proteins. Cell 59, 1061–1070.

Söllner, T., Pfaller, R., Griffith, G., Pfanner, N., & Neupert, W. (1990). A mitochondrial import receptor for the ADP/ATP carrier. Cell 62, 107–115.

Söllner, T., Rassow, J., Wiedmann, M., Schlossmann, J., Keil, P., Neupert, W., & Pfanner, N. (1992). Mapping of the protein import machinery in the mitochondrial outer membrane by crosslinking of translocation intermediates. Nature 355, 84–87.

Steger, H. F., Söllner, T., Kiebler, M., Dietmeier, K. A., Trülzsch, K. S., Tropschug, M., Neupert, W., & Pfanner, N. (1990). Import of ADP/ATP carrier into mitochondria: Two receptors act in parallel. J. Cell Biol. 111, 2353–2363.

Stuart, R. A., Gruhler, A., Van der Klei, I., B, G., Koll, H., & Neupert, W. (1994). The requirement of matrix ATP for the import of precursor proteins into the mitochondrial matrix and intermembrane space. Eur. J. Biochem. 220, 9–18.

Stuart, R. A., & Neupert, W. (1990). Apocytochrome c: An exceptional mitochondrial precursor protein using an exceptional import pathway. Biochimie 72, 115–121.

Steiner, H., Zollner, A., Haid, A., Neupert, W., & Lill, R. (1995). Biogenesis of mitochondrial heme lyases in yeast. Import and folding in the intermembrane space. J. Biol. Chem. 270, 22842–22849.

Tzagoloff, A., & Dieckmann, C. L. (1990). *PET* genes of *Saccharomyces cerevisiae*. Microbiol. Rev. 54, 211–225.

Ungermann, C., Neupert, W., & Cyr, D. M. (1994). The role of Hsp70 in conferring unidirectionality on protein translocation into mitochondria. Science 266, 1250–1253.

Van der Klei, I. J., Veenhuis, M., & Neupert, W. (1994). A morphological view on mitochondrial protein targeting. Microscopy Res. Technique 27, 284–293.

Vestweber, D., Brunner, K. P., Baker, A., & Schatz, G. (1989). A 42K outer membrane protein is a component of the yeast mitochondrial import site. Nature 341, 205–209.

Wachter, C., Schatz, G., & Glick, B. S. (1992). Role of ATP in the intermitochondrial sorting of cytochrome c_1 and the adenine nucleotide translocator. EMBO J. 11, 4787–4794.

Zollner, A., Rödel, G., & Haid, A. (1992). Molecular cloning and characterization of the *Saccharomyces cerevisiae* CYT2 gene encoding cytochrome c_1 heme lyase. Eur. J. Biochem. 207, 1093–1100.

TARGETING AND INSERTION OF PROTEINS INTO THE MITOCHONDRIAL OUTER MEMBRANE

Gordon C. Shore, Heidi M. McBride, Douglas G. Millar, Nancy A.E. Steenaart, and Mai Nguyen

I. Introduction . . . 71
II. Signal Anchor Sequence Selective for the Outer Membrane . . . 72
III. Artificial Signal Anchor Sequence . . . 76
IV. Orientation of the Signal Anchor Sequence . . . 77
Acknowledgments . . . 78
References . . . 78

I. INTRODUCTION

Mitochondria from diverse tissues and sources typically comprise 8–10% of their total protein content as constituents of the outer membrane (OMM). In addition to a number of important enzymes, there exists among this group proteins that

Advances in Molecular and Cell Biology,
Volume 17, pages 71–80.

ISBN: 0-7623-0144-9

facilitate two-way communication between the organelle and the rest of the cytoplasm. These include components of a protein recognition and import machinery (Hartl and Neupert, 1990; Schatz, 1993), a pore complex for export-import of metabolites (porin/VDAC; Mannella, 1993), and receptors that regulate mitochondrial metabolism in response to cytosolic signals (Moreno-Sanchez et al., 1991; Vorobjer and Zurov, 1993; Hirsch et al., 1989).

Until recently, understanding the biogenesis and assembly of many of these proteins has been complicated by a general lack of knowledge concerning their structure and topology in the outer membrane and the extent to which they form higher order complexes. An exception, of course, is porin, for which the high resolution structure of a closely related bacterial homolog has been determined (Weiss et al., 1991). Members of this multi-spanning membrane protein family adopt a trimer of β-barrel structures that exhibit extensive and ordered intra-polypeptide interactions that allow the complex to achieve the necessary amphiphilicity required to form a transbilayer pore (Jap, 1989; Weiss et al., 1991). While β-barrel structures will undoubtedly extend to a large number of outer membrane proteins, specific topogenic information within these proteins may be difficult to dissect. Such difficulties notwithstanding, however, there is evidence from both competition and receptor-binding studies that insertion of porin into the outer membrane is mediated, at least at early steps, by the outer membrane protein import machinery (Pfaller et al.,1990). The finding that another protein with a potential β-barrel structure, monoamine oxidase A, requires ubiquitin for targeting and/or insertion, however, suggests that prior cytosolic events may be critical for presenting the protein to the import machinery in an appropriate conformation (Zhuang and McCauley, 1989).

Recently, a number of integral proteins of the outer membrane have been described whose functions are defined and whose primary structures predict a membrane topology mediated by a single hydrophobic transbilayer segment. These include the components of the protein import machinery in yeast and Neurospora, Mas70p/MOM72 (Hines et al., 1990; Söllner et al., 1990), Mas20p/MOM19 (Ramage et al., 1993; Söllner et al., 1989), and ISP6 (Kassenbrock et al., 1993), and a protein in mammals that functions as a suppressor of programmed cell death, Bcl-2 (Hockenbery et al., 1990; Nguyen et al., 1993). Such proteins provide useful model systems to address not only the problem of targeting and insertion into the outer membrane, but as well the complementary problem of explaining how proteins of the inner membrane that also contain uniformly hydrophobic transmembrane segments are able to pass through the outer membrane without being arrested.

II. SIGNAL ANCHOR SEQUENCE SELECTIVE FOR THE OUTER MEMBRANE

Of the group of single-spanning outer membrane proteins listed previously, the biogenesis of Mas70p has been investigated most extensively. The protein is

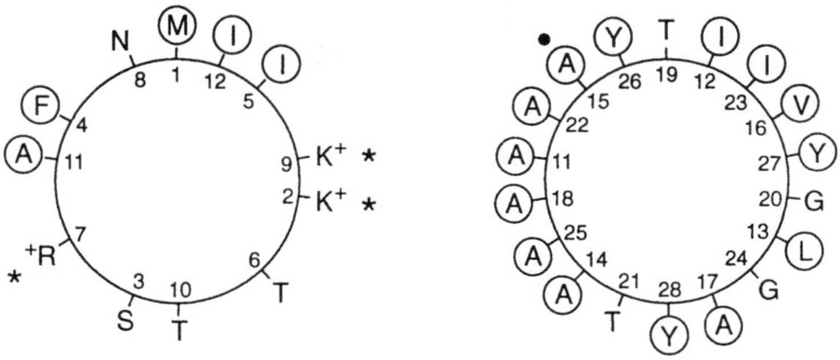

pOMD29
1-12

pOMD29
11-28

Figure 1. Helical wheel projections of amino acids 1–12 and 11–28 of pOMD29. The hybrid protein was created by fusing amino acids 1–29 of Mas70p to amino acids 4–186 of dihydrofolate reductase. Because of the cloning strategy employed (Li and Shore, 1992a), threonine 15 of Mas70p was converted to an alanine in pOMD29. This substitution has no effect on targeting, membrane insertion, and homodimerization of pOMD29 in the OMM.

targeted and inserted into the outer membrane by its NH_2-terminal 29 amino acids (Hase et al., 1984; Nakai et al., 1989; Li and Shore, 1992a), resulting in a predicted Nin-Ccyto orientation that leaves the bulk of the polypeptide facing the cytosol where it functions as a receptor for incoming precursor proteins (Hines et al.,1990). The NH_2-terminal 29 amino acids make up two structural domains: a segment consisting of amino acids 1–10 that is hydrophilic and contains three basic residues (amino acids 2, 7, and 9) and no negatively charged amino acids, followed by a 19 amino acid stretch of uncharged, mostly apolar amino acids which forms the predicted transmembrane segment. A helical wheel display of these two domains reveals interesting features. Amino acids 1–12 can potentially form a short positively-charged amphiphilic helix, whereas the predicted helix made up of amino acids 11–29 is mostly hydrophobic, but has an extensive cluster of alanine residues that line one face of the helix (Figure 1).

We analyzed the topogenic properties of amino acids 1–29 of Mas70p by a combination of fusion protein constructions, deletions, and site-directed mutagenesis. Amino acids 1–29 of Mas70p were fused to a cytosolic reporter protein, dihydrofolate reductase, to create the hybrid protein pOMD29 (Li and Shore, 1992a). pOMD29 inserted into the outer membrane of intact mitochondria *in vitro* with the same orientation as Mas70p; import was dependent on ATP, was temperature-sensitive, and was dependent in part on trypsin-sensitive components on the surface of the organelle (Li and Shore, 1992a, 1992b; McBride et al., 1992). The transmembrane segment (amino acids 11–29) was both necessary and sufficient for targeting and insertion, whereas amino acids 1–10 contributed to an enhanced rate of import due, in large part, to the presence of the basic amino acids at positions 2, 7, and 9 (McBride et al., 1992). Deletion of most of the transmembrane segment of pOMD29 resulted in a protein that did not target to mitochondria *in vitro* (McBride et al., 1992). As with all outer membrane proteins examined to date, pOMD29 (Mas70p) was not proteolytically processed during import, that is, the topogenic domain remained a permanent feature of the protein, and the protein did not require a transmembrane electrochemical potential for insertion (Li and Shore, 1992a,b; McBride et al., 1992).

Taken together, the results suggest a topogenic domain for Mas70p that is remarkably similar, both structurally and functionally, to the well-characterized topogenic sequence of type II membrane proteins that are inserted co-translationally into the endoplasmic reticulum (ER; Wickner and Lodish, 1985; von Heijne, 1988). Typically, a type II signal-anchor sequence is equivalent to a signal sequence except that, unlike a signal sequence, a signal-anchor is not cleaved and remains embedded in the bilayer as a permanent feature of the protein. Structurally, ER signal-anchor sequences are characterized by a hydrophobic core that is responsible for targeting, whereas the extreme amino terminus, which typically contains one or more positively-charged residues, cooperates with the hydrophobic core to increase the efficiency of targeting and membrane translocation, and contributes to the N_{cyto} or C_{cyto} disposition of the protein across the bilayer (von Heijne and Gavel, 1988; Hartmann et al., 1989; High and Dobberstein, 1992). Topogenic sequences, of course, are defined by function, and not structure per se. The hallmark function of an ER signal-anchor sequence is twofold: it initiates translocation of a protein across the membrane, but it is also the sequence that abrogates translocation and results in release of this domain into the surrounding lipid bilayer where it permanently anchors the protein in a transbilayer disposition (Blobel, 1980; Singer, 1990).

By extending this definition of a signal-anchor sequence to Mas70p, insight into the process that causes a protein to be sorted to the OMM may be gained (Shore et al., 1992; McBride, et al., 1992). Because the sequence that triggers protein translocation across the outer membrane is also the sequence that halts such translocation, the Mas70p topogenic sequence may select the outer membrane simply because this is the first membrane that the incoming protein encounters (Figure 2). Consistent with this suggestion is the finding that pOMD29 can be efficiently imported by mitoplasts, where it inserts into the inner membrane with the same orientation that the protein assumes when imported into the outer membrane of intact mitochondria (Li and Shore, 1992a).

Based on the structural and, more importantly, the functional similarities between the Mas70p topogenic sequence and the signal-anchor sequence of type II proteins that are inserted into the ER, we have retained the ER momenclature and refer to the Mas70p topogenic sequence as a signal-anchor. That the Mas70p signal-anchor

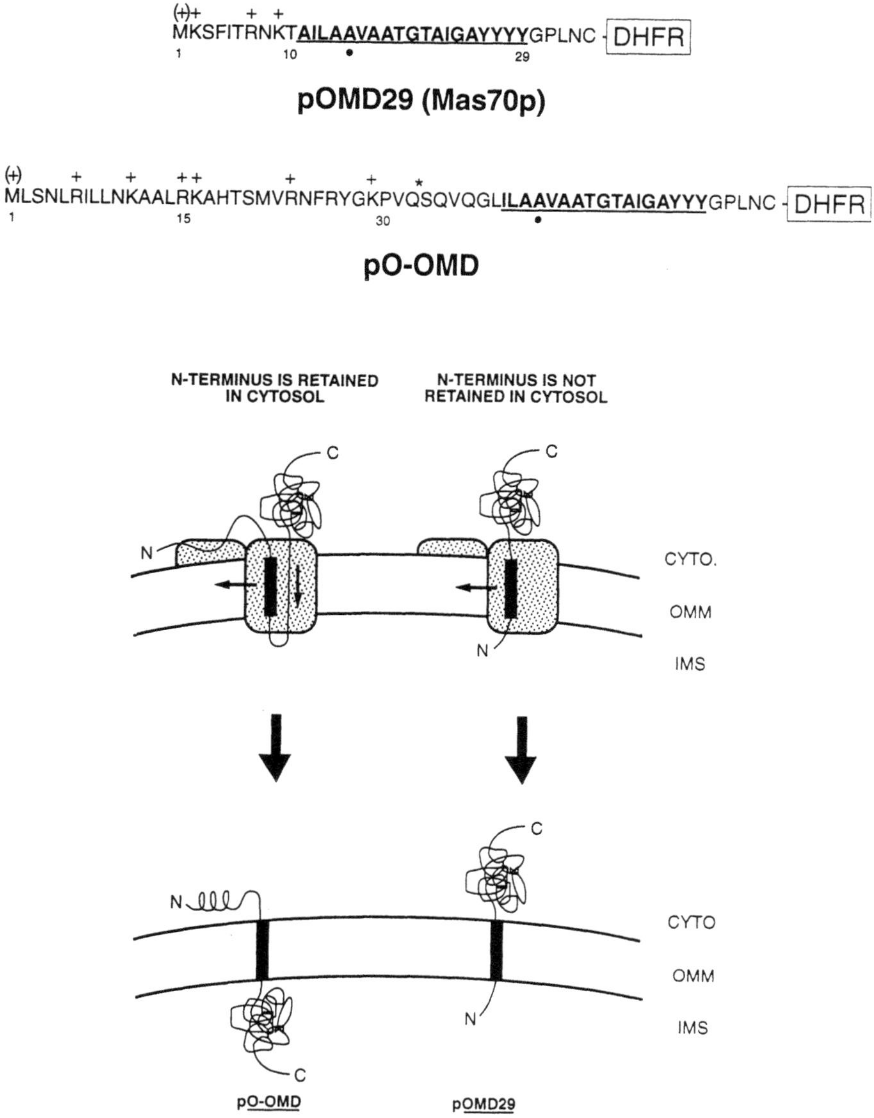

Figure 2. Model for insertion and integration of pOMD29 and pO-OMD into the OMM. See text for details. CYTO, cytosol; OMM, outer mitochondrial membrane; IMS, intermembrane space.

is selective for the mitochondrial outer membrane, rather than for the ER, emphasizes that as yet undetermined features of the OMM signal-anchor sequence distinguish it from that of a type II ER signal anchor.

Because protein import into mammalian mitochondria proceeds by a post-translational mechanism (Shore et al., 1979; Raymond and Shore, 1979), there is no reason a priori to assume that an OMM signal-anchor sequence must be located toward the NH_2-terminus of the protein (Shore et al., 1992). Recently, in fact, we have found that Bcl-2 is targeted to the outer membrane *in vivo* and *in vitro* by its predicted transmembrane segment, which is located at the extreme COOH-terminus of the protein. The transmembrane segment functions as a signal-anchor sequence (Nguyen et al., 1993). Besides being located in the OMM, Bcl-2 can also be found in association with the ER and nuclear envelope in certain cell types (Hockenbery et al., 1990; Chen-Levy et al., 1989; Jacobson et al., 1993; Nguyen et al., 1993,1994). However, whereas Bcl-2 appears to target the OMM by a selective pathway that employs the mitochondrial protein import machinery, its association with ER (and presumably the nuclear envelope) occurs by a different mechanism, for example, a synthetic peptide corresponding to the Mas70p signal-anchor competes for insertion of Bcl-2 into the mitochondrial outer membrane *in vitro*, but has no effect on the association of Bcl-2 with ER microsomes (Nguyen et al., 1993). Bcl-2 is inserted into the outer membrane in the N_{cyto}-C_{in} orientation, that is, an orientation opposite to that of Mas70p but, like Mas70p, the bulk of the polypeptide faces the cytosol. When the Bcl-2 signal-anchor was replaced with that of Mas70p, both targeting and function of the chimeric protein were similar to wild-type bcl-2 (Nguyen et al., 1994).

Finally, a number of proteins found in the plasma membrane form homo- and/or hetero-oligomeric complexes as a consequence of close helix-helix packing between transmembrane segments (Bormann and Engelman, 1992; Sternberg and Gullick, 1990). Characteristics that can promote such helix-helix packing include clusters of small-chain amino acids on one face of the helix (Lemmon et al., 1992; Cosson and Bonifacino, 1992) and a pentapeptide motif that is widespread among members of the tyrosine kinase family of receptors that dimerize in response to signaling by an external ligand (Sternberg and Gullick, 1989, 1990). Interestingly, the hydrophobic domain of the Mas70p signal-anchor exhibits both of these characteristics, which result in homodimerization via the signal-anchor (Millar and Shore, 1993, 1994).

III. ARTIFICIAL SIGNAL ANCHOR SEQUENCE

The motivation to analyze in detail the topogenic domain of native outer membrane proteins came from experiments that had been conducted, paradoxically, on a matrix protein, ornithine carbamyl transferase. This protein, which is made as a pre-protein (pOCT), contains a matrix-targeting signal located at its NH_2-terminus. A heterologous 19 amino acid transmembrane segment, corresponding to that of the G protein of Vesicular stomatitis virus, was engineered into pOCT at either of two positions: immediately downstream of the matrix-targeting signal or far downstream, toward the COOH-terminus (Nguyen et al., 1988). When the membrane-anchor was located toward the COOH-terminus, the protein targeted the mitochondrial inner membrane, where it inserted in the N_{matrix}-C_{out} orientation; toward the NH_2-terminus, the membrane anchor resulted in insertion of the protein into the outer membrane, in the N_{cyto}-C_{in} orientation (i.e., an orientation opposite to that of pOMD29/Mas70p). Of note, when the hydrophilic NH_2-terminal 10 amino acids of pOMD29 were replaced with the pOCT matrix-targeting signal, the modified protein selected the outer membrane for insertion, but in the reverse orientation (N_{cyto}-C_{in}; Li and Shore, 1992b).

These findings suggest an important principle with regard to a functional definition for an OMM signal-anchor sequence. Regardless of the extent of targeting information that immediately flanks the membrane anchor, selection of the outer membrane depends on the membrane anchor gaining access to the protein import apparatus at a very early stage during polypeptide translocation. Otherwise, the membrane anchor may be selective for the inner membrane (Nguyen et al., 1988; Nguyen and Shore, 1987; Shore et al., 1992). Moreover, it must be emphasized that a stretch of hydrophobic amino acids contiguous to a matrix-targeting signal does not necessarily confer a signal-anchor function. For example, certain proteins destined for the mitochondrial intermembrane space contain an apolar segment located toward the amino terminus, immediately downstream of a matrix-targeting signal. However, whereas these sequences contribute to sorting, they do not function as membrane anchor sequences, that is, these proteins do not appear to become embedded into the lipid bilayer at any step during import (Glick et al., 1992; Koll et al, 1992). Characteristics other than simple hydrophobicity alone, therefore, contribute to the signals that specify sorting to the outer membrane and inter membrane space.

IV. ORIENTATION OF THE SIGNAL ANCHOR SEQUENCE

Exchanging the hydrophilic NH_2-terminus of pOMD29 with the matrix-targeting signal of pOCT created the protein pO-OMD, which inserts into the OMM in an orientation opposite to that of pOMD29 (Li and Shore, 1992b; Figure 2). In making this change on the NH_2-terminal side of the hydrophobic core of the Mas70p signal-anchor, three obvious features were altered: length (increased from 10 to 38 amino acids), charge (increased from +4 to +8), and the ability to function independently as a matrix-targeting signal upon deletion of the transmembrane segment. We have now carried out systematic mutagenesis on the NH_2-terminus of pOMD29 and pO-OMD, and find that the determinant in this region that specifies its retention in the cytosol correlates with its ability to function as a strong matrix-targeting signal (Steenaart et al., 1996).

Our original model to account for the retention of the NH_2-terminus of pO-OMD or pOCT-GST35 on the cytosolic side of the outer membrane, that is, insertion in the Ncyto-Cin orientation, suggested that retention resulted from the interaction of the NH_2-terminus with a binding site on the surface of the organelle (Nguyen et al., 1988; Li and Shore, 1992b; Figure 2). More specifically, the rate of release of the NH_2-terminal retention signal from this binding site is slower than the rate at which the hydrophobic core of the signal-anchor enters the outer membrane translocation machinery, which forces the protein to adopt an N_{cyto}-C_{in} orientation (Figure 2). The nature of the putative binding site with which the retention signal interacts remains to be elucidated, but interaction with the lipid-bilayer likely plays an important role (Steenaart et al., 1996).

ACKNOWLEDGMENTS

This research was financed by the Medical Research Council and National Cancer Institute of Canada.

REFERENCES

Blobel, G. (1980). Intracellular protein topogenesis. Proc. Natl. Acad. Sci. USA 77, 1496–1499.

Bormann, B-J., & Engelman, D. M. (1992). Intra-membrane helix-helix association in oligomerization and transmembrane signalling. Annu. Rev. Biophys. Biomol. Struct. 21, 223–242.

Chen-Levy, Z., Nourse, J., & Cleary, M. L. (1989). The Bcl-2 candidate proto-oncogene product is a 24-kD integral-membrane protein highly expressed in lymphoid cell lines and lymphomas carrying the t(14:18). Mol. Cell. Biol. 9, 701–710.

Cosson, P., & Bonifacino, J. S. (1992). Role of transmembrane domain interactions in the assembly of class II MHC molecules. Science 258, 659–662.

Hartl, F-U., & Neupert, W. (1990). Protein sorting to mitochondria. Evolutionary conservations of folding and assembly. Science 247, 930–938.

Hartmann, E., Rapoport, T. A., & Lodish, H. F. (1989). Predicting the orientation of eukaryotic membrane-spanning proteins. Proc. Natl. Acad. Sci. USA 86, 5786–5790.

Hase, T., Mller, U., Riezman, H., & Schatz, G. (1984). A 70-kD protein of the yeast outer membrane is targeted and anchored via its extreme amino terminus. EMBO J. 3, 3157–3164.

High, S., & Dobberstein, B. (1992). Mechanisms that determine the transmembrane disposition of proteins. Curr. Op. Cell Biol. 4, 581–586.

Hines, V., Brandt, A., Griffiths, G., Horstmann, H. Brütsch, H., & Schatz, G. (1990). Protein import into yeast mitochondria is accelerated by the outer membrane protein MAS70. EMBO J. 9, 3191–3200.

Hirsch, J. D., Beyer, C. F., Malkowitz, L., Beer, B., & Blume, A.J. (1989). Mitochondrial benzodiazepine receptors mediate inhibition of mitochondrial respiration control. Mol. Pharmacol. 35, 157–163.

Hockenbery, D. M., Nunez, G., Milliman, C., Schreiber, R. D., & Korsmeyer, S. J. (1990). Bcl-2 is an inner mitochondrial membrane protein that blocks programmed cell death. Nature 348, 334–336.

Jacobson, M. D., Burnett, J. F., King, M. D., Miyashita, T., Reed, J. C., & Raff, M. C. (1993). Bcl-2 blocks apoptosis in cells lacking mitochondrial DNA. Nature 361, 365–369.

Jap, B. K. (1989). Molecular design of PhoE porin and its functional consequences. J. Mol. Biol. 205, 407–419.

Kassenbrock, C. K., Cao, W, & Douglas, M. G. (1993). Genetic and biochemical characterization of ISP6, a small mitochondrial outer membrane protein associated with the protein translocation complex. EMBO J. 12, 3023–3034.

Lemmon, M. A., Flanagan, J. M., Treutlein, H. R., Zhang, J., & Engelman, D. M. (1992). Sequence specificity in the dimerization of transmembrane α-helices. Biochem. 31, 12719–12725.

Li, J-M., & Shore, G. C. (1992a). Protein sorting between mitochondrial outer and inner membranes. Insertion of an outer membrane protein into the inner membrane. Biochim. Biophys. Acta 1106, 233–241.

Li, J-M., & Shore, G. C. (1992b). Reversal of the orientation of an integral protein of the mitochondrial outer membrane. Science 256, 1815–1817.

Mannella, C. A. (1993). The ins and outs of mitochondrial membrane channels. Trends Biochem. Sci. 17, 315–320.

McBride, H. M., Millar, D. G., Li, J-M., & Shore, G. C. (1992). A signal-anchor sequence selective for mitochondrial outer membrane. J. Cell Biol. 119, 1451–1457.

Millar, D. G., & Shore, G. C. (1993). The signal-anchor sequence of mitochondrial Mas70p contains an oligomerization domain. J. Biol. Chem. 268, 18403–18406.

Millar, D. G., & Shore, G. C. (1994). Mitochondrial Mas70p signal-anchor sequence: Mutations in the transmembrane domain that disrupt dimerization but not targeting or membrane insertion. J. Biol. Chem. 269, 12229–12232.

Moreno-Sanchez, R., Hogue, B. A., Bravo, C., Newman, A. H., Basile, A. S., & Chiang, P. K. (1991). Inhibition of substrate oxidation in mitochondria by the peripheral-type benzodiazepine receptor ligand AHN 086. Biochem. Pharmacol. 41, 1479–1484.

Nakai, M., Hase, T., & Matsubara, H. (1989). Precise determination of the mitochondrial import signal contained in the 70-kDa protein of yeast mitochondrial outer membrane. J. Biochem. 105, 513–519.

Nguyen, M., & G. C. Shore. (1987). Import of hybrid Vesicular stomatitis G protein to the mitochondrial inner membrane. J. Biol. Chem. 262, 3929–3931.

Nguyen, M., Bell, A. W., & Shore, G. C. (1988). Protein sorting between mitochondrial membranes specified by position of the stop-transfer domain. J. Cell Biol. 106, 1499–1505.

Nguyen, M., Millar, D. G., Yong, V. W., Korsmeyer, S. J., & Shore, G. C. (1993). Targeting of Bcl-2 to the mitochondrial outer membrane by a COOH-terminal signal anchor sequence. J. Biol. Chem. 268, 25265–25268.

Nguyen, N., Branton, P. E., Walton, P. A., Oltvai, Z. N., Korsmeyer, S. J., & Shore, G. C. (1994). J. Biol. Chem. 269, 16521–16524.

Pfaller, R., Kleene, R., & Neupert, W. (1990). Biogenesis of mitochondrial porin: The import pathway. Experientia 46, 153–161.

Ramage, L., Junne, T., Hahne, K., Lithgow, T., & Schatz, G. (1993). Functional cooperation of mitochondrial protein import receptors in yeast. EMBO J. 12, 4115–4123.

Raymond, Y., & Shore, G. C. (1979). The precursor for carbamyl phosphate synthetase is transported to mitochondria via a cytosolic route. J. Biol. Chem. 254, 9335–9338.

Schatz, G. (1993). The protein import machinery of mitochondria. Protein Science 2, 141–146.

Shore, G. C., Carignan, P., & Raymond, Y. (1979). *In vitro* synthesis of a putative precursor to the mitochondrial enzyme, carbamyl phosphate synthetase. J. Biol. Chem. 254, 3141–3143.

Shore, G. C., Millar, D. G., & Li, J-M. (1992). Protein insertion into mitochondrial outer and inner membranes via the stop-transfer sorting pathway. In: *Membrane Biogenesis and Protein Targeting* (Neupert, W., & Lill, R., Eds.). Elsevier Sci., Amsterdam, pp. 253–263.

Singer, S. J. (1990). The structure and insertion of integral proteins in membranes. Annu. Rev. Cell Biol. 6, 247–296.

Söllner, T., Griffiths, G., Pfaller, R., Pfanner, N., & Neupert, W. (1989). MOM19, an import receptor for mitochondrial precursor proteins. Cell 59, 1061–1070.

Söllner, T., Pfaller, R., Griffiths, G., Pfanner, N., & Neupert, W. (1990). A mitochondrial import receptor for the ADP/ATP carrier. Cell 62, 107–115.

Steenaart, N. A. E., Silvius, J. R., & Shore, G. C. (1996). An amphiphilic lipid-binding domain influences the topology of a signal-anchor sequence in the mitochondrial outer membrane. Biochemistry 35, 3764–3771.

Sternberg, M. J. E., & Gullick, W. J. (1989). Neu receptor dimerization. Nature 339, 587.

Sternberg, M. J. E., & Gullick, W. J. (1990). A sequence motif in the transmembrane region of growth factor receptors with tyrosine kinase activity mediates dimerization. Prot. Eng. 3, 245–248.

von Heijne, G. (1988). Transcending the impenetrable: How proteins come to terms with membranes. Biochim. Biophys. Acta 947, 307–333.

von Heijne, G., & Gavel, Y. (1988). Topogenic signals in integral membrane proteins. Eur. J. Biochem. 174, 671–678.

Vorobjer, I. A., & Zorov, D. B. (1983). Diazepam inhibits cell respiration and induces fragmentation of mitochondrial reticulum. FEBS Lett. 163, 311–314.

Weiss, M. A., Abele, U., Weckesser, J., Welte, W., Schiltz, E., & Schulz, G. E. (1991). Molecular architecture and electrostatic properties of a bacterial porın. Science 254, 1627–1630.

Wickner, W. T. and Lodish, H. F. (1985). Multiple mechanisms of protein insertion into and across membranes. Science 230, 400–407.

Zhuang, Z., & McCauley, R. (1989). Ubiquitin is involved in the *in vitro* insertion of monoamine oxidase B into mitochondrial outer membranes. J. Biol. Chem. 264, 14594–14596.

TARGETING AND TRANSLOCATION OF PREPROTEINS BY THE TOMs OF THE MITOCHONDRIAL RECEPTOR COMPLEX

Petra Keil, Angelika Hönlinger, and Nikolaus Pfanner

I.	Introduction	82
II.	Protein Import Receptors	83
	A. Tom20	84
	B. Tom70	88
	C. Functions of the Receptors in Mitochondrial Protein Import	89
III.	The Mitochondrial Receptor Complex	92
	A. The General Insertion Pore GIP	93
	B. Tom40	93
	C. Tom6 and Tom7	94
	D. The Transfer Component Tom22	95
IV.	Topology of Components of the Outer Membrane Import Machinery	96
V.	Biogenesis of the Import Components	96
VI.	Perspectives	97
	Acknowledgments	98
	References	98

Advances in Molecular and Cell Biology,
Volume 17, pages 81–102.

ISBN: 0-7623-0144-9

INTRODUCTION

Targeting of newly synthesized precursor proteins to mitochondria follows a basic principle of intracellular protein traffic. Signal sequences on the preproteins to be transported across membranes are recognized by complementary organelle-specific structures (Blobel and Dobberstein, 1975; Milstein et al., 1972). In case of mitochondria, the complementary structures (receptors) are integral proteins of the outer membrane and directly bind the preproteins (Pfanner et al., 1991), while in the case of the endoplasmic reticulum, preproteins are recognized by a soluble signal recognition particle that is subsequently targeted to a membrane-bound receptor (Rapoport, 1992).

The precursors of most nuclear-encoded mitochondrial proteins contain typical targeting signals. These signals can be present as amino-terminal cleavable signal sequences, termed presequences, or within the mature part of a protein. The various cleavable signal sequences do not share primary sequence homology, but tend to have common structural motifs, namely a high abundance of positively charged amino acid residues and the potential to form amphipathic α-helical structures (Hurt and van Loon, 1986; von Heijne, 1986; Hartl et al., 1989; Horwich, 1990). Cleavable presequences are found on many proteins destined for the matrix, inner membrane, or the intermembrane space. Outer membrane proteins and also several proteins of the other three mitochondrial subcompartments, however, possess internal targeting information (Pfanner et al., 1987a; Smagula and Douglas, 1988). Only in few cases, these internal targeting signals were characterized in detail (Hurt and van Loon, 1986; Li and Shore, 1992).

The mitochondrial membranes contain specific proteins that are responsible for the selective recognition and translocation of preproteins. In case of the mitochondrial outer membrane, the proteins were termed Toms (*t*ranslocase of the *m*itochondrial *o*uter *m*embrane); in case of the mitochondrial inner membrane, the proteins were termed Tims (*t*ranslocase of the *m*itochondrial *i*nner membrane) (Pfanner et al., 1996). The transport of preproteins across the outer membrane involves two sequential steps: (1) recognition by specific receptors; and (2) translocation through a general insertion pore GIP (Figure 1). The Tom proteins that perform these functions are assembled in a high molecular weight complex. The steps occurring at the outer membrane are independent of a membrane potential $\Delta\Psi$, while the translocation of preproteins across the inner membrane strictly requires a membrane potential. This review describes how the protein import machinery of the mitochondrial outer membrane has been identified and which functions are performed by the Tom proteins. Most of the knowledge about mitochondrial protein import was obtained in experiments with the fungi *Neurospora crassa* and *Saccharomyces cerevisiae*, preferentially with *in vitro* import systems, consisting of isolated mitochondria and preproteins that were synthesized in cell-free translation systems (e.g., rabbit reticulocyte lysate).

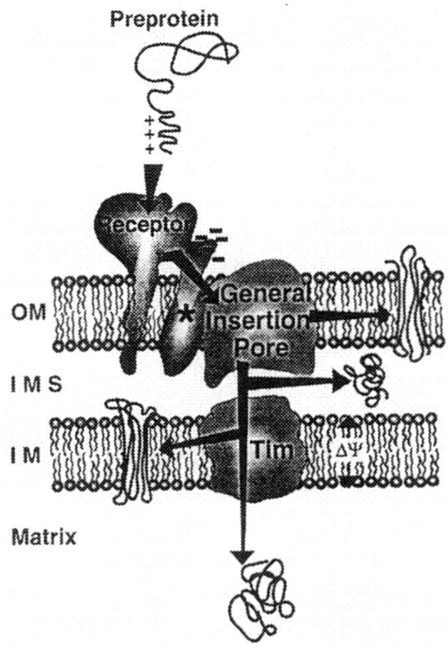

Figure 1. A general pathway for preprotein targeting into mitochondria. Nuclear-encoded mitochondrial preproteins are synthesized in the cytosol. Most of them are recognized by a receptor on the mitochondrial surface. A negatively charged outer membrane protein (*, that is Tom22) may modulate the interaction of preproteins with a receptor and promote the transfer of preproteins to the general insertion pore GIP in the outer membrane (OM). Translocation across the inner membrane (IM) is mediated by the mitochondrial inner membrane import machinery (Tim) and requires a membrane potential $\Delta\Psi$. The entire general import pathway is used by matrix-targeted preproteins. Preproteins destined for mitochondrial outer membrane, inner membrane and in some cases also intermembrane space (IMS) may leave the general import pathway at distinct stages and move to their functional destination (indicated by arrows). Not shown are the complex sorting pathways of some intermembrane space and inner membrane proteins that follow a "detour" via the matrix/inner membrane.

II. PROTEIN IMPORT RECEPTORS

When isolated mitochondria were treated with protease prior to incubation with preproteins, a strong reduction of the amount of bound and imported precursor proteins was observed (Riezman et al., 1983; Zwizinski et al., 1984). Since the protease treatment did not degrade the outer membrane barrier, it was concluded that proteinaceous components (receptors) of the mitochondrial outer membrane were needed for binding and import of precursors. To identify these components, antibodies were generated against most of the proteins found in purified outer membranes of *N. crassa* mitochondria. The antibodies, or Fab-fragments derived thereof, were bound to isolated mitochondria prior to addition of preproteins. Antibodies against two proteins with a molecular mass of 20 kDa and 70 kDa, respectively, led to a strong reduction in protein import (Söllner et al., 1989, 1990). These proteins were termed Tom20 and Tom70 and, as discussed in the following section, were shown to function as import receptors for preproteins. Both receptors have also been found in the yeast *S. cerevisiae* (Moczko et al., 1992, 1993; Ramage et al., 1993; Hines et al., 1990; Steger et al., 1990; Table 1).

A. Tom20

Tom20 is an integral protein of the outer membrane which functions as import receptor for most precursor proteins analyzed so far. It can be co-immunoprecipitated with and crosslinked to bound precursor proteins (Kiebler et al., 1990; Söllner et al., 1992). It is thought that Tom20 is the main receptor for preproteins carrying an amino-terminal presequence. The cleavable presequence has enough information for directing protein import via Tom20 (Becker et al., 1992), although it has to be emphasized that several preproteins without presequence also use Tom20.

Tom20 is anchored in the outer membrane by a hydrophobic sequence at the extreme amino-terminus, while the rest forms a hydrophilic domain exposed to the cytosol (Schneider et al., 1991; Ramage et al., 1993; Moczko et al., 1994). The receptor is equally distributed over the outer membrane, and it is assumed that Tom20 has the ability to laterally move in the outer membrane, collecting precursors from all over the mitochondrial surface and mediating, together with further outer membrane proteins, the delivery of the precursors to translocation sites. Tom20 also binds precursor proteins with internal targeting sequences not resembling presequences (e.g., several precursors of outer membrane proteins; Söllner et al., 1989; Keil et al., 1993; Moczko et al., 1994). Moreover, the proteolytic removal of a short carboxy-terminal segment from Tom20 impairs the import of some, but not of all preproteins. Antibodies directed against specific epitopes of Tom20 inhibit the import of various preproteins to different degrees (Söllner et al., 1989; Keil, unpublished data). We thus speculate that Tom20 carries more than one site for the interaction with preproteins.

Possible candidates for preprotein binding segments of Tom20 are: (1) the negatively charged extreme carboxy-terminus that is predicted to be of α-helical character (targeting sequences are positively charged and are predicted to be of α-helical character); (2) a segment in the middle of Tom20 that represents a tetratricopeptide (TPR) motif. TPR motifs are characterized by a degenerate 34-residue consensus motif and a potential to form an α-helical structure. TPR motifs are assumed to form hole and knob-like structures that promote intra- and intermolecular protein-protein interactions (Goebl and Yanagida, 1991). Interestingly, the second import receptor Tom70, Tom72, a putative peroxisomal import receptor and several other proteins with functions, for example, in cell division

Table 1. Components of the Mitochondrial Receptor Complex

Protein new name (old name)	*Essential for Viability*		*Proposed Function*	*Reference*
	Fermentable Carbon Source	*Non-fermentable Carbon Source*		
Tom 72	No	No	TPR-motif; homologous to Tom70	Bömer et al. (1996)
Tom70 (Mom72, Mas70)	No	No	Receptor preferentially for several preproteins with internal targeting sequences, but also for several cleavable ones; partially overlapping specificity with Tom20	Söllner et al. (1990) Hines et al. (1990) Steger et al. (1990) Hines and Schatz (1993) Moczko et al. (1993, 1994)
Tom40 (Mom38, ISP42)	Yes	Yes	Part of the General Insertion Pore GIP	Kiebler et al. (1990) Baker et al. (1990)
Tom37 (Mas37)	No	No	Receptor with same precursor specificity as Tom70; forms a complex with Tom70	Gratzer et al. (1995)
Tom22 (Mom22, Mas22, Mas17)	Yes	Yes	Transfer of precursor proteins from receptors to GIP (the cytsolic and intermembrane space domains of Tom22 are negatively charged); possibly second subunit of the main receptor Tom20	Kiebler et al. (1993) Lithgow et al. (1994) Hönlinger et al. (1995) Mayer et al. (1995) Nakai and Endo (1995)

(continued)

Table 1. (Continued)

Protein new name (old name)	Essential for Viability		Proposed function	Reference
	Fermentable Carbon Source	Non-fermentable Carbon Source		
Tom20 (Mom19, Mas20)	No	Yes	Receptor for most preproteins, preferentially preproteins with cleavable ones; partially overlapping specificity with Tom70	Söllner et al. (1989) Moczko et al. (1993, 1994) Ramage et al. (1993) Haucke et al. (1995)
Tom7 (Mom7)	No	No	Modulator of dynamics of the receptor complex; role in sorting and accumulation of preproteins at the outer membrane	Hönlinger et al. (1996)
Tom6 (Mom8b, ISP6p)	No	No	Stabilization of Tom40; modulator of dynamics of the receptor complex; promotes efficient transfer of pre-proteins from import receptors	Kassenbrock et al. (1993) Alconada et al. (1995) Hönlinger et al. (1996)
Tom5 (Mom8, Mom8a)	No	at elevated temperature	n.d.	Moczko et al. (1992) Alconada et al. (1995) Hönlinger et al. (1995, 1996) Dietmeier, K., Hönlinger A., Kubrich M., & Pfanner, N. (unpublished)

Notes: GIP, General insertion pore; ISPx, import site protein with approximate molecular mass of x kDa; Mas, mitochondrial assembly (in the case of the outer membrane, the number indicates the approximate molecular mass of a Mas-protein in kDa; in other cases, for example mitochondrial inner membrane or matrix, the Mas-proteins are numbered consecutively in the order of their identification); Mom y, mitochondrial outer membrane protein with approximate molecular mass of y kDa; n.d., not determined; Tomz, subunit of the translocase of the outer mitochondrial membrane with the molecular mass of z kDa.

cycle or neurogenesis contain TPR motifs (McCollum et al., 1993; Moczko et al., 1994; Bömer et al., 1996). The TPR motif in Tom20 may be involved in the interaction with other components of the import machinery like Tom70 (Haucke et al., 1996) or in the interaction with the targeting sequences.

S. cerevisiae mutants lacking Tom20 were constructed. The mutant cells are unable to grow on a nonfermentable carbon source and are impaired in growth on a fermentable carbon source. The cells accumulate mitochondrial precursor proteins in the cytosol. The degree of import inhibition obtained by a deletion of Tom20 is comparable to that obtained by blocking the receptor with specific antibodies or Fab-fragments, demonstrating the validity of the antibody inhibition experiments (Moczko et al., 1994). At a first glance, it may be surprising that cells lacking the main receptor Tom20 are viable. It was previously found that the import receptors Tom20 and Tom70 have a (partially) overlapping specificity and thus can substitute for each other's function to a certain degree (Becker et al., 1992; Steger et al., 1990). Therefore, a deletion of Tom20 is not lethal to yeast cells under all growth conditions.

A human mitochondrial protein with significant homology to Tom20 from *N. crassa* and *S. cerevisiae* was recently identified. The protein shows the three characteristics of Tom20, that is an amino-terminal hydrophobic membrane anchor, a TPR motif in the middle, and a negatively charged carboxy-terminus (Moczko et al., 1994; Goping et al., 1995; Hanson et al., 1995; Seki et al., 1995).

In plant cells, a protein homologous to Tom20 has not been found so far. Though it is reasonable to assume that the mechanisms of protein targeting to mitochondria are similar in all organisms and thus also plant cells should contain a Tom20 homolog, one has to take into account that plant cells have a higher level of complexity due to the presence of chloroplasts. Like mitochondria, chloroplasts are assumed to be derived from prokaryotic ancestors that were taken up into a eukaryotic cell by an endocytosis event (endosymbiont hypothesis). Both organelles are surrounded by two membranes, and several principles of preprotein import seem to be comparable. Mitochondrial and chloroplast targeting sequences are both of hydrophilic character and are mostly devoid of negatively charged amino acid residues. A detailed inspection, however, revealed significant differences between mitochondrial and chloroplast targeting sequences. In particular, chloroplast targeting sequences of higher plants are usually not predicted to form amphipathic α-helical structures (von Heijne et al., 1989). Boutry and colleagues (1987) demonstrated for typical mitochondrial and chloroplast targeting sequences, respectively, that they were selectively transported to the correct organelle, implying that the import receptors of these plant organelles precisely distinguish between the targeting sequences. Surprisingly, however, two preproteins of the chloroplast envelope inner membrane possess targeting sequences with a potential to form amphipathic α-helices, resembling mitochondrial presequences. These preproteins could be efficiently imported into isolated fungal mitochondria in a receptor (Tom20)-dependent manner (Brink et al., 1994). Both preproteins could even be imported into isolated plant mitochondria, although with a low efficiency, suggesting that here the receptor-independent bypass import was used. This implies a higher degree of specificity of the mitochondrial import receptors in plants than in fungi.

B. Tom70

Tom70 functions as receptor for several preproteins without a presequence, including: the inner membrane proteins ADP/ATP carrier, phosphate carrier and probably other members of the inner membrane carrier family (Söllner et al., 1990; Steger et al., 1990; Dietmeier et al., 1993), and the outer membrane proteins Tom40 and Tom22 (Keil and Pfanner, 1993; Keil et al., 1993). Moreover, Tom70 functions as import receptor for several cleavable preproteins, such as the cytochrome c_1 of the inner membrane, although the cleavable preproteins use Tom20 for the major portion of their import (Hines et al., 1990; Steger et al., 1990; Hines and Schatz, 1993; Moczko et al., 1994).

The function of Tom70 as import receptor was demonstrated by several different technics: inhibitory antibodies, co-immunoprecipitation of preproteins with Tom70, crosslinking of preproteins to Tom70, and the use of yeast mutants lacking Tom70 (Söllner et al., 1990, 1992; Hines et al., 1990; Steger et al., 1990). Yeast mutants lacking Tom70 do not only grow well on fermentable carbon sources, but are able to grow on nonfermentable carbon sources, albeit with reduced rate at higher temperature. A deletion of Tom20 thus has a much more pronounced effect on the viability of yeast cells than a deletion of Tom70 (Ramage et al., 1993; Moczko et al., 1994).

The receptors Tom20 and Tom70 do not share an overall primary sequence homology (only the limited segments of the TPR motifs are of significant similarity), yet their topology is similar (Figure 2). Tom70 is anchored in the outer membrane by a hydrophobic amino-terminal sequence and exposes a large 65 kDa-domain to the cytosol. Like Tom20, Tom70 exposes a short amino-terminal segment (roughly 10 amino acids in front of the membrane anchor) to the intermembrane space. Tom70 forms a homodimer (Söllner et al., 1992; Miller and Shore, 1993). It remains to be determined if dimerization is related to the binding of ligand and important for the function of Tom70.

The cytosolic 65 kDa-domain of Tom70 is a stably folded domain which can be removed from the membranes by treatment with protease and is able to bind preproteins. Schlossmann and colleagues (1994) expressed the cytosolic domain in *Escherichia coli* cells and purified it to homogeneity. The soluble 65 kDa-domain showed the same specificity for binding of preproteins as intact Tom70 on the mitochondrial surface. Apparently, the cytosolic domain of Tom70 is sufficient for selective recognition of preproteins, independently of its membrane anchor and other mitochondrial proteins. However, larger amounts of the purified cytosolic domain than of membrane-bound Tom70 were needed to bind a certain amount of

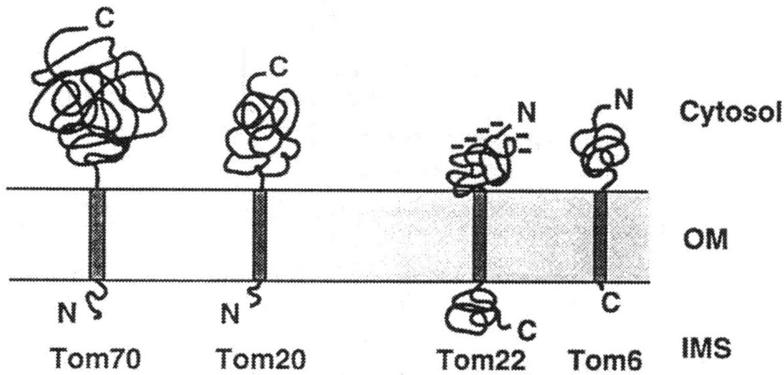

Figure 2. Topology of mitochondrial outer membrane proteins with a single membrane-spanning segment. C, carboxy-terminus; IMS, intermembrane space; Tomx, subunit of the translocase of the mitochondrial outer membrane of x kDa; N, amino-terminus; OM, outer membrane.

preproteins. This may be caused by a partial inactivation of the cytosolic domain during isolation. Alternatively, it cannot be excluded that the membrane anchor sequence or other mitochondrial proteins are involved in stabilization of the cytosolic domain or directly enhance the efficiency of binding.

Denatured preproteins with a tendency for misfolding or aggregation showed a stronger dependence on Tom70 than the nondenatured preproteins (Hines and Schatz, 1993). We propose that, in addition to its receptor function, Tom70 has a chaperone-like role by preventing misfolding or aggregation of bound preproteins and thus helps preproteins to stay in a translocation-competent conformation. Tom70 was found to be in a complex with Tom37, another protein of the outer membrane protein complex. The inactivation of Tom37 leads to an inhibition of import of those preproteins into mitochondria that are also affected in cells with deleted Tom70. Tom37 thus has receptorlike functions and could act in cooperation with Tom70 (Gratzer et al., 1995).

C. Functions of the Receptors in Mitochondrial Protein Import

Tom20 and Tom70 have a preference for distinct classes of preproteins: Tom20 for preproteins with cleavable presequences, and Tom70 for preproteins with internal signal sequences. This distinction, however, is mainly of quantitative, not of qualitative nature, since both receptors possess partially overlapping specificities. Small fractions of cleavable preproteins use Tom70 for their import, and preproteins with internal signal sequences can be imported via Tom20. For example, in the import of members of the inner membrane carrier family, about 20% of the receptor activity is due to Tom20. Moreover, several noncleavable preproteins, such as the porin of the outer membrane, use Tom20 as their main receptor. In summary, the bulk of mitochondrial preproteins is preferentially imported via Tom20. Tom70 functions in parallel to Tom20. The overlapping specificity of both receptors is of particular importance when one of the receptors is nonfunctional or deleted. Then the other receptor ensures the survival of the cells (Moczko et al., 1994). Deletion of both receptors is lethal to yeast cells under all growth conditions (Ramage et al., 1993; Hönlinger et al., 1995).

The receptors appear to directly recognize the preproteins without a strict need for cytosolic cofactors (Becker et al., 1992; Söllner et al., 1992) (the functions of cytosolic cofactors are discussed in the chapters by Estey and Douglas and by Mihara and Omura). The preproteins are then transferred to the general insertion pore GIP for translocation across the outer membrane. It is tempting to speculate that the binding of preproteins induces conformational changes and oligomerization of the receptor molecules that directly or indirectly cause an activation of downstream components of the import machinery such as the general insertion pore or the mitochondrial inner membrane machinery. The short amino-terminal tails of the receptors which reach into the intermembrane space may be involved in such a signal transmission.

The presence of two different protein import receptors on the mitochondrial surface may be explained in view of the endosymbiont hypothesis of mitochondrial origin. Most proteins that were originally encoded in the prokaryotic ancestor of mitochondria received presequences after transfer of their genes to the nucleus of the eukaryotic host. The presequences direct the proteins back to mitochondria via the receptor Tom20. Proteins that were not present in the prokaryotic ancestor, but were probably invented in the eukaryotic cell (e.g., the ADP/ATP carrier; Klingenberg, 1985), were constructed with internal targeting signals, which allowed a high affinity interaction with Tom70. Then an overlapping specificity of the two receptors for the various targeting signals evolved as safety system/back up system in case that one receptor is defective or for an increase of the import efficiency.

Though this view of two import receptors with partially complementary and partially overlapping specificities seems to be complicated, it still represents an oversimplification of mitochondrial protein targeting, as import pathways exist that do not need the import receptors Tom20 or Tom70. (1) Apocytochrome *c*, the precursor of cytochrome *c* of the intermembrane space, is imported without a need for surface receptors (Nicholson et al., 1988). Apocytochrome *c* is a unique preprotein that possesses the ability to spontaneously insert into lipid membranes (Jordi et al., 1992). The specific targeting to the intermembrane space appears to be determined by binding of the imported preprotein to the enzyme cytochrome *c* heme lyase (Dumont et al., 1987; Nargang et al., 1988). It is not clear so far if membrane proteins are involved in the import pathway of apocytochrome *c*. (2) The precursor of yeast cytochrome oxidase subunit Va contains a typical presequence and its import shows several characteristics of the general mitochondrial import pathway, including requirement for a membrane potential across the inner membrane and for a functional heat shock protein hsp70 in the matrix (Miller and Cumsky, 1991; Gärtner et al., 1995). Unexpectedly, import of the preprotein was not affected at all by a removal of the surface receptors. In contrast to apocytochrome *c*, which does not use any of the known components of the general import pathway, pre-subunit Va seems to enter the general pathway, but circumvents the surface receptors. It is unknown which component(s) is/are responsible for the specific targeting of pre-subunit Va. (3) Isolated mitochondria that have been depleted of their surface receptors by a pretreatment with protease are still able to import most preproteins although with a strongly reduced efficiency (about 10 to 20-fold reduction) (Pfaller et al., 1989). This "bypass import" shows several characteristics of mitochondrial protein import, such as dependence on a positively charged targeting sequence and a membrane potential, and is assumed to occur by direct insertion of the preprotein into an outer membrane translocation channel, that is the general insertion pore. The specificity of the bypass import as well as its relevance *in vivo* are unclear. One may speculate that bypass import represents, in evolutionary terms, an early form of mitochondrial protein import. The development of a highly specific and efficient import then required the evolution of specific targeting signals and import receptors. This possibly provided the basis for the massive transfer of genes from the mitochondrion to the nucleus. (4) A complex situation exists for the biogenesis of the components of the mitochondrial outer membrane import machinery itself. As described at the end of the article, a variety of distinct, yet well-controlled, mechanisms are used to ensure a correct targeting of the "targeting components."

In yeast, a third protein, the 32 kDa integral membrane protein p32, has been reported to function as import receptor (Pain et al., 1990). Antibodies recognizing p32 inhibited protein import into isolated mitochondria. p32 was localized to the outer membrane at contact sites by immunoelectron microscopy. Unexpectedly, the primary sequence of p32 (Murakami et al., 1990) was identical to a protein characterized as the mitochondrial phosphate carrier, an inner membrane protein (Phelps and Wohlrab, 1991; Phelps et al., 1991). We found that the bulk of p32/phosphate carrier was indeed located in the mitochondrial inner membrane (Dietmeier et al., 1993) in agreement with a function as phosphate carrier. On the other hand, it was shown that purified p32 binds to mitochondrial presequences, but not to signal sequences required for protein translocation into chloroplasts or endoplasmic reticulum (Murakami et al., 1993). p32 may thus have a dual function as phosphate carrier and as a signal sequence binding subunit of a protein-conducting channel of the inner membrane.

In addition, recent studies indicate that Tom22 has receptor-like activities and can partially substitute for Tom20 and Tom70 (Lithgow et al., 1994; Hönlinger et al., 1995).

III. THE MITOCHONDRIAL RECEPTOR COMPLEX

Upon solubilization of *N. crassa* or *S. cerevisiae* mitochondria with a mild detergent (digitonin), a high molecular weight complex of the outer membrane was identified (Kiebler et al., 1990; Moczko et al., 1992; Hönlinger et al., 1995). It consists of (at least) eight proteins termed Tom70, Tom40, Tom37, Tom22, Tom20, Tom7, Tom6, and Tom5 according to their apparent sizes on an SDS polyacrylamide gel (Table 1, Figure 3). Since it contains the import receptors, the complex is called the mitochondrial receptor complex. As discussed ahead, the complex is responsible for recognition and membrane insertion of precursor proteins.

With antibodies directed against Tom20, Tom22 or Tom40, a co-immunoprecipitation of all eight proteins of the complex is possible. Antibodies against other subunits of the complex, for example, Tom70, preferentially precipitate only a subset of the components (Hönlinger et al., 1996). The complex may be of dynamic nature and may consist of several interacting subcomplexes. The strongest interaction could be found between Tom40, Tom22, Tom20 and the small Toms while Tom37 and Tom70 are more loosely associated with the core complex (Moczko et al., 1992; Hönlinger et al., 1996). At least half of the amount of Tom70 can be found outside the complex; this fraction of Tom70 seems to be responsible for the initial

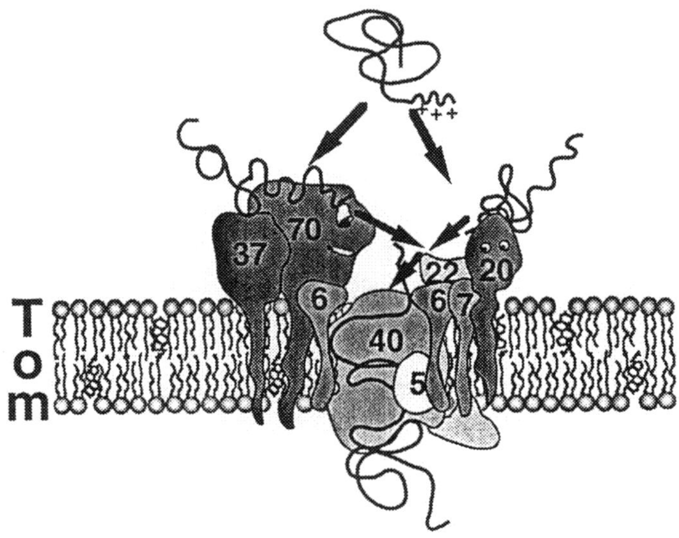

Figure 3. Hypothetical model of the receptor complex in the mitochondrial outer membrane. Tomx = subunit of the translocase of the mitochondrial outer membrane.

binding of the precursor of the ADP/ATP carrier. Binding of the preprotein may then induce the assembly of Tom70 in the complex to allow transfer of the preprotein to the further import components. Moreover, a single receptor complex appears to contain more than one copy of individual subunits. This was suggested by two observations: (1) co-expression of the *Neurospora* Tom20 in yeast and purification of complexes that contained both *N. crassa* and yeast Tom20 (Moczko et al., 1992); (2) the discovery of a homodimer of Tom70 (Söllner et al., 1992; Miller and Shore, 1993).

A treatment of mitochondria with proteases divides the components of the receptor complex into two groups: those exposing major protease-accessible portions to the cytosol (Tom70, Tom37, Tom22, and Tom20); and the others, which are largely protected against proteolytic attack (in particular trypsin) (Figures 2 and 3).

A. The General Insertion Pore GIP

The general insertion pore is responsible for the insertion of preproteins into and translocation across the outer membrane. Its initial characterization was based on the identification of a translocation intermediate of the ADP/ATP carrier. The intermediate was protected against externally added proteases (in contrast to intermediates accumulated at the receptor stage), yet not inserted into the inner mitochondrial membrane. It was concluded that the preprotein was accumulated in the translocation pore of the outer membrane (Pfanner and Neupert, 1987; Pfanner et al., 1987b). The pore was proposed to function as the "general" insertion pore since various precursors analyzed used the same saturable site for insertion into the outer membrane (with the exception of apocytochrome *c*) (Pfaller et al., 1988). A stage-specific crosslinking of the precursor of the ADP/ATP carrier indicated that the proteins Tom40, and Tom5 were in close proximity to the GIP-intermediate (Söllner et al., 1992; Hönlinger, A., Kübrich, M., and Pfanner, N., unpublished). Tom40 was demonstrated to be a central and essential component of the receptor complex (see below).

B. Tom40

Tom40 is an integral membrane protein with at least one membrane spanning region in the amino-terminal third of the protein (Kiebler et al., 1990). After opening of the intermembrane space, Tom40 is digested by protease (Kiebler, 1993), similar to a preprotein accumulated in the general insertion pore GIP (Rassow and Pfanner, 1991). The exact topology of Tom40 in the outer membrane has not been defined so far, the presence of one to four membrane-spanning segments is discussed. The carboxy-terminal half of Tom40 contains two segments with similarities to the A and B-regions of an ATP-binding consensus sequence (Chin et al., 1988), but up to now neither an ATPase activity nor binding of nucleotides to Tom40 could be observed. Two segments of Tom40 with a predicted α-helical conformation are characterized by a high content of negatively charged amino acid residues and the absence of positively charged residues. In both cases, the negative charges are located on one side of the predicted helices. These regions are putative candidates for sites of interaction with positively charged presequences. Moreover, Tom40 contains a segment with similarity to a sequence in the phosphate carrier. The relevance of this similarity has not been analyzed so far.

In *S. cerevisiae* Tom40 was identified by crosslinking to a precursor protein spanning outer and inner membranes at translocation contact sites (Vestweber et al., 1989). Antibodies directed against Tom40 inhibited protein import into mitochondria which had been pretreated with trypsin, that is the antibodies inhibited the bypass import. Antibodies against the carboxy-terminus of Tom40 inhibited import into isolated *N. crassa* mitochondria with opened outer membrane, suggesting that the carboxy-terminal part is located on the intermembrane space side and involved in protein import (Kiebler, 1993). The importance of Tom40 was underscored by the lethal phenotype of disruption of its gene in yeast. Tom40 is essential for the viability of yeast at all growth conditions (Baker et al., 1990).

C. Tom6 and Tom7

The yeast mitochondrial outer membrane contains a small protein of 61 amino acid residues, termed Tom6. Its gene was identified as a high copy number suppressor of a temperature sensitive mutation in the *S. cerevisiae* gene for Tom40. Tom6 has a membrane spanning segment in the carboxy-terminal half and exposes an amino-terminal segment to the cytosol. This amino-terminal domain is accessible for proteases (Figure 2). Kassenbrock and colleagues (1993) showed that Tom6 and several unidentified proteins were co-immunoprecipitated with an antiserum directed against Tom40. Deletion of the *TOM6* gene did not show any detectable effect on the growth of yeast as long as *TOM40* was present. In a temperature-sensitive *tom40*-mutant, however, disruption of *TOM6* was lethal, suggesting a functional or even physical interaction of Tom40 and Tom6.

Alconada and colleagues (1995) showed, that Tom6 promotes the interaction between the receptors and the GIP. In addition it reduces the stability of binding of the preproteins to the receptors. Another small component of the receptor complex is Tom7. It is also an integral membrane protein and consists of 60 amino acid residues. A lack of Tom7 preferentially inhibits import of the mitochondrial outer membrane protein porin, whereas the import of preproteins destined for the mitochondrial interior is impaired only slightly. If the protein import occurred in two steps, that is, preprotein accumulation at the outer membrane in the absence of a membrane potential and subsequent further import after reestablishment of a membrane potential, the protein import into the mitochondrial interior was strongly inhibited. This, Tom7 plays a role in sorting and accumulation of preproteins at the outer membrane. In addition, Tom7 exerts a destabilizing effect on the interaction between Tom20, Tom22, and the GIP protein Tom40 (Hönlinger et al., 1996).

Tom6 and Tom7 have a complementary function in modulating the dynamics of the receptor complex: Tom6 is responsible for stabilizing the interaction of the receptors with the GIP, whereas Tom7 has a destabilizing effect.

D. The Transfer Component Tom22

Tom22 is an integral membrane protein of the mitochondrial receptor complex (Kiebler et al., 1993; Lithgow et al., 1994; Hönlinger et al., 1995; Nakai and Endo, 1995). Depending on the gel system used, Tom22 runs above or below Tom20 on SDS-polyacrylamide gels. The deduced primary sequence shows that Tom22 is by 3 kDa smaller than Tom20. It has two hydrophilic domains, one on either side of the outer membrane, which are connected by a transmembrane segment. The amino-terminal domain comprises half of the molecule and is located on the cytosolic side of the membrane (Figure 2). It is characterized by an exceptional prevalence of negatively charged residues. The accessibility to added proteases groups Tom22 in an intermediate position between the surface-exposed receptors and the membrane-embedded components of the general insertion pore (such as Tom40).

Antibodies directed against the cytosolic domain of Tom22 inhibited the import of both preproteins targeted via Tom20 and preproteins targeted via Tom70. The antibodies did not block the binding of the precursor of the ADP/ATP carrier to its receptor Tom70, but the insertion into the general insertion pore GIP (Kiebler et al., 1993). Therefore, Tom22 seems to function after Tom70, but before the GIP. This suggests that Tom22 acts in the transfer of preproteins from the receptors to GIP; its negatively charged domain may facilitate the insertion of the positively charged presequences into the membrane (Figures 1 and 3). Co-immunoprecipitation experiments suggest that Tom22 is more tightly associated with Tom20 than with Tom70 (Moczko et al., 1992). It is conceivable that Tom22 supports the binding of preproteins to Tom20 by binding or stabilizing the positively charged face of amphophilic presequences.

The possible functional assignments of Tom22 thus comprise a spectrum from the central transfer component of the receptor complex (Kiebler et al., 1993) to a second subunit of the main receptor Tom20. Since the receptor complex is not a static unit, but of dynamic nature (Moczko et al., 1992), it may be possible that the Tom22 molecules in a complex are able to perform both proposed functions. Tom22 is the second protein of the mitochondrial receptor complex found to be essential for viability of yeast, indicating a central importance of Tom22 (Lithgow et al., 1994; Hönlinger et al., 1995). The carboxy-terminal domain of Tom22 in the intermembrane space specifically binds mitochondrial targeting sequences and is required to anchor the N-terminal part of the sequence in the intermembrane space (Bolliger et al., 1995). As discussed for the import receptors, the interaction of a preprotein with the cytosolic domain may cause changes in the conformation or oligomerization state of the intermembrane space part and thereby induce a signal transmission to further components of the import machinery.

IV. TOPOLOGY OF COMPONENTS OF THE OUTER MEMBRANE IMPORT MACHINERY

Four of the outer membrane import components possess one membrane segment each. Their orientation in the outer membrane classifies them into two groups: proteins that expose the carboxy-terminal domain to the cytosol (Tom20 and Tom70) and proteins that have their amino-termini facing the cytosol (Tom22 and Tom6; Figure 2). Interestingly, the two proteins that initially recognize the preproteins expose their carboxy-terminal domain to the outside, while the two proteins acting later on the pathway have the opposite orientation. The characterization of further subunits of the receptor complex will reveal if a relation exists between membrane topology and import function of a component. The mode of targeting and assembly of a subunit into the receptor complex may also be critical for the determination of its orientation in the membrane.

V. BIOGENESIS OF THE IMPORT COMPONENTS

The receptor complex represents the major protein entry gate of mitochondria. Its function is of critical importance to maintain the specific protein composition of mitochondria. The temporally and spatially correct assembly of the receptor complex in the mitochondrial outer membrane seems to be of vital importance for the cell. Mislocalization of the receptor components to another cellular membrane would most likely lead to a deleterious situation for the cell. We therefore asked how the correct targeting and assembly of the "targeting components" (i.e., the subunits of the receptor complex) is assured. All components of the receptor complex are encoded by nuclear genes and have to be imported from the cytosol.

We will have a look at the four components the import pathways of which have been analyzed in some detail. (1) Tom70: A pre-incubation of mitochondria with antibodies against Tom20 or a pretreatment with protease reduce the import of the precursor of Tom70 (Söllner et al., 1990). Tom20 may be involved in the initial recognition or the assembly of the Tom70 precursor. Since mitochondria lacking Tom20 are still able to import the precursor of Tom70 (Ramage et al., 1993; Moczko et al., 1994), further mechanisms controlling the import of Tom20 have to exist. (2) Tom20: The precursor of Tom20 was efficiently imported into mitochondria lacking both surface receptors (after a pretreatment with protease). Antibodies directed against Tom40, however, inhibited the import of Tom20, suggesting that targeting of the precursor of Tom20 does not require a surface receptor, but is controlled by its assembly with Tom40 (Schneider et al., 1991). (3) Tom40: In the assumption that Tom40 is responsible for targeting of the precursor of Tom20, the correct targeting of the precursor of Tom40 to the mitochondrial receptor complex represents a critical step in directing the import of the mitochondrial main receptor. Both receptors Tom20 and Tom70 are involved in the import of the precursor of Tom40 (Keil et al., 1993). The import of Tom40 could be resolved in a two-stage reaction: binding to the mitochondrial surface and assembly into the outer membrane. Both receptors together supported the efficient binding of the precursor to the mitochondrial surface. Inactivation of either one of the receptors led to a reduction of specific binding of about 80%. It has to be noted that this is a new variant of the functions of the surface receptors, since here they are not acting independently of each other, but both cooperate in binding of the preprotein. As is the case with most mitochondrial preproteins, the requirement for surface receptors is not absolute. Mutant mitochondria lacking either one of the receptors were still able to import Tom40, yet with strongly reduced rate. (4) Tom22 Like with the import of Tom40, both surface receptors are required for the import of Tom22 (Keil and Pfanner, 1993). Interestingly, Tom22 was found to be the first preprotein whose import strictly depended on the presence of surface receptors, that is a pretreatment of mitochondria with protease completely blocked its import. It is concluded that Tom22 leaves the typical mitochondrial import pathway already before the general insertion pore GIP and thus is unable to use the bypass import, which is assumed to occur by direct insertion of a preprotein into the GIP.

In summary, none of the subunits of the mitochondrial receptor complex is sufficient for its own targeting. Import of a subunit requires the function of at least one other correctly assembled subunit, in most cases several other subunits may be needed. Therefore, targeting and assembly of the components of the receptor complex can only occur to preexisting complexes, ensuring a formation of receptor complexes only in the mitochondrial outer membrane. We propose that the dynamic nature of the receptor complex provides the molecular basis for a stepwise assembly of newly synthesized components into preformed complexes.

VI. PERSPECTIVES

The mitochondrial receptor complex seems to represent the major site for targeting of preproteins to mitochondria and the translocation across the outer membrane. Subcomplexes may be responsible for initial import steps or the binding of different classes of preproteins. Upon binding of a preprotein, the subcomplexes may assemble to form the entire complex with the translocation channel. The complex dissociates rapidly in most detergents, indicating a labile interaction of the subunits. It is thus conceivable that the receptor complex contains additional subunits that were not detected with the purification procedures used so far. Future studies will include the search for putative further subunits.

Major questions of interest are: What molecular characteristics specify the signal in the targeting sequences and how are the binding sites of the receptors structured? Which components are genuine constituents of the translocation channel and which have only a supportive (indirect) function? It has to be analyzed if the negatively charged regions in several of the Tom-proteins (Tom20, Tom22, and Tom40) are critical for directing the import of the positively charged signal sequences. Does binding of a preprotein induce transfer of information from the receptors to the insertion pore and to components of the inner membrane import machinery, for example to trigger a selective opening of the channel? Further questions are concerned with the driving force of the translocation process, a putative involvement of cytosolic cofactors in the function of the receptor complex, and the regulation of expression and turnover of the subunits of the receptor complex. An important goal is the analysis of protein translocation with the purified receptor complex reconstituted into liposomes.

ACKNOWLEDGMENTS

We are grateful to our colleagues for their contributions to the work reviewed here and for sharing unpublished information with us. We thank Dr. J. Rassow for constructive comments and M. Schmusch for help in preparation of the manuscript. Work of the authors laboratory was supported by the Deutsche Forschungsgemeinschaft and the Fonds der Chemischen Industrie.

REFERENCES

Alconada, A., Kübrich, M., Moczko, M., Hönlinger, A., & Pfanner, N. (1995). The mitochondrial receptor complex: the small subunit Mom8b/Isp6 supports association of receptors with the general insertion pore and transfer of preproteins. Mol. Cell. Biol. 15, 6196–6205.

Baker, K. P., Schaniel, A., Vestweber, D., & Schatz, G. (1990). A yeast mitochondrial outer membrane protein essential for protein import and cell viability. Nature 348, 605–609.

Becker, K., Guiard, B., Rassow, J., Söllner, T., & Pfanner, N. (1992). Targeting of a chemically pure preprotein to mitochondria does not require the addition of a cytosolic signal recognition factor. J. Biol. Chem. 267, 5637–5643.

Blobel, G., & Dobberstein, B. (1975). Transfer of proteins across membranes. I. Presence of proteolytically processed and unprocessed nascent immunoglobulin light chains on the membrane-bound ribosomes of murine myeloma. J. Cell. Biol. 67, 835–851.

Bolliger, L., Junne, T., Schatz, G., & Lithgow, T. (1995). Acidic receptor domains on both sides of the outer membrane mediate translocation of precursor proteins into yeast mitochondria. EMBO J. 14, 6318–6326.

Bömer, U., Pfanner, N., & Dietmeier, K. (1996). Identification of a third yeast mitochondrial Tom protein with tetratricopeptide repeats. FEBS Lett. 382, 153–158.

Boutry, M., Nagy, F., Poulsen, C., Aoyagi, K., & Chua, N. H. (1987). Targeting of bacterial chloramphenicol acetyltransferase to mitochondria in transgenic plants. Nature 328, 340–342.

Brink, S., Flügge, U. I., Chaumont, F., Boutry, M., Emmermann, M., Schmitz, U., Becker, K., & Pfanner, N. (1994). Preproteins of chloroplast envelope inner membrane contain targeting information for receptor-dependent import into fungal mitochondria. J. Biol. Chem. 269, 16478–16485.

Chin, D., Goff, A., Webster, T., Smith, T., & Goldberg, A. (1988). Sequence of the lon gene in *E. coli* (a heat shock gene that encodes the ATP dependent protease La). J. Biol. Chem. 263, 11718–11728.

Dietmeier, K., Zara, V., Palmisano, A., Palmieri, F., Voos, W., Schloßmann, J., Moczko, M., Kispal, G., & Pfanner, N. (1993). Targeting and translocation of the phosphate carrier/p32 to the inner membrane of yeast mitochondria. J. Biol. Chem. 268, 25958–25964.

Dumont, M. E., Ernst, J. F., Hampsey, D. M., & Sherman, F. (1987). Identification and sequence of the gene encoding cytochrome *c* heme lyase in the yeast *Saccharomyces cerevisiae*. EMBO J. 6, 235–241.

Gärtner, F., Voos, W., Querol, A., Miller, B., Craig, E. A., Cumsky, M. G., & Pfanner, N. (1995). Mitochondrial import of COXVa characterized with yeast mutants: Independence from receptors, but requirement for matrix hsp70 translocase function. J. Biol. Chem. 270, 3788–3795.

Goebl, M., & Yanagida, M. (1991). The TPR snap helix: A novel protein repeat motif from mitosis to transcription. Trends Biochem. Sci. 16, 173–177.

Goping, I. S., Miller, D. G., & Shore, G. C. (1995). Identification of human mitochondrial protein import receptor, huMAS20p. Complementation of Δ*mas20* in yeast. FEBS Lett. 373, 45–50.

Gratzer, S., Lithgow, T., Bauer, R. E., Lamping, E., Paltauf, F., Kohlwein, S. D., Haucke, V., Junne, T, Schatz, G., & Horst, M. (1995). Mas37p, a novel subunit for protein import into mitochondria. J. Cell. Biol. 129, 25–34.

Hanson, B., Nuttall, S., & Hoogenraad, N. (1996). A receptor for the import of proteins into human mitochondria. Eur. J. Biochem. 235, 750–753.

Hartl, F. U., Pfanner, N., Nicholson, D. W., & Neupert, W. (1989). Mitochondrial protein import. Biochim. Biophys. Acta 988, 1–45.

Haucke, V., Lithgow, T., Rospert, S., Hahne, K., & Schatz, G. (1995). The yeast mitochondrial protein import receptor Mas20p binds precursor proteins through electrostatic interaction with the positively charged presequence. J. Biol. Chem. 270, 5565–5570.

Haucke, V., Horst, M., Schatz, G., & Lithgow, T. (1996). The Mas20p and Mas70p subunits of the protein import receptor of yeast mitochondria interact via the tetratricopeptide repeat motif in Mas20p: Evidence for a single hetero-oligomeric receptor. EMBO J. 15, 1231–1237.

Hines, V., Brandt, A., Griffiths, G., Horstmann, H., Brütsch, H., & Schatz, G. (1990). Protein import into yeast mitochondria is accelerated by the outer membrane protein MAS70. EMBO J. 9, 3191–3200.

Hines, V., & Schatz, G. (1993). Precursor binding to yeast mitochondria: A general role for the outer membrane protein Mas70p. J. Biol. Chem. 268, 449–454.

Hönlinger, A., Kübrich, M., Moczko, M., Gärtner, F., Mallet, L., Bussereau, F., Eckerskorn, C., Lottspeich, F., Dietmeier, K., Jacquet M., & Pfanner, N. (1995). The mitochondrial receptor complex: Mom22 is essential for cell viability and directly interacts with preproteins. Mol. Cell. Biol. 15, 3382–3389.

Hönlinger, A., Bömer, U., Alconada, A., Eckerskorn, C., Lottspeich, F., Dietmeier, K., & Pfanner, N. (1996). Tom7 modulates the dynamics of the mitochondrial outer membrane translocase and plays a pathway related role in protein import. EMBO J. 15, 2125–2137.

Horwich, A. (1990). Protein import into mitochondria and peroxisomes. Curr. Op. Cell Biol. 2, 625–633.

Hurt, E. C., & van Loon, A. P. G. M. (1986). How proteins find mitochondria and intramitochondrial compartments. Trends Biochem. Sci. 11, 204–207.

Jordi, W., Hergersberg, C., & de Kruijff, B. (1992). Bilayer-penetrating properties enable apocytochrome *c* to follow a special import pathway into mitochondria. Eur. J. Biochem. 204, 841–846.

Kassenbrock, C. K., Cao, W., & Douglas, M. G. (1993). Genetic and biochemical characterization of ISP6, a small mitochondrial outer membrane protein associated with the protein translocation complex. EMBO J. 12, 3023–3034.

Keil, P., & Pfanner, N. (1993). Insertion of MOM22 into the mitochondrial outer membrane strictly depends on surface receptors. FEBS Lett. 321, 197–200.

Keil, P., Weinzierl, A., Kiebler, M., Dietmeier, K., Söllner, T., & Pfanner, N. (1993). Biogenesis of the mitochondrial receptor complex: Two receptors are required for binding of Tom40 to the outer membrane surface. J. Biol. Chem. 268, 19177–19180.

Kiebler, M. (1993). Untersuchungen zur Struktur und Funktion des mitochondrialen Rezeptorkomplexes. Thesis, Universität München, Germany.

Kiebler, M., Pfaller, R., Söllner, T., Griffiths, G., Horstmann, H., Pfanner, N., & Neupert, W. (1990). Identification of a mitochondrial receptor complex required for recognition and membrane insertion of precursor proteins. Nature 348, 610–616.

Kiebler, M., Keil, P., Schneider, H., van der Kleij, I., Pfanner, N., & Neupert, W. (1993). The mitochondrial receptor complex: A central role of MOM22 in mediating preprotein transfer from receptors to the general insertion pore. Cell 74, 483–492.

Klingenberg, M. (1985). Principles of carrier catalysis elucidated by comparing two similar membrane translocators from mitochondria, the ADP/ATP carrier and the uncoupling protein. Ann. N.Y. Acad. Sci. 456, 279–288.

Li, J-M., & Shore, G. C. (1992). Reversal of orientation of an integral protein of the mitochondrial outer membrane. Science 256, 1815–1817.

Lithgow, T., Junne, T., Suda, K., Gratzer, S., & Schatz, G. (1994). The mitochondrial outer membrane protein Mas22p is essential for protein import and viability of yeast. Proc. Natl. Acad. Sci. USA. 91, 11973–11977.

Mayer, A., Nargang, F. E., Neupert, W., & Lill, R. (1995). MOM22 is a receptor for mitochondrial targeting sequences and cooperates with MOM19. EMBO J. 14, 4204–4211.

McCollum, D., Monosov, E., & Subramani, S. (1993). The *pas8* mutant of *Pichia pastoris* exhibits the peroxisomal protein import deficiencies of Zellweger syndrome cells—the PAS8 protein binds to the COOH-terminal tripeptide peroxisomal targeting signal, and is a member of the TPR protein family. J. Cell Biol. 121, 761–774.

Miller, B. R., & Cumsky, M. G. (1991). An unusual mitochondrial import pathway for the precursor to yeast cytochrome *c* oxidase subunit Va. J. Cell Biol. 112, 833–841.

Miller, D. G., & Shore, G. C. (1993). The signal anchor sequence of mitochondrial Mas70p contains an oligomerization domain. J. Biol. Chem. 268, 18403–18406.

Milstein, C., Brownlee, G. G., Harrison, T. M., & Mathews, M. D. (1972). A possible precursor of immunoglobulin light chains. Nature 239, 117–120.

Moczko, M., Dietmeier, K., Söllner, T., Segui, B., Steger, H. F., Neupert, W., & Pfanner, N. (1992). Identification of the mitochondrial receptor complex in *Saccharomyces cerevisiae*. FEBS Lett. 310, 265–268.

Moczko, M., Gärtner, F., & Pfanner, N. (1993). The protein import receptor Tom20 of yeast mitochondria. FEBS Letters 326, 251–254.

Moczko, M., Ehmann, B., Gärtner, F., Hönlinger, A., Schäfer, E., & Pfanner, N. (1994). Deletion of the receptor Tom20 strongly impairs import of cleavable preproteins into *Saccharomyces cerevisiae* mitochondria. J. Biol. Chem. 269, 9045–9051.

Murakami, H., Blobel, G., & Pain, D. (1990). Isolation and characterization of the gene for a yeast mitochondrial receptor. Nature 347, 488–491.

Murakami, H., Blobel, G., & Pain, D. (1993). Signal sequence region of mitochondrial precursor proteins binds to mitochondrial import receptor. Proc. Natl. Acad. Sci. USA 90, 3358–3362.

Nakai, M., & Endo, T. (1995). Identification of yeast *MAS17* encoding the functional counterpart of the mitochondrial receptor complex protein MOM22 of *Neurospora crassa*. FEBS Lett. 357, 202–206.

Nargang, F. E., Drygas, M. E., Kwong, P. L., Nicholson, D. W., & Neupert, W. (1988). A mutant of *Neurospora crassa* deficient in cytochrome *c* heme lyase activity cannot import cytochrome *c* into mitochondria. J. Biol. Chem. 263, 9388–9394.

Nicholson, D. W., Hergersberg, C., & Neupert, W. (1988). Role of cytochrome *c* heme lyase in the import of cytochrome *c* into mitochondria. J. Biol. Chem. 263, 19034–19042.

Pain, D., Murakami, H., & Blobel, G. (1990). Identification of a receptor for protein import into mitochondria. Nature 347, 444–449.

Pfaller, R., Steger, H. F., Rassow, J., Pfanner, N., and Neupert, W. (1988). Import pathways of precursor proteins into mitochondria: Multiple receptor sites are followed by a common membrane insertion site. J. Cell Biol. 107, 2483–2490.

Pfaller, R., Pfanner, N., & Neupert, W. (1989). Mitochondrial protein import: Bypass of proteinaceous surface receptors can occur with low specificity and efficiency. J. Biol. Chem. 264, 34–39.

Pfanner, N., & Neupert, W. (1987). Distinct steps in the import of the ADP/ATP carrier into mitochondria. J. Biol. Chem. 262, 7528–7536.

Pfanner, N., Hoeben, P., Tropschug, M., & Neupert, W. (1987a). The carboxyl-terminal two-thirds of the ADP/ATP carrier polypeptide contains sufficient information to direct translocation into mitochondria. J. Biol. Chem. 262, 14851–14854.

Pfanner, N., Tropschug, M., & Neupert, W. (1987b). Mitochondrial protein import: Nucleoside triphosphates are involved in conferring import-competence to precursors. Cell 49, 815–823.

Pfanner, N., Söllner, T., & Neupert, W. (1991). Mitochondrial import receptors for precursor proteins. Trends Biochem. Sci. 16, 63–67.

Pfanner, N., Douglas, M. G., Endo, T., Hoogenraad, N. J., Jensen, R. E., Meijer, M., Neupert, W., Schatz, G., Schmitz, U., & Shore, G. C. (1996). Uniform nomenclature for the protein transport machinery of the mitochondrial membranes. Trends Biochem. Sci. 21, 51–52.

Phelps, A., & Wohlrab, H. (1991). Mitochondrial phosphate transport: The *Saccharomyces cerevisiae* (threonine 43 to cysteine) mutant protein explicitly identifies transport with genomic sequence. J. Biol. Chem. 266, 19882–19885.

Phelps, A., Schobert, C. T., & Wohlrab, H. (1991). Cloning and characterization of the mitochondrial phosphate transport protein gene from the yeast *Saccharomyces cerevisiae*. Biochemistry 30, 248–252.

Ramage, L., June, T., Hahnke, K., Lithgow, T., & Schatz, G. (1993). Functional cooperation of mitochondrial protein import receptors in yeast. EMBO J. 12, 4115–4123.

Rapoport, T. A. (1992). Transport of proteins across the endoplasmic reticulum membrane. Science 258, 931–936.

Rassow, J., & Pfanner, N. (1991). Mitochondrial preproteins en route from the outer membrane to the inner membrane are exposed to the intermembrane space. FEBS Lett. 293, 85–88.

Riezman, H., Hay, R., Witte, C., Nelson, N., & Schatz, G. (1983). Yeast mitochondrial outer membrane specifically binds cytoplasmically-synthesized precursors of mitochondrial proteins. EMBO J. 2, 1113–1118.

Schlossmann, J., Dietmeier, K., Pfanner, N., & Neupert, W. (1994). Specific recognition of mitochondrial preproteins by the cytosolic domain of the import receptor Tom70. J. Biol. Chem. 269, 11893–11901.

Schneider, H., Söllner, T., Dietmeier, K., Eckerskorn, C., Lottspeich, F., Trülzsch, B., Neupert, W., & Pfanner, N. (1991). Targeting of the master receptor Tom20 to mitochondria. Science 254, 1659–1662.

Seki, N., Moczko, M., Nagase, T., Zufall, N., Ehmann, B., Dietmeier, K., Schäfer, E., Nomura, N., & Pfanner, N. (1995). A human homolog of the mitochondrial protein import receptor Mom19 can assemble with the yeast mitochondrial receptor complex. FEBS Lett. 375, 307–310.

Smagula, C. S., & Douglas, M. G. (1988). ADP-ATP carrier of *Saccharomyces cerevisiae* contains a mitochondrial import signal between amino acids 72 and 111. J. Cell. Biochem. 36, 323–328.

Söllner, T., Griffiths, G., Pfaller, R., Pfanner, N., & Neupert, W. (1989). Tom20, an import receptor for mitochondrial precursor proteins. Cell 59, 1061–1070.

Söllner, T., Pfaller, R., Griffiths, G., Pfanner, N., & Neupert, W. (1990). A mitochondrial import receptor for the ADP/ATP carrier. Cell 62, 107–115.

Söllner, T., Rassow, J., Wiedmann, M., Schlossmann, J., Keil, P., Neupert, W., & Pfanner, N. (1992). Mapping of the protein import machinery in the mitochondrial outer membrane by crosslinking of translocation intermediates. Nature 355, 84–87.

Steger, H. F., Söllner, T., Kiebler, M., Dietmeier, K. A., Pfaller, R., Trülzsch, K. S., Tropschug, M., Neupert, W., & Pfanner, N. (1990). Import of the ADP/ATP carrier into mitochondria: Two receptors act in parallel. J. Cell Biol. 111, 2353–2363.

Vestweber, D., Brunner, J., Baker, A., & Schatz, G. (1989). A 42K outer-membrane protein is a component of the yeast mitochondrial protein import site. Nature 341, 205–209.

von Heijne, G. (1986). Mitochondrial targeting sequences may form amphophilic helices. EMBO J. 5, 1335–1342.

von Heijne, G., Steppuhn, J., & Herrmann, R. G. (1989). Domain structure of mitochondrial and chloroplast targeting peptides. Eur. J. Biochem. 180, 535–545.

Zwizinski, C., Schleyer, M., & Neupert, W. (1984). Proteinaceous receptors for the import of mitochondrial precursor proteins. J. Biol. Chem. 259, 7850–7856.

MITOCHONDRIAL IMPORT OF CYTOCHROME *C*

Mark E. Dumont

I. Introduction 104
II. Heme Attachment to Cytochrome *c* 105
    A. Role in Transport 105
    B. Subcellular Location of Heme Lyase 106
    C. Enzymology of Heme Attachment. 109
III. Transport Across the Outer Mitochondrial Membrane 110
IV. Additional Cellular Components Involved in Cytochrome *c* Import 111
    A. Cytoplasmic Factors 111
    B. Membrane Receptors and Translocators 112
    C. The *CYC2* Gene Product 113
    D. Other Components 114
V. Targeting Signals in Cytochrome *c* 115
VI. Relation to Import Pathways of Other Mitochondrial Proteins. 118
VII. Some Evolutionary Relationships of Cytochrome *c* Targeting 120
VIII. Summary 122
    Acknowledgment 122
    References 123

Advances in Molecular and Cell Biology,
Volume 17, pages 103–126.

ISBN: 0-7623-0144-9

I. INTRODUCTION

The *c*-type cytochromes comprise one of the most extensively studied families of proteins. The availability of large quantities of these small, highly soluble proteins has made them the subject of many biochemical and biophysical studies. Amino acid sequences of cytochromes *c* from over 100 different species are known. The structures of cytochromes *c* from several species have been solved by x-ray diffraction and NMR techniques. The ease of genetic manipulation of the genes encoding cytochrome *c* in the yeast *Saccharomyces cerevisiae* have made this system the subject of many genetic analyses (see Rothstein and Sherman, 1980; Hampsey et al., 1988).

The best-studied example of the *c*-type cytochromes is found in mitochondria, where it participates in electron transport. Mitochondrial cytochrome *c* binds as a peripheral protein to the outer surface of the inner mitochondrial membrane, where it can associate with the other electron-transporting enzymes of the inner membrane. It may also spend a significant fraction of the time as a soluble protein of the intermembrane space (Cortese and Hackenbrock, 1993). Like most mitochondrial proteins, cytochrome *c* is encoded by nuclear genes, synthesized in the cytoplasm, and then imported into mitochondria. However, the import pathway of cytochrome *c* appears to be distinct in important respects from that followed by other commonly studied imported mitochondrial proteins (Stuart and Neupert, 1990). In particular, the available evidence suggests that the cytochrome *c* pathway does not involve the same targeting sequences, receptors, or membrane insertion machinery as proteins targeted to the matrix or inner membrane, or several other proteins as targeted to the mitochondrial intermembrane space. Transport of cytochrome *c* into mitochondria appears to take place directly across the outer membrane, in contrast to other proteins of the intermembrane space which appear, first, to be transiently targeted to the inner membrane or matrix. Unlike most other imported mitochondrial proteins, cytochrome *c* does not undergo post-translational proteolytic processing. Furthermore, cytochrome *c* does not use the same mechanisms as other mitochondrially imported proteins to regulate protein folding during transport.

Given the biochemical tractability of the cytochromes *c*, and the wealth of information available about them, it might be expected that it would be possible to obtain a detailed picture of the mechanisms by which their mitochondrial import occurs. However, major questions about the mechanisms of cytochrome *c* import into mitochondria remain unanswered. This review is an attempt to present the information that is currently available about cytochrome *c* import in the framework of these unanswered questions.

II. HEME ATTACHMENT TO CYTOCHROME *C*

A. Role in Transport

Cytochrome *c* contains a covalently attached protoheme IX prosthetic group that is linked via two thioether bonds to cysteine residues in the protein. The protein is initially synthesized as apocytochrome *c*, lacking heme. This precursor appears to be an unfolded protein under normal physiological conditions (see below). Formation of the covalent linkage to heme is catalyzed by the enzyme cytochrome *c* heme lyase (also referred to as holocytochrome *c* synthase; EC 4.4.1.17). Attachment of heme is stereospecific. The vinyl group at the protoheme C3 carbon is attached to the cysteine at position 17 of the protein and the vinyl group at the C8 carbon is attached to the cysteine at position 14 (Takano and Dickerson, 1981). (All references to positions in amino acid sequences in this review are numbered with reference to the sequence of horse cytochrome *c*.)

Like much of the fundamental information on cytochrome *c* import, a relationship between heme attachment and import of cytochrome *c* was first uncovered in the laboratory of Walter Neupert. Hennig and Neupert (1981) found that the presence of an excess of deuteroheme, a heme analog that contains hydrogen atoms in place of the vinyl groups, inhibited the formation of holocytochrome *c* from radiolabeled apocytochrome *c* by isolated *Neurospora* mitochondria. Furthermore, this blockage of holocytochrome *c* formation caused apocytochrome *c* to remain accessible to proteases added external to the mitochondria. Similar protease accessibility of apocytochrome *c* was seen when *in vitro* import reactions were carried out in the absence of reducing agents, since heme lyase is believed to only be capable of acting on heme in the reduced state (Nicholson et al., 1988).

The identification and cloning of the genes encoding cytochrome *c* heme lyase in yeast (Dumont et al., 1987), and *Neurospora* (Drygas et al., 1989), made it possible to examine the role of the enzyme in import. In yeast, no detectable import of cytochrome *c* occurs *in vivo* in strains containing a deletion of *CYC3*, the gene encoding heme lyase (Dumont et al., 1991). Similar results were obtained by examining import of radiolabeled precursor into mitochondria isolated from strains of yeast (Dumont et al., 1988) or *Neurospora* (Nargang et al., 1988) containing heme lyase mutations. Thus, the heme-attaching enzyme appears to play a critical role in import or accumulation of cytochrome *c* in mitochondria.

To further define the role of heme lyase, mitochondrial import of altered forms of apocytochrome *c* that could not have heme covalently attached was studied *in vitro* and *in vivo*. In the *in vitro* studies, altered *cyc1* genes, encoding yeast iso-1-cytochrome *c* with amino acid substitutions at one or the other cysteine residues involved in heme attachment, were used for *in vitro* transcription and translation. The altered apocytochromes *c* could not be converted to holocytochrome *c*. They could, however, be imported into isolated yeast mitochondria at low efficiency, compared with the normal precursor (Dumont et al., 1988). In the *in vivo* studies, the subcellular distribution of apocytochrome *c* was examined in a yeast strain containing an altered *CYC7* gene, encoding a yeast iso-2-cytochrome *c* in which serine residues had been substituted for both cysteines involved in heme attachment. Following subcellular fractionation, most of this altered apocytochrome *c* was found in the cytoplasm. However, about 5% was imported into a mitochondrial compartment that was protected against proteases added to the mitochondrial fraction. Furthermore, the amount of protease-protected apocytochrome *c* varied with the gene dosage of heme lyase. In a strain containing a deletion of the *CYC3* gene, encoding heme lyase, the amount of the altered apocytochrome *c* inside mitochondria was no greater than the level of cross-contamination by cytoplasmic markers. On the other hand, in a strain containing about twenty times the normal level of heme lyase, expressed from a multi-copy plasmid under control of an active promoter, about 40% of the total cellular complement of altered apocytochrome *c* was found in mitochondria (Dumont et al., 1991). These experiments produced two unexpected conclusions: (1) Apocytochrome *c* can be transported into mitochondria without having heme attached, although the precise intramitochondrial location of the imported apocytochrome *c* has not yet been determined. (2) Mitochondrial import of an altered form of apocytochrome *c* that can not have heme attached depends on heme lyase.

B. Subcellular Location of Heme Lyase

Based on subcellular fractionation of its activity, cytochrome *c* heme lyase was found to be associated with mitochondria (Basile et al., 1980; Korb and Neupert, 1978). Further submitochondrial fractionations indicated that the protein was membrane associated, facing the intermembrane space (Enosawa and Ohashi, 1986; Nicholson et al., 1988).

Anti-peptide antibodies recognizing cytochrome *c* heme lyase were prepared using synthetic peptides based on the sequence of the yeast *CYC3* gene. The protein recognized by these antibodies fractionates with inner mitochondrial membrane, facing the intermembrane space (Dumont et al., 1991). Heme lyase appears not to be an integral membrane protein, based on the lack of hydrophobic residues in the predicted amino acid sequences, although detergents have been used to remove the protein from membrane-containing subcellular fractions (Dumont et al., 1988; Nicholson et al., 1988).

Localization of heme lyase to the inner mitochondrial membrane is difficult to reconcile with the demonstrated involvement of the enzyme in mitochondrial accumulation of apo- and holocytochrome *c*, unless the role of heme lyase is actually the trapping of cytochrome *c* in the intermembrane space, rather than direct involvement in translocation across the outer membrane. If this is the case, transport of normal apocytochrome *c* into mitochondria would be a reversible process. Modification by heme lyase, followed by cytochrome *c* folding would prevent re-export of apocytochrome *c* molecules that had been transported across the outer membrane and diffused through the intermembrane space to the inner membrane (see Figure 1). Results consistent with reversible import have also been obtained in *in vitro* import experiments (Hakvoort et al., 1990, see below).

This reversible model of cytochrome *c* import provides an explanation for the results of the *in vivo* studies of import of altered apocytochromes *c* containing amino acid substitutions of the cysteine residues involved in heme attachment. These altered precursors appear to retain the ability to bind to heme lyase, trapping the apocytochrome *c* inside the outer membrane as a nonproductive enzyme-substrate complex. Because cytochrome *c* normally appears to be present in great excess over heme lyase, such trapping could only result in low levels of apocytochrome *c* inside mitochondria. However, when the heme lyase level is increased by introducing multiple copies of the *CYC3* gene, larger amounts of altered apocytochrome *c* could be trapped, consistent with the results described previously (Dumont et al 1991). The presence of the nonproductive apocytochrome *c*-heme lyase complex in these strains has been confirmed by cross-linking and co-immunoprecipitation under nondenaturing conditions (Lapan, K.A., Nelson, D., Testa, C.A., and Dumont, M.E., unpublished). Apocytochrome *c* has also been imported *in vitro* into vesicles derived from outer mitochondrial membranes under conditions where the vesicles contained internally trapped anti-cytochrome antibodies (Mayer et al., 1995). In this case, antibody binding appears to substitute for the action of heme lyase in providing a driving force for mitochondrial uptake.

While the reversible import model explains much of the data, some issues remain unresolved. It has been reported that there are binding sites for apocytochrome *c* on the outer surface of mitochondria (see below) and that one class of these sites is missing in mitochondria derived from *Neurospora* containing mutations in the heme lyase gene (Nargang et al., 1988). A mutant form of apocytochromes *c* is stalled during import in partially-translocated state that remains accessible to external proteases. Formation of this state is dependent on the presence of heme lyase (Wang et al., 1996). These results suggest that at least a fraction of the population of heme lyase molecules is associated with the outer mitochondrial membrane where it can interact with apocytochrome *c* during translocation across the outer membrane. One way to reconcile these results with the immunologic localization of heme lyase to the inner membrane would be if the enzyme were to be enriched at sites where the inner and outer mitochondrial membranes are in contact or are closely apposed. When yeast mitochondria are fractionated by a technique that has been reported to yield a fraction enriched in sites of contact between the inner and outer membranes (Pon et al., 1989), this fraction was not found to contain any higher concentration of heme lyase than is found in the inner membrane (Dumont et al., 1991). On the other hand, contact sites may be transient structures (Pfanner et al., 1992). Thus, failure to find enrichment of heme lyase in this fraction does not rule out a temporary association of the enzyme with these sites *in vivo*.

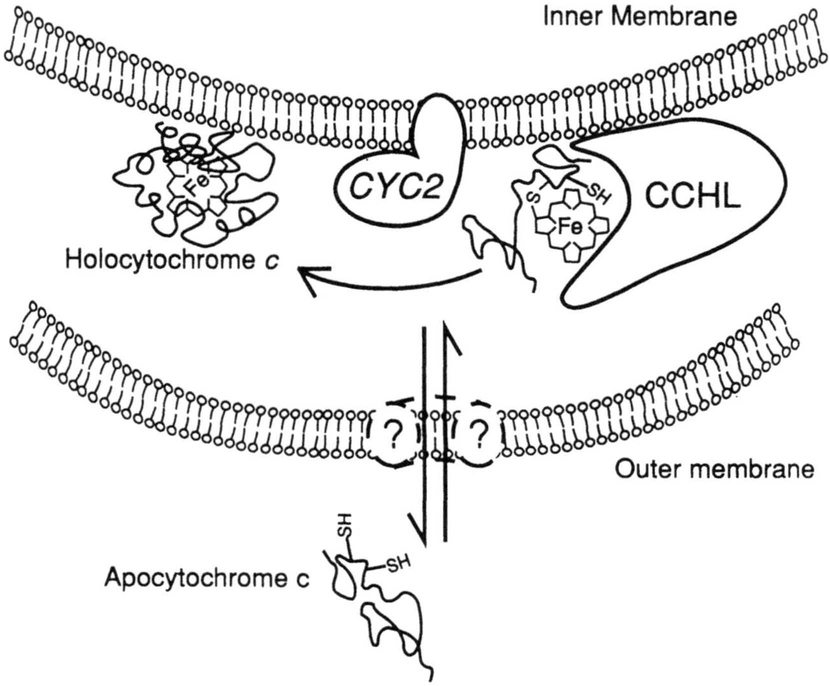

Figure 1. Schematic model of cytochrome *c* import into mitochondria. Apocytochrome *c*, synthesized in the cytoplasm, traverses the outer membrane, either directly through the lipid phase or through a proteinaceous pore, indicated by question marks. The initial translocation across the outer membrane may be reversible. Once inside the outer membrane, apocytochrome *c* binds to cytochrome *c* heme lyase (CCHL) and heme is covalently attached. Protein folding associated with heme attachment causes trapping of holocytochrome *c* in the intermembrane space. Cytochrome *c* heme lyase is primarily associated with the inner membrane, but could be transiently localized to the outer membrane or sites of contact between the inner and outer membranes, leading to direct coupling between heme attachment and membrane translocation (not shown). The *CYC2* gene product increases the rate or efficiency of heme attachment or mitochondrial import of cytochrome *c* but also appears to have effects on proteins other than cytochrome *c*. It is shown associated with the inner membrane based on its identification as a mitochondrial protein that is resistant to protease digestion in intact mitochondria and on the presence of potential transmembrane segments in the sequence.

C. Enzymology of Heme Attachment

Although cytochrome *c* heme lyase has only been partially purified, the enzyme activity has been studied in mitochondrial extracts and *E. coli* containing the yeast *CYC3* gene. The mechanism of the reaction is still not known. At least three possibilities can be envisioned: (1) Heme lyase first binds separately to heme and apocytochrome *c* and then catalyzes the covalent linkage. It has not yet been demonstrated that heme binds directly to heme lyase; (2) Heme lyase binds first to apocytochrome *c*, causing a conformational change that then allows the heme to bind spontaneously, as has recently been reported for an altered form of cytochrome b_5 containing an extra cysteine in the heme-binding site (Barker et al., 1993); and (3) Heme binds first to apocytochrome *c* to form a partially folded complex that is subsequently acted upon by heme lyase. In contrast to earlier reports (Fisher et al., 1973; Parr and Taniuchi, 1980), heme binds to purified apocytochrome *c* with a dissociation constant of about 10^{-7} M, even in the presence of high concentrations of cyanide (Dumont et al., 1994). Upon binding heme, apocytochrome *c* undergoes a transition from a relatively extended conformation (see below) to a compact state. However, it is not known whether the affinity of apocytochrome *c* for heme is high enough to allow formation of the noncovalent complex as an intermediate in holocytochrome *c* formation *in vivo*.

In addition to deuteroheme (see above), mesoheme, another heme analog with ethyl groups in place of the vinyl groups, also inhibits holocytochrome *c* formation in isolated mitochondria. On the other hand, hematoheme, which contains hydroxyethyl side chains instead of vinyl groups, does not compete for holocytochrome *c* production, presumably because the larger hydrophilic groups cannot fit in the necessary site on the heme lyase or on apocytochrome *c* (Hennig et al., 1983).

Maintenance of heme in the reduced state appears to be necessary for heme attachment. In *Neurospora*, this reduction is apparently achieved through the action of NADH and a flavin-mediated electron transfer (Nicholson and Neupert, 1989). Early reports in the yeast system suggested that NADPH, rather than NADH, is the optimal source of electrons for the reduction (Basile et al., 1980). Interpretation of some of the experiments demonstrating the need for reduction of heme is complicated by the fact that free heme is unstable in the presence of reducing agents and oxygen (Lamberg and Legge, 1949).

Sequences of the genes encoding cytochrome *c* heme lyases from yeast and *Neurospora*, as well that encoding the yeast cytochrome c_1 heme lyase, which attaches heme to the nuclear-encoded cytochrome c_1 of the cytochrome bc_1 complex, are now available (Zollner et al., 1992). The predicted amino acid sequences show regions of similarity, some of which are likely to be important for catalysis. All three amino acid sequences contain regions that are similar to a consensus sequence seen in other heme-binding proteins (Fytlovich et al., 1993). There is also a limited similarity between yeast cytochrome *c* heme lyase and a consensus sequence found in thioredoxins and related proteins, suggesting a role for this particular heme lyase in reduction of protein disulfides or of heme (see below).

III. TRANSPORT ACROSS THE OUTER MITOCHONDRIAL MEMBRANE

Cytochrome *c* is a hydrophilic protein that contains approximately 30% charged amino acid residues and lacks contiguous hydrophobic sequences longer than six residues. Nonetheless, both apo- and holocytochrome *c* bind to membranes. In the case of holocytochrome *c*, the interaction appears to consist primarily of electrostatic binding to membranes containing acidic lipids, possibly leading to partial destabilization of the protein structure (see Muga et al., 1991; de Jongh et al., 1992). The interaction of apocytochrome *c* with membranes, on the other hand, appears to go beyond what is typical of peripheral membrane proteins. Apocytochrome *c* binds to lipid vesicles and detergents at physiological ionic strengths. Binding to vesicles is accompanied by at least partial insertion into the membrane, leading to exposure of the apocytochrome *c* to the aqueous interior (Dumont and Richards, 1984; Rietveld and de Kruijff, 1984). Binding and insertion depend on the lipid composition of the vesicles. The membrane-active properties of apocytochrome *c* raise the possibility that cytochrome *c* transport into mitochondria occurs by a mechanism that depends on direct physical-chemical interaction of the precursor with the membrane lipids, and not on facilitation of transport by any proteins of the outer membrane. Additional support for this idea is derived from the following results: (1) Digestion of isolated mitochondria by proteolytic enzymes does not diminish the efficiency of *in vitro* cytochrome *c* import, as long as the conditions used do not lead to disruption of the permeability barrier of the outer membrane (Nicholson et al., 1988). (2) Negatively charged lipid vesicles effectively compete with isolated mitochondria for binding of apocytochrome *c*, preventing formation of holocytochrome *c* in the mitochondria (Jordi et al., 1992). (3) Fusion of apocytochrome *c* to the amino terminus of cytochrome c_1, another imported mitochondrial protein, allows bypass of the normal requirement for receptors on the mitochondrial surface for import of cytochrome c_1 (Stuart et al., 1990a). This suggests that the direct interaction of apocytochrome *c* with the outer membrane can fulfill the same function that is provided by receptors for other imported proteins.

A principal argument against the idea that cytochrome *c* translocation across the outer membrane is mediated by a direct protein-lipid direction is the insensitivity of the translocation step to low temperatures (Nicholson et al., 1988; Hakvoort et al., 1990). Any process that involved direct passage through the lipid fatty acyl chains would be expected to be strongly slowed by low temperature, which should greatly increase the viscosity of this environment. The outer mitochondrial membrane appears to be permeable to molecules up to about 5,000 molecular weight. While cytochrome *c* exceeds this weight, and the outer membrane is known to be an effective barrier for holocytochrome *c* (Wojtczak and Zaluska, 1969), it remains possible that the unusual folding properties of apocytochrome *c*, or its unusual shape (Fisher et al., 1973; see below), allow it to pass through relatively small pores.

IV. ADDITIONAL CELLULAR COMPONENTS INVOLVED IN CYTOCHROME *C* IMPORT

A. Cytoplasmic Factors

The absence of a cleaved signal sequence in cytochrome *c* makes it possible to obtain large amounts of apocytochrome *c* by chemically removing the heme from holocytochrome *c*. Removal of heme involves denaturing treatment of the protein, followed by re-equilibration to more mild conditions. While this treatment could be affecting the conformation of the protein, "synthetic" apocytochrome *c* prepared in this way has similar electrophoretic and chromatographic behavior to authentic apocytochrome *c* produced in cells (Matner and Sherman, 1982). Furthermore, this "synthetic" apocytochrome *c* made by removing the heme from holocytochrome *c* is capable of serving as a substrate for cytochrome *c* heme lyase (Basile et al., 1980). Thus, if no cytoplasmic components are involved in mitochondrial import of cytochrome *c*, it might be possible to observe import of "synthetic" apocytochrome *c* into isolated mitochondria. While apocytochrome *c* made in this way competes with newly synthesized apocytochrome *c* from a cell free translation system for import into mitochondria, actual import of purified apocytochrome *c* into mitochondria has not been reported. One problem with such an experiment is the difficulty of finding a structurally benign way of labeling the purified apocytochrome *c* to the high specific activity usually used for assaying import.

Another way of addressing the requirement for cytoplasmic components is to use apocytochrome *c* synthesized in a cell-free translation system and then purified by reverse phase HPLC. The import of purified *Drosophila* cytochrome *c* produced in this way into isolated mouse mitochondria was found to be enhanced by a soluble factor derived from the wheat germ translation extract (Hakvoort et al. 1990). However, even in the presence of this factor, levels of heme attachment were very low in these experiments.

A second type of cytoplasmic factor implicated in cytochrome *c* import consists of cellular components that use NADH to keep the heme reduced prior to attachment to cytochrome *c*. These components were initially identified as a cytoplasmic extract from *Neurospora* that could be substituted for by flavin nucleotide (Nicholson and Neupert, 1989). In contrast to the requirements for ATP and an electrochemical potential across the inner membrane that are seen for import of other proteins destined for the intermembrane space, inner membrane, and matrix, there is no known additional requirement for nucleotides or other energy sources for import of cytochrome *c*.

Cytoplasmic heat shock-related proteins appear to play a role in unfolding many imported mitochondrial proteins or maintaining them in a translocation-competent conformation. One demonstration of importance of these proteins was the observation that many mitochondrial precursors synthesized in reticulocyte cell-free translation systems can be translocated into isolated mitochondria, whereas the same precursors synthesized in wheat germ systems can not. This difference can be explained in part by the presence of higher levels of Hsp70 heat shock-related proteins in the reticulocyte systems. Consistent with this, import from the wheat germ extract can be reconstituted upon addition of purified members of the Hsp70 family and a cytoplasmic factor from yeast (Murakami et al., 1988). However, cytochrome *c* can be imported at nearly 100% efficiency from the same wheat germ lysate that fails to import other mitochondrial proteins (Dumont et al., 1988). This suggests that cytoplasmic heat shock-related proteins are not as important for import of cytochrome *c* as they are for the import of other mitochondrial proteins.

An explanation for the lack of dependence of cytochrome *c* import on heat shock proteins may lie in the unusual conformational properties of apocytochrome *c*. When prepared by removal of heme from holocytochrome *c*, apocytochrome *c* appears to be unfolded. The hydrodynamic properties of apocytochrome *c* are indicative of an anomalously large radius of gyration for a protein of this molecular weight (Stellwagen et al., 1972; Fisher et al., 1973). The NMR spectrum of apocytochrome *c* is characteristic of a protein without a high degree of structural order (Cohen et al., 1974; de Jongh et al., 1992). The circular dichroism spectrum does not change significantly after heating to high temperatures (Fisher et al., 1973). Given these properties, it is surprising that apocytochrome *c* is also quite soluble and is monomeric, if treated carefully.

The unusual conformational properties of apocytochrome *c* may circumvent the need for cellular factors to modulate the folding of the precursor protein. The stable, monomeric, unfolded structure could, in itself, constitute translocation competence. In spite of the apparent lack of requirement for heat shock proteins *in vivo*, incubation of purified apocytochrome *c* with the cytoplasmic Hsp70 homolog, Hsc70, leads to stimulation of the ATPase activity of Hsc70, consistent with the idea that this activity is related to the binding and unbinding of Hsp70s to unfolded proteins (Sadis and Hightower, 1992).

B. Membrane Receptors and Translocators

At least two classes of binding sites for apocytochrome *c* have been found on mitochondria (Hennig et al., 1983). Based on their abundance, the lower affinity class of sites could represent binding to a subpopulation of lipids. The number of the higher affinity sites was decreased in *Neurospora* strains with mutations in the gene encoding heme lyase, suggesting that these sites are associated with heme lyase (Nargang et al., 1988). On the other hand, proteolytic digestion of mitochondria with disrupted outer membranes led to destruction of heme lyase activity without a corresponding loss of apocytochrome *c* binding (Nicholson et al., 1988). The higher affinity sites can be destroyed by mild proteolytic treatments of intact mitochondria that do not affect the efficiency of mitochondrial import, suggesting that these sites may not be critical for import (Nicholson et al., 1988). Furthermore, apocytochrome *c* binding to isolated mitochondria and to purified outer mitochondrial membranes exhibits a strong dependence on ionic strength, decreasing rapidly with increasing salt concentrations, even at levels below physiological ionic strength (Hennig et al., 1983; Mayer et al., 1995). Thus, the role of membrane receptors in cytochrome *c* import remains unclear. If the higher affinity binding sites are on heme lyase, or are on a molecule associated with heme lyase, this would imply that apocytochrome *c* binding to mitochondria actually represents an insertional event allowing the precursor access to heme lyase in the intermembrane space. As discussed previously, such events would be difficult to reconcile with the apparent localization of heme lyase at the inner mitochondrial membrane (Dumont et al., 1991) unless the enzyme is localized to sites of contact or close approach between the inner and outer membranes.

If apocytochrome *c* transport across the outer membrane occurs via a proteinaceous pore, this would be an important component to detect. Yeast strains that lack the major pore-forming protein of the outer membrane are partially, but not completely, deficient in several cytochromes, including cytochrome *c* (Dihanich et al., 1987). Interpretation of this result is complicated by compensatory changes that accumulate in porin-deficient strains. Apocytochrome *c* has been reported to bind to outer membrane porin, based on changes observed in electron micrographs of ordered arrays of the protein (Mannella et al., 1987, 1992).

C. The *CYC2* Gene Product

Over the past 30 years, an extensive search for genes affecting cytochrome *c* biosynthesis in yeast has been conducted by Fred Sherman and co-workers. This search uncovered four genes that, when mutated, specifically reduce cellular levels of holocytochrome *c* (Rothstein and Sherman, 1980). Two of these, *CYC1* and *CYC7*, are the structural genes encoding iso-1- and iso-2-cytochromes *c*, the two isozymes found in yeast. A third gene, *CYC3*, encodes cytochrome *c* heme lyase. Mutations in the fourth gene, *CYC2*, were shown by Matner and Sherman (1982) to lead to the accumulation of apocytochrome *c* in certain strains, indicating that the gene product acts at a post-translational step in cytochrome *c* synthesis. According to recent genomic sequencing of yeast chromosome XV containing the *CYC2* locus (de Haan, M., Maarse, A.C., and Grivell, L.A., unpublished; accession number X87331) this gene encodes a 404 amino acid protein without significant similarity to any other known protein. Due to a single-base sequencing error this reading frame is 236 amino acids longer at its amino terminus than was originally reported (Dumont et al., 1993). A role for the protein in mitochondrial import or accumulation of cytochrome *c* was inferred based on immunologic analyses of subcellular fractions derived from isogenic pairs of *cyc2*$^-$ and *CYC2*$^+$ strains (Dumont et al., 1993). Mitochondrial levels of cytochrome *c* were decreased, and cytoplasmic levels were increased in the *cyc2*$^-$ strains. No effect of *cyc2* mutations on the level or subcellular localization cytochrome *c* heme lyase, has been detected.

The exact role of the *CYC2* gene product in mitochondrial accumulation of cytochrome *c* is still not known. Although point mutations and a partial deletion of the gene caused only a diminution of mitochondrial import of cytochrome *c* (Dumont et al., 1993), deletion of the complete *CYC2* reading frame leads to general defects in the ability to grow on non-fermentable substrates, in addition to the effects of cytochrome *c* (Cardillo, T.S., Tzagoloff, A., Pearce, D.A., and Sherman, F., unpublished). Anti-*CYC2* antibodies recognize a protein that is found in a compartment of mitochondria that is protected from externally added proteases (Dumont et al., 1993).

While the *CYC2* gene product appears to have a role in cytochrome *c* import or accumulation, it is not strictly required for this process. Yeast strains with a complete deletion of the gene always contain at least 10% of the normal level of holocytochrome *c* in mitochondria. Furthermore, the extent of the cytochrome *c* deficiency in *cyc2*$^-$ strains depends on which isozyme of cytochrome *c* is being studied. *cyc2* mutations have more drastic effects on iso-1-cytochrome *c*, which normally comprises about 95% of cellular cytochrome *c*, than on the less abundant isozyme, iso-2-cytochrome *c*. This can be explained by the difference in the stabilities of the apo-forms of the two isozymes. Apo-iso-1-cytochrome *c* is rapidly degraded in the cytoplasm if import is inhibited. Apo-iso-2-cytochrome *c*, on the other hand, has a half life on the order of hours (Dumont et al., 1990). Thus, a partial defect in mitochondrial import could lead to a loss of iso-1-cytochrome *c*, whereas apo-iso-2-cytochrome *c* should be able to accumulate in the cytoplasm, eventually resulting in partial import.

Thus, the *CYC2* gene product appears to have multiple roles in mitochondrial synthesis or function. The effects on cytochrome *c* can be explained by involvement of the *CYC2* protein either in translocation of cytochrome *c* into mitochondria, or as an accessory factor in attachment of heme. The partial nature of the blockage of cytochrome *c* maturation in *cyc2*$^-$ mutants can be explained in two ways: (1) The *CYC2* gene product increases the efficiency of import or accumulation but is not strictly necessary for these processes or (2) the function performed by the *CYC2* gene product is a necessary one, but there is an additional yet-to-be identified gene product that is capable of taking over in the absence of *CYC2*.

D. Other Components

The genetic procedures used to identify the *CYC2* and *CYC3* genes have been saturated without uncovering evidence for involvement of any additional cellular components in import. However, certain features of previous searches for genes affecting cytochrome *c* biosynthesis make it possible that not all the relevant genes have been uncovered: (1) Previous searches have focused on genes with specific effects on cytochrome *c* that do not affect other mitochondrial proteins. However, some cellular components could be involved in the import of cytochrome *c* as well as other proteins; (2) It may be necessary to seek temperature sensitive mutations. Some mutations that destroy the function of mitochondrial proteins are lethal (Baker and Schatz, 1991), and thus would have to be studied as conditional alleles; and (3) There may be multiple genes capable of carrying out similar functions. Mutations affecting one of a set of such genes may only be detectable in genetic backgrounds where other members of the set are deleted. For example, if there is another gene with a function similar to *CYC2*, it would be most convenient to detect it in a *cyc2⁻* background.

V. TARGETING SIGNALS IN CYTOCHROME *C*

While the patterns of amino acid sequence mediating the import of many mitochondrial proteins are now characterized, the sequences involved in targeting cytochrome *c* to mitochondria have yet to be clearly defined. One reason for this is the lack of an obvious cleaved signal sequence. Another reason is the difficulty of separating the requirements for targeting to the organelle from the requirements for the covalent heme attachment. If, as proposed above, cytochrome *c* import occurs via a nonspecific interaction of the precursor with the outer membrane, followed by trapping in the intermembrane space when heme is attached, it may prove difficult to uncover simple separable sequence requirements for these two processes.

The large number of evolutionarily related cytochromes *c* that have been sequenced, as well as the hundreds of mutated forms of cytochrome *c* that have been expressed in yeast, have provided initial information about targeting sequences for import. Cytochromes *c* from horse, pigeon, tuna, and rat have all been expressed in yeast, with high yields of holocytochrome *c* (see Hickey et al., 1991). This indicates that there is substantial conservation of the import pathway. The amino terminal region of cytochrome *c* varies considerably in comparing different species. The first nine residues of yeast iso-1-cytochrome *c* (corresponding to the first four residues of horse cytochrome *c*) can be deleted without affecting cellular levels of holocytochrome *c* (see Hampsey et al., 1988). Thus, the extreme amino terminal of the protein does not contain critical targeting information.

Among the many single amino acid substitutions that have been studied in yeast cytochromes *c*, only three positions have been identified at which substitutions cause a complete loss of holocytochrome *c* from cells. These positions are the two cysteine residues that are the sites of covalent heme attachment and His 18, which is one of the ligands of the heme iron and is immediately adjacent to one of the two cysteines. Cytochromes *c* containing single nonconservative amino acid substitutions at these positions showed greatly reduced, but detectable, import into isolated mitochondria *in vitro*. However, chemical modification of the cysteines with N-ethyl maleimide completely abolished import (Dumont et al., 1988). On the other hand, cytochrome *c* containing serine residues substituted for both of the two cysteine residues could be imported with relatively high efficiency into mitochondria containing elevated levels of heme lyase, though it could not be converted to holocytochrome *c* (Dumont et al., 1991). These results suggest that the cysteine residues, although they play a crucial role in heme attachment, may not be critical for targeting to mitochondria and translocation across the outer membrane. Amino acid substitutions for methionine 80, the second ligand of the heme iron, allow production of holocytochrome *c* with reduced yield *in vivo* (see Hampsey et al., 1988). Cytochrome *c* with a substitution at this position could be imported into isolated mitochondria with normal efficiency (Dumont et al., 1988).

The results of three systematic studies of the sequences responsible for targeting cytochrome *c* to mitochondria have been published. All of these were carried out in systems where sequences required for mitochondrial targeting could be distinguished from sequences required for heme attachment. The first of the studies made use of gene fusions between cytochrome *c* sequences and the chloramphenicol acetyltransferase (CAT) gene in yeast (Nye and Scarpulla 1990a,b). The presence of the large, folded, CAT protein caused a partial block in the ability to have heme attached *in vivo* and prevented the protein from being functional in respiration, presumably because it was arrested in a partially translocated form in the outer membrane. Introduction of a small deletion in the CAT domain that interfered with folding increased the efficiency of heme attachment and allowed the protein to function in respiration. Substitution of serine residues for the cysteines in cytochrome *c* that are the sites of heme attachment did not interfere with initial targeting to mitochondria, but rendered the fusion protein with the CAT deletion incompetent to function in respiration. By fusing different portions of the cytochrome *c* sequence to the intact CAT gene, two large regions capable of targeting the fusions to mitochondria were defined. These comprised amino acids 1–63 and amino acids 63–80 (numbered relative to horse cytochrome *c*). The presence of only one of these two sequences was sufficient to target the fusion to mitochondrial membranes.

The second systematic analysis of cytochrome *c* targeting was carried out *in vitro*, assaying import of *Drosophila* cytochrome *c* into isolated mouse mitochondria (Hakvoort et al., 1990; Sprinkle et al., 1990). The fact that heme is not attached to cytochrome *c* to any significant extent in this system, for unknown reasons, allows separation of heme lyase action from translocation. A deletion from the carboxyl terminal of cytochrome *c*, extending back to residue 58, had no effect on uptake. Deletion of residues from the carboxyl terminal extending beyond residue 58, diminished uptake, implicating the region just to the amino terminal of this residue in targeting. However, an internal deletion extending from residue 36 to 66 could be imported. This means that a carboxyl terminal sequence can substitute for deleted information in the amino terminal region of the sequence, in agreement with the evidence for functional redundancy from the CAT fusion system. Substitution of serine residues for the cysteines involved in heme attachment did not affect uptake, although treatment of these residues with N-ethyl maleimide did have a significant effect.

The third systematic study of the sequences involved in cytochrome *c* targeting to mitochondria was based on a sequential series of approximately 10 amino acid deletions extending throughout the *CYC7* gene, with additional amino acid replacements (Wang et al., 1996). Mitochondrial accumulation of mutated iso-2-cytochromes *c* was measured in a yeast strain that overexpressed heme lyase, allowing import in the absence of heme attachment (see above). Most deletions in the protein prevented heme attachment, but still allowed accumulation in mitochondria. No role in this accumulation was detected for carboxyl terminal sequences of cytochrome c or for the cysteines involved in heme attachment.

Modification of the following regions of iso-2-cytochrome *c* interfered with accumulation of apoprotein inside mitochondria: (1) The amino terminal—a minimal length of polypeptide chain at the amino terminal, rather than any specific sequence, was found to be required to stable association with mitochondria; (2) His 27 (corresponding to His 18 of horse cytochrome *c*)—substitution of alanine at this position led to accumulation of precursor bound to mitochondria but not completely translocated across the outer membrane. Such binding was dependent on the overexpression of heme lyase. (3) Phe 19 (corresponding to Phe 10 of horse cytochrome *c*) and several additional amino acids to the amino terminal side of this phenylalanine residue—replacement or deletion of these residues prevented association of precursor with mitochondria. Surprisingly, these altered apocytochromes *c* retained the ability to strongly compete the import of normal cytochrome *c*, suggesting an interaction with a saturable cytoplasmic component.

The propensity of different regions of apocytochrome *c* to interact with membranes has been studied by de Kruijff and co-workers, who found that fragments derived from both the amino and carboxyl termini exhibit membrane-active properties, although only the amino terminal fragments were able to insert across the bilayer of lipid vesicles (Jordi et al., 1989).

A role for the carboxyl terminal of cytochrome *c* in import was initially inferred from the observation that a cyanogen bromide fragment consisting of the carboxyl terminal half of apocytochrome *c* (but not smaller carboxyl terminal fragments) could complete the import of newly-synthesized labeled apocytochrome *c* into isolated mitochondria (Matsuura et al., 1981). Similarly, a mutation changing the last eight amino acids and appending 19 extra amino acids to the carboxyl terminal of *Neurospora* cytochrome *c* prevents import of the altered protein (Stuart et al., 1987). However, the systematic studies of cytochrome *c* import described above found little effect of deletions of the extreme carboxyl terminal on mitochondrial targeting, under conditions where effects on targeting could be separated from effects on covalent attachment of heme. This suggests that the carboxyl terminal is actually important for heme attachment, rather than initial targeting and translocation. Although the carboxyl terminal is far from the site of heme attachment in the primary sequence, it is actually rather close in the tertiary structure of the folded protein.

Only limited information is available regarding the sequence requirements for recognition of cytochrome *c* by heme lyase. Yeast cytochrome *c* heme lyase is capable of attaching heme to horse apocytochrome *c* and to a peptide corresponding to residues 1–25 of horse cytochrome *c* (Veloso et al., 1984). This suggests that the recognition signals for heme lyase lie in the region encompassed by this peptide, and not in the carboxyl terminal. However, it is difficult to compare this reaction with conditions *in vivo*, where short deletions throughout the molecule prevent heme attachment, even in situations where transport across the outer membrane can be shown to occur (Wang et al., 1996).

The cysteine residues involved in bonding to heme are obvious candidates for recognition sites; however, the apparent ability of the enzyme to bind to apocytochrome *c* in which the cysteines are replace by serines (Wang et al., 1996) suggests that these sites are not critical for the initial enzyme-substrate recognition. Some protozoan species contain heme that is only attached via one cysteine residue, and certain mutant cytochromes *c* containing only one cysteine can be very inefficiently converted to holocytochrome *c* at low levels in yeast (Sorrell and Martin, 1989; Tanaka et al., 1990).

There are two distinct heme lyases in mitochondria. One of these acts on cytochrome *c*, the other on cytochrome c_1, a nuclear-encoded protein that is imported by a pathway distinct from that of cytochrome *c*. The complete lack of cytochrome *c* in cells with mutations in cytochrome *c* heme lyase indicates that cytochrome *c* can not be acted upon *in vivo* by the cytochrome c_1 heme lyase. This could be a result of different submitochondrial locations or of different signals for recognition by the enzyme. If cytochrome *c* is fused at its amino terminus to targeting signals derived from cytochrome c_1, the fusion protein can be imported into mitochondria via the cytochrome c_1 pathway, but can not subsequently be acted upon by either of the heme lyases (Stuart et al., 1990b). In the case of the cytochrome c_1 heme lyase, this is probably because the proper recognition elements are lacking. In the case of the cytochrome *c* heme lyase, this suggests that for the reaction to proceed, the apocytochrome *c* must be presented to the enzyme in the proper topological orientation.

It has been suggested that trimethylation of lysine 72 of cytochrome *c* is an important signal for import in plants and fungi, where this modification is found. However, substitution of arginine, which cannot be methylated, for this lysine residue did not lead to any detectable decrease in the cellular levels or activity of the resulting holocytochrome *c* (Holzschu et al., 1987).

VI. RELATION TO IMPORT PATHWAYS OF OTHER MITOCHONDRIAL PROTEINS

The import pathways of several proteins of the mitochondrial intermembrane space, cytochrome c_1, cytochrome b_2, and the Rieske Fe/S protein of cytochrome *c* reductase complex, have been studied in considerable detail (see Glick and Schatz,

1991; Kiebler et al., 1993). These proteins appear to be imported first to the matrix or the inner membrane through contact sites, a step that depends on an electrochemical potential across the inner membrane, and then to be exported from the matrix or set free from the inner membrane by proteolytic processing. The bipartite targeting signals of cytochrome c_1 and cytochrome b_2 consist of an amino-terminal "matrix" targeting sequence followed by a second sequence specifying translocation to the intermembrane space. These two signals are sequentially cleaved during targeting. The Rieske Fe/S protein also contains a doubly-cleaved targeting signal, but both cleavage events appear to take place in the matrix prior to re-export.

Several fundamental differences between the characteristics of cytochrome *c* import and that of other mitochondrial proteins, including the lack of an ATP requirement, the lack of evidence for involvement of heat shock-related proteins, and the lack of a cleaved bipartite intermembrane space targeting signal have been presented previously. In addition, import of cytochrome *c*, unlike that of many other mitochondrial proteins, does not depend on a potential across the inner mitochondrial membrane (Zimmerman et al., 1981). Furthermore, while the presence of excess cold apocytochrome *c* inhibits the import of newly-synthesized radiolabeled apocytochrome *c* into isolated *Neurospora* mitochondria, excess apocytochrome *c* does not affect import of other mitochondrial precursors in this system (Zimmerman et al., 1981). The *Neurospora* protein MOM19 is a mitochondrial receptor that has been implicated in the import of a number of proteins other than cytochrome *c*. Anti-MOM19 antibodies that inhibit the import of these other mitochondrial proteins do not affect cytochrome *c* import into isolated mitochondria and cytochrome *c* import is not affected by mild proteolytic treatments that destroy MOM19 binding sites (Söllner et al., 1989).

Thus, the import pathway of cytochrome *c* clearly differs from the paradigm defined by cytochrome c_1, cytochrome b_2, and the Fe/S protein pathways. However, cytochrome *c* is not the only protein that fails to conform to this paradigm: (1) Cytochrome *c* heme lyase, which also ends up in the intermembrane space, appears to be imported via a pathway that uses the same outer membrane receptors as these other proteins, but does not involve the inner membrane, since heme lyase does not have typical amino-terminal targeting sequences and its transport does not require ATP hydrolysis or a potential across the inner membrane (Lill et al., 1992); (2) Adenylate kinase, a soluble protein of the intermembrane space, also lacks a cleavable targeting sequence, and thus, seems unlikely to be targeted via the inner membrane or matrix (Magdolen et al., 1992); (3) While the exact submitochondrial location of the *CYC2* gene product is not yet clear, the protein is presumed to face the intermembrane space because of its role in cytochrome *c* import and its inaccessibility to protease digestion in intact mitochondria. However, the predicted sequence of the *CYC2* protein lacks conventional amino-terminal mitochondrial targeting signals; and (4) Cytochrome *c* peroxidase, a protein of the intermembrane space that contains a typical bipartite targeting sequence, may under some circumstances be capable of being translocated directly across the outer membrane (Kaput et al., 1989). Thus, import pathways of some of these proteins of the intermembrane space may have more in common with proteins destined for the outer membrane than with the pathway followed by cytochrome c_1, cytochrome b_2, and the Rieske Fe/S protein.

VII. SOME EVOLUTIONARY RELATIONSHIPS OF CYTOCHROME *C* TARGETING

C-type cytochromes are found in a number of prokaryotic organisms, either as soluble proteins of the periplasmic space, or as membrane proteins of the cytoplasmic membrane, facing the periplasm. In every known case, these cytochromes *c* are synthesized as precursors, with cleavable sequences targeting them for secretion. In *Parococcus denitrificans*, *c*-type cytochromes can be transported to the periplasm as apoproteins, indicating that heme attachment occurs outside the cytoplasmic membrane and is not strongly coupled to secretion (Page and Ferguson, 1990), however, a *c*-type cytochrome from the hydrogen-oxidizing thermophilic bacterium *Hydrogenobacter thermophilus* could be converted to holocytochrome *c* in the cytoplasm of *E. coli* when expressed in the absence of a targeting sequence (Sanbongi et al., 1991)

The conservative sorting hypothesis, as presented by Hartl and Neupert (1990), explains mitochondrial import pathways in terms of the endosymbiotic evolutionary history of mitochondria. According to this view, as mitochondria adapted to intracellular existence, nuclear genes could be targeted to the intermembrane space by first entering the matrix, and then making use of the original prokaryotic secretion pathway to be re-exported. Of the two mitochondrial *c*-type cytochromes, cytochrome c_1 appears to be imported via this "conservative" pathway, while cytochrome *c* is not.

Cytochrome *c* may have escaped the need to follow an evolutionarily conserved pathway by virtue of its interaction with membranes, allowing it to directly cross the outer membrane and follow a direct route to the intermembrane space. The interaction with cytochrome *c* heme lyase, leading to heme attachment and protein folding, may have also freed cytochrome *c* from the requirement to become exposed to the matrix-localized heat shock proteins involved in folding other mitochondrial proteins. The differences between the pathways followed by cytochromes *c* and c_1 could reflect differences in their interactions with membranes, different properties of the two heme lyases, or the fact that only cytochrome c_1 has to be assembled as part of the multi-protein cytochrome bc_1 complex.

Several genetic searches have uncovered bacterial genes that are specifically involved in post-translational steps of the biosynthesis of *c*-type cytochromes. A number of these genes have recently been cloned and sequenced. They include: (1) The cycV, cycW, and cycX genes from *Bradyrhizobium japonicum*, a nitrogen-fixing root bacterium, and the similar helA, helB, and helD genes of *Rhodobacter capsulatus*, a photosynthetic bacterium. The cycV and helA genes show sequence similarity to the family of ABC-type ATP-driven membrane transporters. An example of this type of transporter has also recently been identified and shown to have effects on levels of various cytochromes, including cytochrome *c*, in yeast mitochondria (Leighton and Schatz, 1995). The cycW, cycX, helB, and helD genes appear from the hydrophobicity profiles to be membrane proteins. The three gene products in each organism may form a complex that is involved in heme transport to the periplasm (Beckman et al., 1992; Ramseier et al., 1991); (2) The cycH gene of *B. japonicum*, which causes a deficiency of all *c*-type cytochromes except cytochrome c_1. Topological analyses, using gene fusions, indicate that this encodes a transmembrane protein that is primarily exposed to the periplasmic space (Ritz et al., 1993); (3) The *ccl1* and *ccl2* genes of *R. capsulatus*. These genes appear to encode periplasmic proteins. The sequence of ccl1 is similar to that of several chloroplast and mitochondrial open reading frames of unknown function (Beckman et al., 1992); and (4) ORF132 of *B. japonicum* and the related helX gene of *R. capsulatus*. Both of these genes show rather limited sequence similarity to thioredoxins, but share a consensus sequence found in the active site of thioredoxins, protein disulfide isomerase, and DsbA, a periplasmic protein that is involved in disulfide bond formation in *E. coli* (Ramseier et al., 1991; Beckman and Kranz, 1993).

None of the genes that have been implicated to date in bacterial cytochrome *c* biosynthesis appear to affect targeting of apoprotein to the periplasmic space. Furthermore, none of these genes shows extensive similarity to the three cytochrome *c* heme lyase sequences. However, there is very limited similarity between the cytochrome *c* heme lyase of *Saccharomyces cerevisiae* and the putative active site region of the thioredoxins and protein disulfide isomerases, in particular that of HelX (see Figure 2). This similarity, though weak, involves most of the highly conserved residues of the consensus sequence. It could be indicative of a role for yeast cytochrome *c* heme lyase in control of the redox state of either heme or the cytochrome *c* sulfhydryl groups, in addition to its obvious role in formation of the thioether linkages of holocytochrome *c*. However, since this sequence similarity is not shared with the cytochrome *c* heme lyase from *Neurospora* or the cytochrome

```
CYC3    160 MGKPGVLSPRARWMHLCGLLFPS 182
             ||  ::   | |:  | :  |
HelX     62 DGKIKLVNFWASWCAPCRVEHPN 84
```

Figure 2. Sequence relationship between *CYC3* and the HelX gene of *Rhodobacter capsulatus* (Dumont, M.E. unpublished). Solid lines indicate identical residues in the two sequences. Dots indicate valine to leucine or methionine to cysteine substitutions. Underlined letters in the HelX sequence indicate positions that are identical in the four following sequences: HelX, thioredoxin from *Rhodobacter spheroides*, and both the amino and carboxyl consensus sequences of protein disulfide isomerase from rat (see Beckman and Kranz, 1993).

c_1 heme lyase from yeast, the second of these functions may be unique to yeast cytochrome *c* heme lyase.

VIII. SUMMARY

Cytochrome *c* is synthesized in the cytoplasm as apocytochrome *c*, lacking heme, and subsequently transported to the mitochondrial intermembrane space via a pathway that is substantially different from that followed by other well-studied mitochondrial proteins. Heme is covalently attached to the protein by the enzyme cytochrome *c* heme lyase during or after translocation across the outer mitochondrial membrane. Attachment of heme appears to be coupled to mitochondrial accumulation of apocytochrome *c* through a mechanism in which apocytochrome *c* is reversibly translocated across the outer membrane and then trapped in the intermembane space through the action of heme lyase. Heme lyase is predominantly localized at the inner mitochondrial membrane. The ability of apocytochrome *c* to insert into artificial lipid membranes suggests that the precursor might be able to translocate itself at least partially across the outer mitochondrial membrane, but participation of a specific mitochondrial protein pore cannot currently be excluded. The unique properties of apocytochrome *c* as a stably unfolded protein, with folding triggered by attachment of heme, may allow transport to occur without involvement of heat-shock-related proteins. The *CYC2* gene product is a mitochondrial protein that renders import or accumulation of cytochrome *c* more efficient. However, low levels of mitochondrial accumulation are seen in strains with *cyc2* deletions. The signals in the cytochrome *c* sequence that are responsible for targeting the protein to mitochondria appear to differ from other known mitochondrial targeting sequences. Under conditions that allow measurement of mitochondrial import in the absence of heme attachment, several sequence elements near the amino terminal of the protein appear to be involved in particular stages of binding to mitochondria and translocation across the outer membrane. In contrast, alterations throughout the sequence of cytochrome *c* interfere with heme attachment. Cytochrome *c* is just one of several examples of proteins of the mitochondrial intermembrane space that are imported via pathways that appear not to involve the inner membrane. A number of genes whose products are required for post-translational steps in the synthesis of *c*-type cytochromes in bacterial cells have been identified. The relationship of these gene products to cellular components involved in the synthesis of mitochondrial cytochrome *c* remains to be clarified.

ACKNOWLEDGMENT

I thank Fred Sherman for his comments on this manuscript and for many helpful discussions.

REFERENCES

Baker, K. P., & Schatz, G. (1991). Mitochondrial genes essential for viability mediate protein import into yeast mitochondria. Nature 348, 605–609.

Barker, P. D., Ferrer, J. C., Muthusamy, M., Loehr, T. M., Feng, R., Konishi, Y., Funk, W., MacGillivray, R. T. A., & Mauk, A. G. (1993). Transmutation of a heme protein. Proc. Natl. Acad. Sci. USA 90, 6542–6546.

Basile, G., Di Bello, C., & Taniuchi, H. (1980). Formation of and iso-1-cytochrome *c*-like species containing a covalently bonded heme group from apoprotein by a yeast cell-free system in the presence of hemin. J. Biol. Chem. 255, 7181–7191.

Beckman, D. L., Trawick, D. R., & Kranz, R. G. (1992). Bacterial cytochromes *c* biogenesis: Involvement of a putative heme transporter and other novel components. Genes and Dev. 6, 268–283.

Beckman, D. L., & Kranz, R. G. (1993). Cytochromes *c* biogenesis in a photosynthetic bacterium requires a periplasmic thioredoxin-like protein. Proc. Natl. Acad Sci. USA 90, 2179–2183.

Bedwell, D. M., Klionsky, D. L., & Emr, S. (1987). The yeast F1-ATPase β-subunit contains functionally redundant mitochondrial protein import information. Mol. Cell. Biol. 7, 4038–4047.

Cohen, J. S., Fisher, W., & Schechter, A. N. (1974). Spectroscopic studies on the conformation of cytochrome *c* and apocytochrome *c*. J. Biol. Chem. 249, 1113–1118.

de Jongh, H. H. J., Killian, J. A., & de Kruijff, B. (1992). A water-lipid interface induces a highly dynamic folded state in apocytochrome *c* and cytochrome *c*, which may represent a common folding intermediate. Biochemistry 31, 1636–1643.

Cortese, J. G., & Hackenbrock, C. R. (1993). Motional dynamics of functional cytochrome *c* delivered by low pH fusion into the intermembrane space of intact mitochondria. Biochim. Biophys. Acta 1142, 194–202.

Dihanich, M., Suda, K., & Schatz, G. (1987). A yeast mutant lacking mitochondrial porin is respiratory-deficient but can recover respiration with simultaneous accumulation of an 86-kd extramitochondrial protein. EMBO J. 6, 723–728.

Drygas, M. E., Lambowitz, A. M., & Nargang, F. E. (1989). Cloning and analysis of the *Neurospora crassa* gene for cytochrome *c* heme lyase. J. Biol. Chem. 264, 17897–17906.

Dumont, M. E., & Richards, F. M. (1984). Insertion of apocytochrome *c* into lipid vesicles. J. Biol. Chem. 259, 4147–4156.

Dumont, M. E., Ernst, J. F., Hampsey, D. M., & Sherman, F. (1987). Identification and sequence of the gene encoding cytochrome *c* heme lyase in the yeast *Saccharomyces cerevisiae*. EMBO J. 6, 235–241.

Dumont M. E., Ernst J. F., & Sherman, F. (1988). Coupling of heme attachment to import of cytochrome *c* into yeast mitochondria: Studies with heme lyase-deficient mitochondria and altered apocytochromes *c*. J. Biol. Chem. 263, 15928–15937.

Dumont, M. E., Mathews, A. J., Nall, B. T., Baim, S. B., Eustice, D. C., & Sherman, F. (1990). Differential stability of two apo-isocytochromes *c* in the yeast *Saccharomyces cerevisiae*. J. Biol. Chem. 265, 2733–2739.

Dumont, M. E., Cardillo, T. S., Hayes, M. K., & Sherman, F. (1991). Role of cytochrome *c* heme lyase in mitochondrial import and accumulation of cytochrome *c* in *Saccharomyces cerevisiae*. Mol. Cell. Biol. 11, 5487–5496.

Dumont, M. E., Schlichter, J., Cardillo, T. S., Hayes, M. K., Bethlendy, G., & Sherman, F. (1993). *CYC2* encodes a factor involved in mitochondrial import of yeast cytochrome *c*. Mol. Cell. Biol. 13, 6442–6451.

Dumont, M. E., Corin, A. F., & Campbell, G. A. (1994). Noncovalent binding of heme induces a compact apocytochrome *c* structure. Biochemistry 33, 7368–7378.

Enosawa, S., & Ohashi, A. (1986). Localization of enzyme for heme attachment to apocytochrome *c* in yeast mitochondria. Biochem. Biophys. Res. Commun. 141, 1145–1150.

Fisher, W. R., Taniuchi, H., & Anfinson, C. B. (1973). On the role of heme in the formation of the structure of cytochrome *c*. J. Biol. Chem. 248, 3188–3195.

Fytlovich, S., Gervais, M., Agrimonti, C., & Guiard, B. (1993). Evidence for an interaction between the CYP1(HAP1) activator and a cellular factor during heme-dependent transcritptional regulation in the yeast *Saccharomyces cerevisiae*. EMBO J. 12, 1209–1218.

Glick, B., & Schatz, G. (1991). Import of proteins into mitochondria. Annu. Rev. Genet. 25, 21–44.

Hakvoort, T. B. M., Sprinkle, J. R., & Margoliash, E. (1990). Reversible import of apocytochrome *c* into mitochondria. Proc. Natl. Acad. Sci. USA 87, 4996–5000.

Hampsey, D. M., Das, G., & Sherman, F. (1988). Yeast iso-1-cytochrome *c*: Genetic analysis of structural requirements. FEBS Lett. 231, 275–283.

Hartl, F-U., & Neupert, W. (1990). Protein sorting to mitochondria: Evolutionary conservation of folding and assembly. Science 247, 930–938.

Hennig, B., & Neupert, W. (1981). Assembly of cytochrome *c*: Apocytochrome *c* is bound to specific sites on mitochondria before its conversion to holocytochrome *c*. Eur. J. Biochem. 121, 203–212.

Hennig, B., Koehler, H., & Neupert, W. (1983). Receptor sites involved in post-translational transport of apocytochrome *c* into mitochondria: Specificity, affinity, and number of sites. Proc. Natl. Acad. Sci. USA 80, 4963–4967.

Hennig, B., Koehler, H., & Neupert, W. (1983). Transport of cytochrome *c* into mitochondria: Involvement of specific receptors and of cytochrome *c* heme lyase. In: *Mitochondria* 1983, 30th ed. Walter de Guyer & Co., Berlin.

Hickey, D. R., Jayaraman, K., Goodhue, C. T., Shah, J., Clements, J. M., Tsunasawa, S., & Sherman, F. (1991). Synthesis and expression of genes encoding tuna, pigeon, and horse cytochromes *c* in the yeast *Saccharomyces cerevisiae*. Gene 105, 73–81.

Holzschu, D., Principio, L., Conklin, K. T., Hickey, D. R., Short, J., Rao, R., McLendon, G., & Sherman, F. (1987). Replacement of the invariant lysine 77 by arginine in yeast iso-1-cytochrome *c* results in enhanced and normal activities *in vitro* and *in vivo*. J. Biol. Chem. 262, 7125–7131.

Jordi, W., Li-Xin, Z., Pilon, M., Demel, R., & de Kruijff, B. (1989). The importance of the amino terminus of the mitochondrial precursor protein cytochrome *c* for translocation across model membranes. J. Biol. Chem. 264, 2292–2301

Jordi, W., Hergersberg, C., & de Kruijff, B. (1992). Bilayer-penetrating properties enable apocytochrome *c* to follow a special import pathway into mitochondria. Eur. J. Biochem. 204, 841–846.

Kaput, J., Brandriss, M., & Prussak-Wieckowska, T. (1989). *In vitro* import of cytochrome *c* peroxidase into the intermembrane space: Release of the processed form by intact mitochondria. J. Cell Biol. 109, 101–112.

Kiebler, M., Brecker, K., Pfanner, N., & Neupert, W. (1993). Mitochondrial protein import: Specific recognition and membrane translocation of preproteins. J. Membr. Biol. 135, 191–207.

Korb, H., & Neupert, W. (1978). Biogenesis of cytochrome *c* in *Neurospora crassa*. Eur. J. Biochem. 91, 609–620.

Lamberg, R., & Legge, J. W. (1949). In: *Hematin Compounds and Bile Pigments*. Interscience Publishers, New York, pp. 456–470.

Leighton, J., & Schatz, G. (1995). An ABC transporter in the mitochondrial inner membrane is required for normal growth of yeast. EMBO J. 14, 188–195.

Lill, R., Stuart, R. A., Drygas, M. E., Nargang, F. E., & Neupert, W. (1992). Import of cytochrome *c* heme lyase into mitochondria: A novel import pathway into the intermembrane space. EMBO J. 11, 449–456.

Magdolen, V., Schricker, R., Strobel, G., Germaier, H., & Bandlow, W. (1992). *In vivo* import of yeast adenylate kinase into mitochondria affected by site-directed mutagenesis. FEBS Lett. 299, 262–272.

Mannella, C. A., Guo, X. W., & Dias, J. (1992). Binding of a synthetic peptide to a mitochondrial channel protein. J. Bioenergetics and Biomembranes 24, 55–61.

Mannella, C. A., Ribeiro, A. J., & Frank, J. A. (1987). Cytochrome *c* binds to lipid domains in arrays of mitochondrial outer membrane channels. Biophys. J. 51, 221–226.

Matner, R. R., & Sherman, F. (1982). Differential accumulation of two apo-iso-cytochromes *c* in processing mutants of yeast. J. Biol. Chem. 257, 9811–9821.

Matsuura, S., Arpin, M., Hannum, C., Margoliash, E., Sabatini, D.D., & Morimoto, T. (1981). *In vitro* synthesis and posttranslational uptake of cytochrome *c* into isolated mitochondria, role of a specific addressing signal in the apocytochrome. Proc. Natl. Acad. Sci. USA 78, 4368–4372.

Mayer, A., Neupert, W., & Lill, R. (1995). Translocation of apocytochrome *c* across the outer mitochondrial membrane of mitochondria. J. Biol. Chem. 270, 12390–12397.

Muga, A., Mantsch, H. H., & Surewicz, W. K. (1991). Membrane binding induces destabilization of cytochrome *c* structure. Biochemistry 30, 7219–7224.

Murakami, H., Pain, D., & Blobel, G. (1988). 70 kD heat shock related protein is one of at least two distinct cytosolic factors stimulating protein import into mitochondria. J. Cell Biol. 107, 2051–2057.

Nargang, F. E., Drygas, M. E., Kwong, P. L., Nicholson, D. W., & Neupert, W. (1988). A mutant of *Neurospora crassa* deficient in cytochrome *c* heme lyase activity can not import cytochrome *c* into mitochondria. J. Biol. Chem. 263, 9388–9394.

Nicholson, D. W., Hergersberg, C., & Neupert, W. (1988). Role of cytochrome *c* heme lyase in the import of cytochrome *c* into mitochondria. J. Biol. Chem. 263, 19034–19042.

Nicholson, D. W., & Neupert, W. (1989). Import of cytochrome *c* into mitochondria: Reduction of heme, mediated by NADH and flavin nucleotides is obligatory for its covalent linkage to apocytochrome *c*. Proc. Natl. Acad. Sci. USA 86, 4340–4344.

Nye, S. H., & Scarpulla, R. C. (1990a). *In vivo* expression and mitochondrial targeting of yeast apoiso-1-cytochrome *c* fusion proteins. Mol. Cell. Biol. 10, 5753–5762.

Nye, S. H., & Scarpulla, R. C. (1990b). Mitochondrial targeting of yeast apoiso-1-cytochrome *c* is mediated through functionally independent structural domains. Mol. Cell. Biol. 10, 5763–5771.

Page, M. D., & Ferguson, S. J. (1990). Apo forms of cytochrome *c*550 and cytochrome *c*d1 are translocated to the periplasm of paracoccus denitificans in the absence of haem incorporation caused by either mutation or inhibition of heme synthesis. Mol. Microbiol. 4, 1181–1192.

Parr, G. R., & Taniuchi, H. (1980). An investigation of ordered complex formation with chemically modified fragments of cytochrome *c*. J. Biol. Chem. 255, 2616–2623.

Pfanner, N., Rasow, J., van der Klei, I., & Neupert, W. (1992). A dynamic model of the mitochondrial protein import machinery. Cell 68, 999–1002.

Pon, L., Moll, T., Vestweber, D., Marshallsay, B., & Schatz, G. (1989). Protein import into mitochondria: ATP-dependent protein translocation activity in a submitochondrial fraction enriched in membrane contact sites and specific proteins. J. Cell Biol. 109, 2603–2616.

Ramseier, T. M., Winteler, H. V., & Hennecke, H. (1991). Discovery and sequence analysis of bacterial genes involved in the biogenesis of *c*-type cytochromes. J. Biol. Chem. 266, 7793–7803.

Rietveld, A., & De Kruijff, B. (1984). Is the mitochondrial precursor protein apocytochrome *c* able to pass a lipid barrier? J. Biol. Chem. 259, 6704–6707.

Ritz, D., Bott, M., & Hennecke, H. (1993). Formation of several bacterial *c*–type cytochromes requires a novel membrane-anchored protein that faces the periplasm. Mol. Microbiol. 9, 729–740.

Rothstein, R. J., & Sherman, F. (1980). Genes affecting the expression of cytochrome *c* in yeast: Genetic mapping and genetic interactions. Genetics 94, 871–889.

Sadis, S., & Hightower, L. E. (1992). Unfolded proteins stimulate molecular chaperone Hsc70 ATPase by accelerating ADP/ATP exchange. Biochemistry 31, 9406–9412.

Söllner, T., Griffiths, G., Pfaller, R., Pfanner, N., & Neupert, W. (1989). MOM19, an import receptor for mitochondrial precursor proteins. Cell 59, 1061–1070.

Sanbongi, Y., Yang, J-H., Igarashi, Y., & Kodoma, T. (1991). Cloning and nucleotide sequence of the cytochrome *c*-551 gene from Hydrogenobacter thermopilus., Eur. J. Biochem. 198, 7–12.

Sorrell, T. N., & Martin, P. K. (1989). A novel, functional variant of cytochrome *c*: Replacement of the histidine ligand with arginine via site-directed mutagenesis. J. Am. Chem. Soc. 111, 766–767.

Sprinkle, J. R., Hakvoort, T. B. M., Koshy, T. I., Miller, D. D., & Margoliash, E. (1990). Amino acid sequence requirements for the association of apocytochrome *c* with mitochondria. Proc. Natl. Acad. Sci. USA 87, 5729–5733.

Stellwagen, E., Rysavy, R., & Babul, G. (1972). The conformation of horse heart apocytochrome *c*. J. Biol. Chem. 247, 8074–8077.

Stuart, R. A., & Neupert, W. (1990). Apocytochrome *c* an exceptional mitochondrial precursor protein using an exceptional import pathway. Biochemie. 72, 115–121.

Stuart, R. A., Neupert, W., & Tropschug, M. (1987). Deficiency in mRNA splicing in a cytochrome *c* mutant of *Neurospora crassa*: Importance of carboxy terminus for import of apocytochrome *c* into mitochondria. EMBO J. 6, 2131–2137.

Stuart, R. A., Nicholson, D. W., Wienhues, U., & Neupert, W. (1990a). Import of apocytochrome *c* into the mitochondrial intermembrane space along a cytochrome c_1 sorting pathway. J. Biol. Chem. 265, 20210–20219.

Stuart, R. A., Nicholson, D. W., & Neupert, W. (1990b). Early steps in mitochondrial protein import: Receptor functions can be substituted by the membrane insertion activity of apocytochrome *c*. Cell 60, 31–43.

Takano, T., & Dickerson, R. E. (1981). Conformational change of cytochrome *c*. 1. Ferrocytochrome *c* structure refined at 1.5 A resolution. J. Mol. Biol. 153, 79–94.

Tanaka, Y., Kubota, I., Amachi, T., Yoshizumi, H., & Matsubara, H. (1990). Site-directed mutated human cytochrome *c* which retains heme *c* via only one thioether bond. J. Biochem. 108, 7–8.

Veloso, D., Juillerat, M., & Taniuchi, H. (1984). Synthesis of a heme fragment of horse cytochrome *c* which forms a productive complex with native apofragment. J. Biol. Chem. 259, 6067–6073.

Wang, X., Dumont, M. E., & Sherman, F. (1996). Sequence requirements for mitochondrial import of yeast cytochrome *c*. J. Biol. Chem. 271, 6594–6604.

Wotczak, L., & Zaluska, H. (1969). On the impermeability of the outer mitochondrial membrane to cytochrome *c*. I. Studies on whole mitochondria. Biochim. Biophys. Acta 193, 64–72.

Zimmerman, R., Henning, B., & Neupert, W. (1981). Different transport pathways of individual precursor proteins in mitochondria. Eur. J. Biochem. 116, 455–460.

Zollner, A., Rodel, G., & Haid, A. (1992). Molecular cloning and characterization of the *Saccharomyces cerevisiae CYT2* gene encoding cytochrome-c_1-heme lyase. Eur. J. Biochem. 207, 1093–1100.

TRANSLOCATION OF PREPROTEINS ACROSS THE MITOCHONDRIAL INNER MEMBRANE:
TIMs AND HSP70

Michiel Meijer, Ammy Maarse,
Michael Kübrich, and Nikolaus Pfanner

I. Introduction 128
II. Energetics of Protein Translocation Across the Inner Membrane 129
A. The Membrane Potential $\Delta\Psi$ 129
B. ATP 132
III. Tims Form the Import Machinery of the Inner Membrane 132
A. Genetic Identification of the Tims 134
B. The Integral Inner Membrane Proteins Tim17 and Tim23 as Parts of a Protein Translocation Channel 134
C. Direct Interaction of the Peripheral Inner Membrane Protein Tim44 with Preproteins 137
D. Interactions of Tim-Proteins 139
IV. Role of Hsp70 in the Matrix in Unfolding, Translocation, and Folding of Preproteins 139

Advances in Molecular and Cell Biology,
Volume 17, pages 127–148.

ISBN: 0-7623-0144-9

V. Components Interacting with Mitochondrial Hsp70 143
VI. Perspectives 143
Acknowledgments 144
References 144

I. INTRODUCTION

The mitochondrial inner membrane is a sealed membrane that is impermeable to many small ions, unless they are transported by specific carrier proteins (Krämer and Palmieri, 1989). However, this membrane is able to translocate a large number of different preproteins. Even during the translocation of a polypeptide chain, the membrane maintains its transmembrane proton gradient. This was shown by Vestweber and Schatz (1988) and Wienhues and colleagues (1991) who accumulated a preprotein as membrane-spanning intermediate and demonstrated that at the same time a membrane potential was present. This implies that the inner membrane possesses a specific translocation channel, which permits the passage of many different polypeptides, but is selective enough to prevent the leakage of ions. In the past years, it was a major goal in the field of mitochondrial protein import to identify components of the import machinery of the inner membrane.

Before we discuss the current knowledge about the inner membrane import machinery, let us have a look at the historical development of the research in this field. R. Butow and colleagues proposed that mitochondrial preproteins were preferentially imported at sites of close contact between mitochondrial outer and inner membranes (Kellems et al., 1975). Electron micrographs of isolated mitochondria showed areas of the outer membrane (about 5% of the total outer membrane) that seemed to be stably associated with the inner membrane. It was thus concluded that these morphological contact sites served as main import sites for mitochondrial preproteins. A major breakthrough in the field was the discovery by Schleyer and Neupert (1985) of translocation intermediates of preproteins that were spanning both mitochondrial membranes at the same time. This promoted the view that the transport machineries of the outer and inner membranes were permanently linked at contact sites and that a preprotein passing through the tightly connected channels never gets into contact with the intermembrane space. More recent findings, however, did not fit to the proposal of stable and permanently sealed translocation contact sites. (1) Schatz and colleagues opened the outer membrane of isolated yeast mitochondria and demonstrated that preproteins could be directly translocated across the inner membrane without using the receptors and other components of the outer membrane (Ohba and Schatz, 1987; Hwang et al., 1989). (2) We found that insertion of the precursor of the ADP/ATP carrier into the outer membrane was independent of the membrane potential, while transport of the precursor into the inner membrane strictly depended on a $\Delta\Psi$ (Pfanner and Neupert, 1987; Pfanner et al., 1987b). (3) The final proof that the import machineries of both mitochondrial membranes were not stably linked by a sealed channel came from the identification of bona fide translocation intermediates in the intermembrane space (Hwang et al., 1991; Rassow and Pfanner, 1991). This suggests that the import machineries of the mitochondrial outer and inner membranes are independent entities that can be transiently linked by a polypeptide chain in transit across both membranes ("translocation contact sites"). Since *in vivo*, most of the mitochondrial outer membrane is in close contact to parts of the inner membrane (the inner boundary membrane), translocation contact sites may be formed all over the mitochondrial surface (Pfanner et al., 1992). The translocation contact sites do not seem to be identical with the permanent morphological contact sites.

The view that the inner membrane possesses a structurally independent protein import system has been substantiated by the recent identification of three mitochondrial inner membrane proteins which are essential components of the transport machinery. This review focuses on the properties of these Tim-proteins. We will also discuss the energetics of translocation, that is, the role of the membrane potential $\Delta\Psi$ and of ATP. The importance of ATP seems to be largely due to the ATP-dependent function of the heat shock protein Hsp70 in the matrix, the fourth essential protein of the import apparatus.

II. ENERGETICS OF PROTEIN TRANSLOCATION ACROSS THE INNER MEMBRANE

The process of mitochondrial protein uptake appears to be unidirectional and irreversible, that is, an imported protein is unable to move back into the cytosol. The conformational changes accompanying the import, folding, and assembly of proteins are by themselves usually not sufficient to drive an unidirectional transport. This is particularly obvious in the case of protein translocation across the mitochondrial inner membrane. The supply of two external energy sources, ATP and the membrane potential $\Delta\Psi$ across the inner membrane, is needed to drive protein import (Figure 1). We will discuss that with some preproteins short segments may move back across the inner membrane when the function of the matrix Hsp70 is strongly impaired, for example, by ATP-depletion of the matrix, supporting the view of an energy-driven unidirectional transport.

A. The Membrane Potential $\Delta\Psi$

The mitochondrial respiratory chain generates a proton electrochemical gradient across the inner membrane which consists of two components, a proton concentration difference ΔpH and an electrical potential difference, the membrane potential $\Delta\Psi$ (Nicholls and Ferguson, 1992). Energization of the inner membrane is an essential prerequisite for the import of preproteins. Investigations with translocation intermediates indicated that the initial insertion of the amino-terminal presequence required an energized membrane (Figure 1, stage *a* to stage *b*). Once the

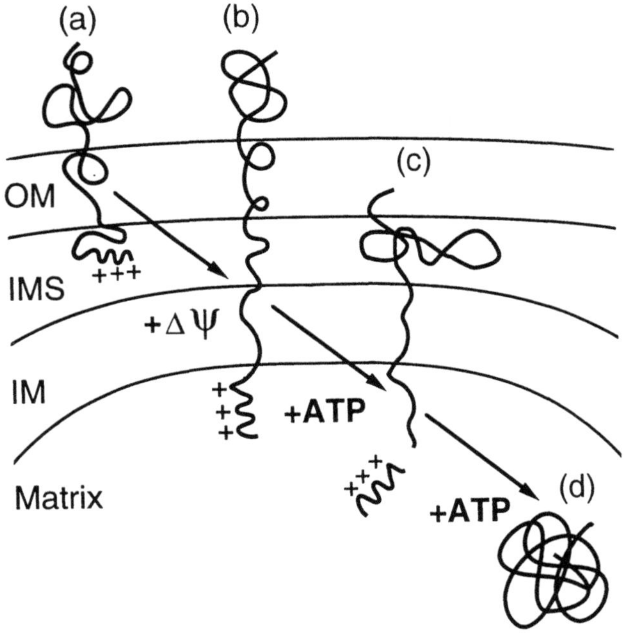

Figure 1. Energetics of preprotein translocation across the mitochondrial inner membrane. The membrane potential ΔΨ drives the initial translocation of the positively charged presequence. ATP in the matrix is needed to complete the translocation of the presequence and to promote the unfolding and translocation of the mature part of the preprotein. IM, inner membrane; IMS, intermembrane space; OM, outer membrane.

presequence had been transferred across the membrane, translocation of the mature protein portion was independent of a membrane potential (Schleyer and Neupert, 1985). The efficiency (yield) of preprotein translocation depends on the magnitude of the potential gradient, as found with an artificially generated potassium diffusion potential. For the import of a few preproteins it was possible to experimentally distinguish between the contributions of ΔpH and ΔΨ. In those cases, only the ΔΨ but not the ΔpH, appeared to be required, implying that protein import was not driven by a movement of protons, but depended on the electrical potential gradient (Pfanner and Neupert, 1985; Martin et al., 1991).

Different models can be envisaged that explain the role of the membrane potential in protein import. (1) The presequences are positively charged and one might imagine that the ΔΨ (negative on the matrix side) exerts an electrophoretic effect on the presequence. This is supported by the observation that different presequences require different threshold membrane potentials to initiate translocation. Presequences with a high number of positive charges can be translocated already at a low membrane potential (Martin et al., 1991). Nicholls and Ferguson (1992) discussed that, based on energetic considerations, it is unlikely that a $\Delta\Psi$ alone is the driving force for translocation of a polypeptide. It may be speculated that $\Delta\Psi$ helps to bring the presequence into an orientation compatible with insertion into the membrane and the translocation channel. The membrane potential may facilitate the initial entrance of the presequence into the lipid membrane (Maduke and Roise, 1993), that is rapidly followed by contact of the presequence to and insertion into the translocation channel. (2) The activity of a protein component of the translocation channel of the inner membrane may be modulated by the membrane potential. This component should be involved in the initial import step, that is, binding of the preprotein or opening of the channel, and should be activated only in the presence of a $\Delta\Psi$. (3) Eventually, it cannot be excluded that the import of some preproteins requires the transport of protons driven by the total protonmotive force. In this case, both $\Delta\Psi$ and ΔpH would be effective in driving protein import. The movement of protons might induce conformational changes of a channel protein that is required for initial import steps.

Some polytopic membrane proteins, such as the ADP/ATP carrier of the inner membrane, carry several internal signal sequences (Pfanner et al., 1987a; Smagula and Douglas, 1988). It is unknown in which sequential order the polypeptide segments of the preprotein are inserted into the membrane and which steps require a membrane potential.

According to the endosymbiont hypothesis of mitochondrial origin, the mitochondrial inner membrane is derived from a prokaryotic plasma membrane. However, the direction of protein movement into mitochondria (towards the negatively charged face of the membrane) is opposite to that of protein export in bacteria (toward the positively charged face of the membrane). It is, therefore, not surprising that the protein import system of mitochondria and the export system of bacteria are different in many aspects, including the types of signal sequences and the components involved. In bacteria, the total protonmotive force, that is, a movement of protons, promotes protein export, but an energization of the plasma membrane is not absolutely essential to drive translocation. Moreover, the protonmotive force is not required for the initial insertion of the preprotein into the plasma membrane, yet for later translocation steps (Driessen and Wickner, 1991; Schiebel et al., 1991). Protein transport into the endoplasmic reticulum (ER) is similar to bacterial protein transport in both signal sequences and components involved, and protein transport across the chloroplast envelope shares basic principles with mitochondrial protein import (Rapoport, 1992; Soll and Alefsen, 1993). Although a requirement for an energized membrane has not been found for protein transport into the ER or chloroplasts, it cannot be excluded that these transport systems require the presence of a low membrane potential that has not been detected in the studies performed so far.

B. ATP

The transport of preproteins into mitochondria typically requires hydrolysis of ATP both in the cytosol and in the matrix. There are multiple roles of ATP in the import process. The ATP-requiring steps defined so far are mostly those involving the functions of heat shock proteins. (1) ATP in the cytosol seems to be involved in maintaining preproteins in a transport-competent (nonaggregated) conformation (Chen and Douglas, 1987; Pfanner et al., 1987b, 1988, 1990; Verner and Schatz, 1987; Murakami et al., 1988; Skerjanc et al., 1990; Hachiya et al., 1993). However, several (folded or unfolded) preproteins were found that were imported in the absence of cytosolic ATP and Hsp70s. (2) Transport of preproteins into the mitochondrial matrix requires the hydrolysis of matrix ATP (Hwang and Schatz, 1989). As discussed in the text that follows, the mitochondrial Hsp70, which is essential for binding and translocation of preproteins, functions in an ATP-dependent cycle (Manning-Krieg et al., 1991; Gambill et al., 1993). The membrane potential $\Delta\Psi$ and matrix ATP are needed for protein import in a sequential order (Figure 1). $\Delta\Psi$ is needed for the initial insertion of the presequence into the inner membrane, while matrix ATP is needed to complete the translocation of the presequence and to drive the unfolding and translocation of the mature protein part (Cyr et al., 1993; Gambill et al., 1993). The ATP-dependent roles of Hsp70s in protein translocation appear to be of general importance, not only for mitochondria and chloroplasts, but also for the endoplasmic reticulum (Sanders and Schekman, 1992). (3) The Hsp60 in the matrix operates in an ATP-dependent fashion. While its dominant role in folding of imported proteins is well documented (Cheng et al., 1989; Ostermann et al., 1989), there is no evidence so far that Hsp60 is required for membrane translocation of preproteins.

It is conceivable that the requirement of import for ATP is not limited to the steps involving heat shock proteins. Components of the translocation channels of the outer membrane or the inner membrane may be actively involved in the translocation process and require ATP. Currently, however, evidence for an ATP-dependent function of membrane components of the inner membrane import machinery is lacking.

III. TIMs FORM THE IMPORT MACHINERY OF THE INNER MEMBRANE

While many of the components of the outer membrane import machinery have been identified and are already partly characterized, identification of components of the mitochondrial inner membrane import machinery has started only recently. Biochemical and genetic approaches were used to analyze the inner membrane machinery. The major advance leading to the discovery of three proteins of the mitochondrial inner membrane import machinery (TIM) was made by the characterization of yeast mutants defective in protein import. In contrast, the identification of outer membrane components was mainly achieved by biochemical approaches.

Table 1. Components of the Inner Membrane Import Machinery of Yeast Mitochondria

Protein[*]	*Tim17* (*MIM17, Sms1p*)	*Tim23* (*MIM23, as6p*)	*Tim44* (*MIM44, ISP45*)	*mt-Hsp70* (*Ssc1p*)
Gene	*Tim17 (MIM17, SMS1)*	*Tim23 (MIM23, MAS6)*	*Tim44 (MIM44, MPI1)*	*SSC1*
Essential for viability	Yes	Yes	Yes	Yes
Location	Inner membrane (integral)	Inner membrane (integral)	Inner membrane (peripheral)	Matrix/partially inner membrane
Synthesized with presequence	No	No	Yes	Yes
Import requires $\Delta\Psi$	Yes	Yes	Yes	Yes
Involvement in import	Part of channel	Part of channel	Binds to preprotein	Binds to preprotein
References	Dekker et al. (1993) Kübrich et al. (1994) Maarse et al. (1994) Ryan et al. (1994)	Dekker et al. (1993) Emtage and Jensen (1993) Ryan and Jensen (1993) Kübrich et al. (1994)	Maarse et al. (1992) Scherer et al. (1992) Blom et al. (1993) Horst et al. (1993)	Craig et al. (1989) Kang et al. (1990) Manning-Krieg et al. (1991) Gambill et al. (1993) Rassow et al. (1994)

Notes: [*] In the "MIM," "Tim," and "Hsp" terminologies, the number indicates the molecular mass of a protein in kDa.
Hsp, heat shock protein; *MAS*, mitochondrial assembly; *MIM*, mitochondrial inner membrane import machinery; *MPI*, mitochondrial protein import; *mt*, mitochondrial; *SSC*, stress seventy subfamily C; *Tim*, translocase of the inner mitochondrial membrane.

A. Genetic Identification of the Tims

To identify new components of the mitochondrial protein import machinery we have devised a positive selection procedure for *Saccharomyces cerevisiae* mutants with partial defects in translocation of mitochondrial preproteins across the inner membrane (Maarse et al., 1992). This genetic approach is based on the efficient mislocalization of a cytosolic enzyme into the mitochondrial matrix. A chimeric protein was constructed, consisting of the mitochondrial targeting sequence of the matrix enzyme manganese superoxide dismutase and the complete cytosolic enzyme orotidine 5′-phosphate (OMP) decarboxylase (encoded by the *URA3* gene). A yeast strain was used that carried a deletion of the chromosomal *URA3* locus and thus required the addition of uracil to the growth medium. When the chimeric protein was expressed in the yeast strain, growth of transformed cells was still dependent on added uracil, despite the fact that significant OMP decarboxylase activity was measured in cellular extracts. These observations indicated that the mitochondrial targeting signal efficiently directed the fusion protein into mitochondria, thereby preventing it from participation in cytosolic pyrimidine biosynthesis. Mutations that block import of the chimeric protein into mitochondria are expected to allow growth of transformed cells in the absence of uracil. Upon chemical mutagenesis of the transformed cells, such mutations were obtained as cis-acting forms in the mitochondrial targeting sequence and trans-acting forms in components of the mitochondrial protein import machinery. Sixteen trans-acting recessive nuclear mutations were classified into four complementation groups (Table 1; Dekker et al., 1993).

Mutants of one complementation group carry mutations in *SSC1*, the gene encoding mitochondrial Hsp70, an essential matrix protein required for protein translocation across the mitochondrial inner membrane (Craig et al., 1989). The other three complementation groups represent mutants of new essential genes, named *MIM44 (MPI1)*, *MIM17* (*SMS*1), and *MIM23* (*MAS*6). (Maarse et al., 1992, 1994; Dekker et al., 1993). As described in the text following, all three genes were found to encode proteins of the protein import machinery of the *m*itochondrial *i*nner *m*embrane which were thus termed MIM44, MIM17, and MIM23 (proteins with molecular masses of 44 kDa, 17 kDa, and 23 kDa. According to a recently introduced uniform nomenclature for subunits of translocases of both mitochondrial membranes, these *t*ranslocases of the *i*nner mitochondrial *m*embrane are renamed as Tim44, Tim17, and Tim23, respectively (Pfanner et al., 1996).

B. The Integral Inner Membrane Proteins Tim17 and Tim23 as Parts of a Protein Translocation Channel

Tim17

Tim17 (Sms1p) contains four hydrophobic regions, almost equally distributed over the protein sequence, which may function as membrane anchors (Figure 2).

In particular, the carboxy-proximal region is predicted to function as a membrane spanning sequence. Fractionation experiments showed that Tim17 is an integral protein of the mitochondrial inner membrane (Kübrich et al., 1994; Maarse et al., 1994; Ryan et al., 1994). Experiments with an epitope-tagged Tim17-derivative suggest that the extreme carboxy-terminus of the protein is oriented towards the intermembrane space side (Figure 2).

Import of the precursor of Tim17 into mitochondria requires the presence of a membrane potential $\Delta\Psi$. While many inner membrane proteins are synthesized with a cleavable presequence, Tim17 is apparently not processed upon import into mitochondria, indicating that its targeting information resides in the mature protein.

Deletion of the *Tim17* gene is lethal to yeast cells not only on nonfermentable, but also on fermentable carbon sources (Maarse et al., 1994; Ryan et al., 1994). Hence, Tim17 is one of the few mitochondrial proteins that are essential for viability of yeast under all growth conditions. The majority of mitochondrial proteins, such as the components of the respiratory chain or the mitochondrial protein synthesis apparatus, are not essential for growth of yeast on fermentable carbon sources (Attardi and Schatz, 1988). All essential mitochondrial proteins identified so far are components of the machinery for import or folding of nuclear encoded proteins:

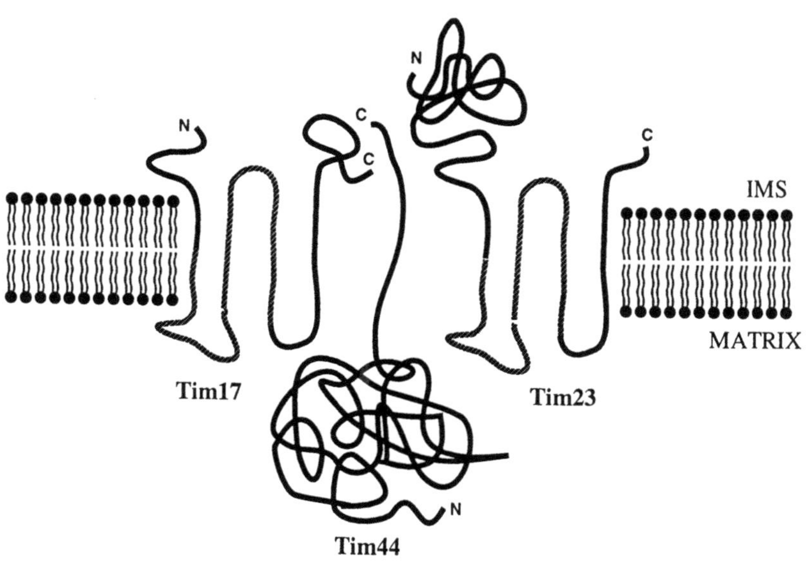

Figure 2. Speculations about the topology of the Tim-proteins in the mitochondrial inner membrane. The region of homology between the mitochondrial inner membrane proteins Tim17 and Tim23 is hatched. While Tim17 and Tim23 directly interact with each other it is unknown which components mediate the stable association of Tim44 with the inner membrane. C, carboxy-terminal end; IMS, intermembrane space; N, amino-terminal end.

the outer membrane proteins Tom40 (ISP42; Baker et al., 1990) and Tom22 (Mas22p; Lithgow et al., 1994; Hönlinger et al., 1995); the matrix proteins Hsp70, Hsp60, Mge1p (the homolog of the prokaryotic heat shock protein GrpE; Craig et al., 1989; Cheng et al., 1989; Laloraya et al., 1994), and the two subunits of the mitochondrial processing peptidase (Jensen and Yaffe, 1988; Pollock et al., 1988; Witte et al., 1988); the inner membrane proteins Tim17 and, as described following, Tim23, and Tim44 (Maarse et al., 1992, 1994; Emtage and Jensen, 1993).

Mitochondria isolated from a mutant yeast strain with a temperature-sensitive tim17 allele are defective in the import of several preproteins when the import reaction is performed at the restrictive temperature. To obtain direct evidence that Tim17 is part of a putative protein translocation channel of the inner membrane, crosslinking studies were carried out with a preprotein that was accumulated as a translocation intermediate spanning both mitochondrial membranes (Figure 1, stage *b*). Tim17 was crosslinked to the translocation intermediate, whereas it could not be crosslinked to preproteins bound to the mitochondrial surface or to fully imported proteins. This suggests that Tim17 is in close contact with preproteins in transit across the mitochondrial inner membrane (Kübrich et al., 1994).

Tim23

Tim23 has many properties in common with Tim17. It is a mitochondrial inner membrane protein which appears to be synthesized without a cleavable targeting sequence. Import of the precursor of Tim23 into mitochondria requires a membrane potential (Table 1). While the amino-terminal half of Tim23 has an overall hydrophilic character, the carboxy-terminal half contains four hydrophobic regions. At least two of the hydrophobic sequences are of sufficient length to function as membrane anchor sequences in agreement with the observation that imported Tim23 behaves as an integral membrane protein. Both the amino-terminal half and the extreme carboxy-terminus of Tim23 seem to face the intermembrane space (Figure 2; Emtage and Jensen, 1993; Kübrich et al., 1994). Tim17 and Tim23 share a domain of 90 amino acids with a high similarity, hydrophobic regions at corresponding positions, and a similar distribution of charged residues. This domain contains most of the potential membrane spanning sequences of Tim17 and Tim23 (Figure 2; Dekker et al., 1993; Kübrich et al., 1994). It is therefore tempting to speculate that both proteins have a comparable membrane topology and that their homologous domains play a crucial role in translocation of preproteins across the mitochondrial inner membrane. This hypothesis is supported by the observation that a deletion of Tim23 sequences at either the amino-terminal or the carboxy-terminal flank of this domain does not cause a drastic loss of Tim23-function (Ryan and Jensen, 1993; Dekker, unpublished data).

The yeast mutant *mas6*, which has a temperature-sensitive defect in mitochondrial protein import, has been characterized in detail. The *MAS6* gene was found to be identical to the Tim23 gene. Deletion of the gene was lethal to yeast cells on all carbon sources (Emtage and Jensen, 1993). Mitochondria isolated from the temperature-sensitive *mas6* mutant were defective in import of several preproteins. When the expression of Tim23 was turned off, the cells accumulated mitochondrial precursor proteins. Moreover, antibodies raised against Tim23 inhibited protein import into mitochondria whose outer membrane had been disrupted, and Tim23 could be crosslinked to preproteins in transit across the mitochondrial membranes (Emtage and Jensen, 1993; Ryan and Jensen, 1993; Kübrich et al., 1994). These observations suggest that Tim23 is an essential part of the protein import machinery of the mitochondrial inner membrane.

We speculate that Tim17 and Tim23 represent (part of) the protein translocation channel of the inner membrane. It is shown that Tim17 and Tim23 are associated with each other in a protein complex, which may form the putative translocation channel. This protein complex also contains 2–4 other protein subunits, but their characterization has yet not been performed (Blom et al., 1995; Berthold et al., 1995). The role of the peripheral membrane protein Tim44 and the soluble protein Hsp70 are discussed following. Moreover, Tim17 or Tim23 may be involved in the recognition of preproteins to be translocated across the inner membrane, that is, function as receptor(s) of the inner membrane.

C. Direct Interaction of the Peripheral Inner Membrane Protein Tim44 with Preproteins

Tim44 is associated with the mitochondrial inner membrane. Its precursor is synthesized with an amino-terminal presequence that is cleaved off after the membrane potential-dependent import into mitochondria. The primary sequence of Tim44 predicts a rather hydrophilic protein without obvious hydrophobic stretches (Maarse et al., 1992). Consistently, Tim44 does not behave as an integral membrane protein embedded into the lipid phase of the membrane. It is, however, stably associated with the inner membrane and is not released from the membranes by sonication even at high salt concentrations. Therefore, Tim44 can be defined as a firmly associated peripheral inner membrane protein (Blom et al., 1993). The major portion of Tim44 is probably exposed to the matrix space, but the extreme carboxy-terminus of the protein seems to face the intermembrane space (Figure 2). We assume that Tim44 is associated with the inner membrane via interaction with unidentified integral membrane proteins (Rassow et al., 1994). Tim17 and Tim23 are possible partners but their weak and probably only transient interaction with Tim44 indicates that other proteins are involved in the stable association of Tim44 with the inner membrane.

Tim44 is an essential yeast protein at all growth conditions. In yeast cells with strongly reduced levels of Tim44, the steady-state level and rate of maturation of imported mitochondrial proteins was drastically reduced (Maarse et al., 1992). Tim44 could be crosslinked to a preprotein accumulated in the mitochondrial membranes (Blom et al., 1993). Interestingly, Scherer and colleagues (1992) identified a mitochondrial protein with a size of about 45 kDa by crosslinking to a partially translocated preprotein. They used a polyspecific antiserum recognizing this import site protein ISP45 and showed inhibition of protein import into inner membrane vesicles. Horst and colleagues (1993) reported that Tim44 and ISP45 are identical proteins. The combined data thus provide strong evidence that Tim44 is a further component of the protein import machinery of the mitochondrial inner membrane.

Protein translocation across the mitochondrial inner membrane can be divided into distinct steps. In a relatively early stage the preprotein spans both the outer membrane and inner membrane and extends from the cytosol into the matrix space (Figure 1, stage *b*). In a later stage, the preprotein has traversed the outer membrane and spans from the intermembrane space to the matrix (Figure 1, stage *c*). Tim44 was crosslinked with high efficiency to the translocation intermediate *b*, suggesting that Tim44 is in close proximity to preproteins in an early step of transfer across the mitochondrial inner membrane. Crosslinking to the late intermediate *c* occurred only with low efficiency. It remains to be determined whether this was due to the technical conditions of the crosslink approach used or whether it indicates a preferential role of Tim44 in the early import stage. Co-immunoprecipitation

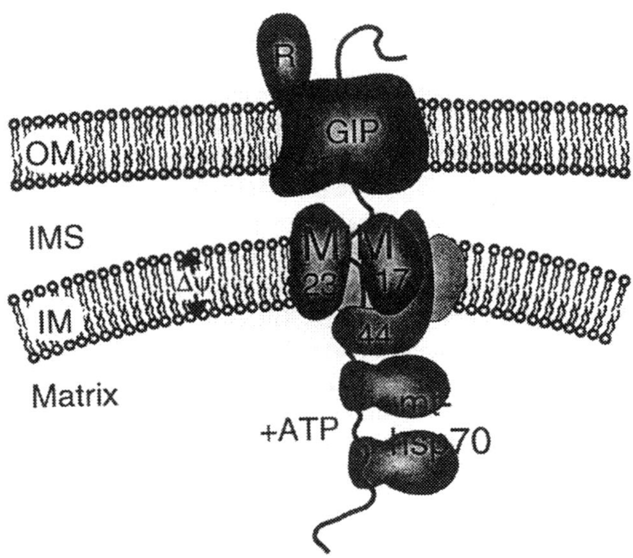

Figure 3. Hypothetical model of the protein import machinery of the mitochondrial inner membrane. A preprotein is shown that spans across the import machineries of the outer membrane (OM) and inner membrane (IM). ΔΨ, membrane potential; GIP, general insertion pore; IMS, intermembrane space; Timx, mitochondrial inner membrane protein of x kDa; mt-Hsp70, heat shock protein of 70 kDa in the mitochondrial matrix; R, receptor.

experiments indicate that solubilized Tim44 interacts with *in vitro* synthesized preproteins. The mechanism and specificity of the interaction of Tim44 and preproteins have not been elucidated in detail. A first result is that the presence of the presequence enhances the binding efficiency of a protein to Tim44 (Blom et al., 1993).

D. Interactions of Tim-Proteins

Although it has only been possible to demonstrate a physical interaction of Tim17 and Tim23, genetic evidence indicates that Tim17 and Tim23 functionally cooperate with Tim44 (Blom et al., 1995). Furthermore, all three Tims were found as crosslink-products with the same preprotein that was accumulated as translocation intermediate in the mitochondrial membranes (Kübrich et al., 1994), suggesting that they act in close vicinity of the same inner membrane import site (Figure 3).

Further advanced are the studies on an interaction of Tim44 and the mitochondrial Hsp70 of the matrix. First, the protein import defect of yeast cells with mutated Tim44 can be suppressed by overexpression of mitochondrial Hsp70. Second, a fraction of mitochondrial Hsp70 was found in a complex with Tim44. The binding of mt-Hsp70 to Tim44 explains the observation that part of this matrix protein seems to be bound to the inner membrane (Kronidou et al., 1994; Rassow et al., 1994; Schneider et al., 1994). Binding of ATP induces a conformational change in nt-Hsp70 and seems to induce dissociation of the mt-Hsp70-Tim44 complex (von Ahsen et al., 1995; Horst et al., 1996).

IV. ROLE OF Hsp70 IN THE MATRIX IN UNFOLDING, TRANSLOCATION, AND FOLDING OF PREPROTEINS

The heat shock protein of 70 kDa in the mitochondrial matrix belongs to the multiprotein family of Hsp70s. Hsp70s have a high binding affinity for unfolded or partially folded polypeptide segments and release the polypeptides in an ATP-dependent manner (summarized in Rothman, 1989). Yeast cells contain at least eight different Hsp70s, six in the cytosol (Ssa1p-Ssa4p, Ssb1p, and Ssb2p), one in the endoplasmic reticulum (Kar2p), and one in the mitochondrion (Ssc1p, *s*tress *s*eventy subfamily *c*; Craig et al., 1989, 1993). Ssa1p - Ssa4p form an essential subfamily and appear to be involved in preventing misfolding and aggregation of precursors of organellar proteins. Ssb1p and Ssb2p are not essential; they seem to interact with the nascent polypeptide chain emerging from the ribosome. Deletion of the gene encoding mitochondrial Hsp70, as well as that encoding the endoplasmic reticulum Hsp70, is lethal to yeast cells. The experimental dissection of protein transport into isolated mitochondria by use of translocation and folding intermediates (Hartl et al., 1989; Söllner et al., 1991) promoted a detailed analysis of the function of mt-Hsp70. Thus, the function of Hsp70s in translocation or folding of preproteins is currently best understood for the mt-Hsp70. The Hsp70 in the endoplasmic reticulum may play similar roles in driving protein import (Sanders and Schekman, 1992), representing one of the few cases of similarities on the molecular level between protein transport into mitochondria and the endoplasmic reticulum.

We used two temperature-sensitive yeast mutants, each carrying a single amino acid substitution in *SSC1*, to study the role of mt-Hsp70 (Kang et al., 1990; Ostermann et al., 1990; Gambill et al., 1993; Voos et al., 1993). The mutant phenotypes can be induced by the shift of isolated mitochondria to the nonpermissive temperature of 37°C for a few minutes. Since a shift of the cell culture to the non-permissive temperature is not necessary, indirect effects of the mutations can be minimized. In the mutant *ssc1-2*, mt-Hsp70 is still able to bind preproteins, yet is impaired in release of preproteins. In the mutant *ssc1-3*, mt-Hsp70 is already unable to bind preproteins. In *ssc1-2* mitochondria, preproteins were partially imported. The presequence was completely translocated into the matrix and cleaved off by the processing peptidase, while the mature portion of the protein spanned across the mitochondrial membranes. Artificial unfolding of the preprotein in 8 M urea prior to addition to isolated mitochondria allowed a complete import of the polypeptide chain into the matrix, where it was found associated with the mutant Hsp70. This indicated that mt-Hsp70 plays a role in the unfolding of polypeptide chains during translocation across the mitochondrial membranes (Table 2).

Table 2. Functions of Mitochondrial Hsp70 in Protein Import

Function	*Preproteins affected*	*Proposed mechanism*	*References*
Unfolding	Preproteins for matrix, inner membrane and (in some cases) intermembrane space; the preproteins have (partially) folded structures on the cytosolic side	Step-wise binding of mt-Hsp70 molecules to an extended preprotein chain emerging on the matrix side; thereby trapping of preprotein segments in the matrix and facilitating unfolding on the cytosolic side	Kang et al. (1990) Gambill et al. (1993) Voos et al. (1993)
Translocation	Preproteins destined for the matrix	Binding of mt-Hsp70 is an essential step to drive complete import of a preprotein into the matrix (independent of the folding state of the preprotein)	Gambill et al. (1993) Voos et al. (1993)
Folding	Matrix proteins	Keeping preproteins in a loosely folded (folding-competent) conformation and transfer to Hsp60	Kang et al. (1990) Manning-Krieg et al. (1991)

To test the transport defect in *ssc1-3* mitochondria, a precursor with the presequence of F_0-ATPase subunit 9 was used. This highly positively charged presequence is cleaved twice by the matrix processing peptidase (Schmidt et al., 1984). Only the first half of the presequence could be transported far enough into the matrix of *ssc1-3* mitochondria to be cleaved. Unfolding of the preprotein did not promote translocation of the complete presequence or of the mature protein part into the matrix (Gambill et al., 1993). Unfolding of the polypeptide chain, however, promoted the translocation of the mature part across the outer membrane to the intermembrane space of *ssc1-3* mitochondria. We conclude that mt-Hsp70 has a dual role in membrane translocation of preproteins: (1) facilitation of unfolding of preproteins for translocation across outer and inner membranes (unfoldase function); and (2) mt-Hsp70 is a genuine component of the inner membrane import machinery, that is, binding of Hsp70 to the preprotein is essential to drive complete transport of the polypeptide chain across the inner membrane into the matrix, independent of the folding state of the preprotein (translocase function; Table 2).

The mutation in Ssc1-3p is located in the ATPase domain of the Hsp70. Lowering of the levels of matrix ATP in wild-type or *ssc1-2* mitochondria led to a similar phenotype as the *ssc1-3* mutation, indicating that both ATP and a functional ATPase domain are needed for binding of the preprotein to mt-Hsp70 (Gambill et al., 1993). Presequences that were not as positively charged as the subunit 9 presequence were not cleaved at all in *ssc1-3* mitochondria (or after depletion of the matrix ATP in wild-type mitochondria) and seemed to move back out of the inner membrane transport channel. In the absence of binding to functional Hsp70, the membrane potential alone is obviously not sufficient to drive those presequences far enough across the inner membrane or to stably keep the presequences in the transport channel (Cyr et al., 1993; Glick et al., 1993; Voos et al., 1993; Ungermann et al., 1994).

The proposed dual role of mt-Hsp70 as unfoldase and translocase could be confirmed by using a series of fusion proteins between the precursor of the intermembrane space protein cytochrome b_2 and the cytosolic protein dihydrofolate reductase (DHFR; Voos et al., 1993). (1) Induction of a folded structure in the preprotein on the cytosolic side, such as stabilization of the noncovalent heme binding domain of cytochrome b_2 by the addition of heme caused a dependence on the unfoldase function of mt-Hsp70 (Voos et al., 1993; Glick et al., 1993). (2) When the preprotein was completely directed into the mitochondrial matrix by a typical matrix targeting signal, it depended strictly on the translocase function of mt-Hsp70. The presence of a hydrophobic intermembrane space sorting signal beyond the matrix targeting signal, as is the case in the authentic cytochrome b_2 presequence, rendered the import independent of the translocase function of mt-Hsp70 (Voos et al., 1993). A yet to be identified sorting component seems to bind to the hydrophobic sorting signal, substitute for the translocase function of mt-Hsp70 and direct the preprotein to the intermembrane space. Strikingly, a preprotein containing the hydrophobic sorting signal, but lacking stably folded structures, was completely independent of functional mt-Hsp70 and matrix ATP for translocation into mitochondria. Induction of a folded structure, for example, in the heme binding domain, caused a dependence on the unfoldase function of mt-Hsp70. Thus, import and sorting of the precursor of wild-type cytochrome b_2 to the intermembrane space seem to be directed by the translocase function of the hypothetical sorting component; mt-Hsp70 is needed to promote the unfolding of the preprotein.

Under normal import conditions, the unfoldase and translocase functions of mt-Hsp70 are not separated, but it is the binding of Hsp70 molecules to the polypeptide chain that promotes unfolding and translocation at the same time. Different preproteins, however, have different requirements depending on their folding properties and sorting signals, and thus their import may for example depend on only one of the functions of mt-Hsp70 (Gärtner et al., 1995).

At a first glance, it seems to be surprising that Hsp70 in the matrix facilitates the unfolding of a preprotein on the cytosolic side. The following minimal hypothesis provides an explanation (Kang et al., 1990). The initial translocation and unfolding of the amino-terminal portion of a preprotein is promoted by the membrane potential and components of the inner membrane import system. Mt-Hsp70 then binds to the unfolded polypeptide segment emerging in the matrix. A major part of the preprotein is still (partially) folded and on the cytosolic side. Limited polypeptide segments may be released out of this part in a possibly spontaneous and reversible manner and may thus be able to move across the mitochondrial membranes. These segments may then move back again unless mt-Hsp70 molecules bind to them and trap them in the matrix. A step-wise binding of mt-Hsp70s to incoming polypeptide segments thus facilitates a step-wise unfolding of the preprotein on the outside (ratchet mechanism). However, it is also possible that mt-Hsp70 plays an even more active role in the translocation process besides the trapping of translocated polypeptide segments. Its dynamic interaction with Tim44 has led to the proposal that Tim44-bound mt-Hsp70 utilizes the ATP hydrolysis-dependent conformational change to generate a unidirectional pulling force that can promote translocation of even folded domains of preproteins (translocation motor model; Pfanner and Meijer, 1995; Glick, 1995). It is a matter of debate whether mt-Hsp70 exerts it unfoldase and translocase function merely by trapping or by a combination of trapping and pulling preprotein segments (Ungermann et al., 1996; Voos et al., 1996).

The release of polypeptides from Hsp70 requires ATP. Hydrolysis of ATP seems to be needed for the recycling of Hsp70 for new rounds of transport. Mt-Hsp70 keeps the preproteins in a loosely folded (unfolded) conformation before the transfer to Hsp60 where the actual folding occurs (Ostermann et al., 1989; Kang et al., 1990; Manning-Krieg et al., 1991). In the *ssc1-2* mitochondria, imported proteins remain associated with Hhsp70 in an unfolded conformation. Mt-Hsp70 may thus be seen as an initial component of the (re)folding pathway in the mitochondrial matrix. It helps to maintain a folding-competent conformation of preproteins.

V. COMPONENTS INTERACTING WITH MITOCHONDRIAL HSP70

The transfer of a preprotein from mt-Hsp70 to Hsp60 might occur via a free pool of preproteins as suggested by Manning-Krieg and colleagues (1991). Alternatively, it cannot be excluded that a preprotein in transit from mt-Hsp70 to Hsp60 is in physical contact with both chaperones at the same time. It is likely that a series of further components is involved, as found in the prokaryotic system where an ordered action of Hsp70 (DnaK) and the heat shock proteins DnaJ and GrpE is needed for the transfer of preproteins to Hsp60 (GroEL; Langer et al., 1992). Recently, mitochondrial homologs of DnaJ and GrpE, termed Mdj1p and Mge1p, were identified (Stuart et al., 1994; Laloraya et al., 1994; Rowley et al., 1994). We found a protein complex in the mitochondral matrix that contains mt-Hsp70 and Mge1p. The complex is dissociated in the presence of ATP (Voos et al., 1994). Surprisingly, Mge1p is already involved in the translocation of preproteins across the membranes (Laloraya et al., 1994; Voos et al., 1994), while Mdj1p is required later in the folding of imported proteins (Rowley et al., 1994).

As discussed previously, a fraction of mt-Hsp70 was found to interact with Tim44. We conclude that mt-Hsp70 is a component of two machineries, that for protein translocation across the mitochondrial membranes and that for folding of imported proteins. The sharing of an essential component promotes a tight interaction between the machineries and an efficient transfer of preproteins from translocation to folding reactions.

VI. PERSPECTIVES

Each of the four components found to be required for protein translocation across the mitochondrial inner membrane is essential for the viability of yeast cells. This emphasizes the high importance of Tim17, Tim23, Tim44, and mt-Hsp70. Besides the homology between Tim17 and Tim23, the three Tim-proteins do not show homology to any known protein, in particular not to components of other transport machineries. The protein import apparatus of the mitochondrial inner membranes seems to represent a novel transport system.

The identification of the Tim-proteins now opens a broad area for characterization of the import apparatus. This will include the topology of the Tims, their mode of interaction with preproteins, and the specificity of the transport machinery. Of particular importance will be the analysis of the interaction of the Tims with each other and with mt-Hsp70 and partner proteins, as well as the identification of putative further components required for inner membrane transport. An example of an unidentified component is the hypothetical sorting component that is assumed to bind to intermembrane space sorting signals and thereby partially substitutes for the function of mt-Hsp70. It will be investigated if matrix proteins and integral inner membrane proteins use the same channel and if the inner membrane proteins are able to leave the channel in a lateral direction. Our aim is to unravel the molecular architecture and function of the protein-conducting channel(s) in the mitochondrial inner membrane. It is possible that the view of a signal peptide-gated channel, which, besides the selective opening, plays a more passive role in protein translocation, is too simple. It cannot be excluded that the Tims in conjunction with mt-Hsp70 play a more active role in driving polypeptide chains across the inner membrane.

ACKNOWLEDGMENTS

We are grateful to our colleagues for their contributions to the work reviewed here and for sharing unpublished information with us. We thank M. Schmusch for help in preparation of the manuscript.

REFERENCES

Attardi, G., & Schatz, G. (1988). Biogenesis of mitochondria. Annu Rev. Cell Biol. 4, 289–333.

Baker, K. P., Schaniel, A., Vestweber, D., & Schatz, G. (1990). A yeast mitochondrial outer membrane protein essential for protein import and cell viability. Nature 348, 605–609.

Berthold, J., Bauer, M. F., Schneider, H-C., Klaus, C., Dietmeier, K., Neupert, W., & Brunner, M. (1995). The MIM complex mediates preprotein translocation across the mitochondrial inner membrane and couples it to the mt-Hsp70/ATP driving system. Cell 81, 1085–1093.

Blom, J., Dekker, P. J. T., & Meijer, M. (1995). Functional and physical interactions of components of the yeast mitochondrial inner membrane import machinery (MIM). Eur. J. Biochem. 232, 309–314.

Blom, J., Kübrich, M., Rassow, J., Voos, W., Dekker, P. J. T., Maarse, A. C., Meijer, M., & Pfanner, N. (1993). The essential yeast protein MIM44 (encoded by *MPI1*) is involved in an early step of preprotein translocation across the mitochondrial inner membrane. Mol. Cell. Biol. 13, 7364–7371.

Chen, W.-J., & Douglas, M. G. (1987). Phosphodiester bond cleavage outside mitochondria is required for the completion of protein import into the mitochondrial matrix. Cell 49, 651–658.

Cheng, M. Y., Hartl, F.-U., Martin, J., Pollock, R. A., Kalousek, F., Neupert, W., Hallberg, E. M., Hallberg, R. L., & Horwich, A. L. (1989). Mitochondrial heat-shock protein hsp60 is essential for assembly of proteins imported into yeast mitochondria. Nature 337, 620–625.

Craig, E. A., Gambill, B. D., & Nelson, R. J. (1993). Heat shock proteins: Molecular chaperones of protein biogenesis. Microbiol. Rev. 57, 402–414.

Craig, E. A., Kramer, J., Shilling, J., Werner-Washburne, M., Holmes, S., Kosic-Smithers, J., & Nicolet, C. M (1989). *SSC1*, an essential member of the yeast HSP70 multigene family, encodes a mitochondrial protein. Mol. Cell. Biol. 9, 3000–3008.

Cyr, D. M., Stuart, R. A., & Neupert, W. (1993). A matrix ATP requirement for presequence translocation across the inner membrane of mitochondria. J. Biol. Chem. 268, 23751–23754.

Dekker, P. J. T., Keil, P., Rassow, J., Maarse, A. C., Pfanner, N., & Meijer, M. (1993). Identification of MIM23, a putative component of the protein import machinery of the mitochondrial inner membrane. FEBS Lett. 330, 66–70.

Driessen, A. J. M., & Wickner, W. (1991). Proton transfer is rate-limiting for translocation of precursor proteins by the *Escherichia coli* translocase. Proc. Natl. Acad. Sci. USA 88, 2471–2475.

Emtage, J. L. T., & Jensen, R. E. (1993). *MAS6* encodes an essential inner membrane component of the yeast mitochondrial protein import pathway. J. Cell Biol. 122, 1003–1012.

Gambill, D., Voos, W., Kang, P. J., Miao, B., Langer, T., Craig, E. A., & Pfanner, N. (1993). A dual role for mitochondrial heat shock protein 70 in membrane translocation of preproteins. J. Cell Biol. 123, 109–117.

Gärtner, F., Voos, W., Querol, A., Miller, B., Craig, E. A., Cumsky, M. G., & Pfanner, N. (1995). Mitochondrial import of COXVa characterized with yeast mutants: independence from receptors, but requirement for matrix hsp70 translocase function. J. Biol. Chem. 270, 3788–3795.

Glick, B. S. (1995). Can hsp70 proteins act as force-generating motors? Cell 80, 11–14.

Glick, B. S., Wachter, C., Reid, G. A., & Schatz, G. (1993). Import of cytochrome b_2 to the mitochondrial intermembrane space: The tightly folded heme-binding domain makes import dependent upon matrix ATP. Protein Science 2, 1901–1917.

Hachiya, N., Alam, R., Sakasegawa, Y., Sakaguchi, M., Mihara, K., & Omura, T. (1993). A mitochondrial import factor purified from rat liver cytosol is an ATP-dependent conformational modulator for precursor proteins. EMBO J. 12, 1579–1586.

Hartl, F-U., Pfanner, N., Nicholson, D. W., & Neupert, W. (1989). Mitochondrial protein import. Biochim. Biophys. Acta 988, 1–45.

Hönlinger, A., Kübrich, M., Moczko, M., Gärtner, F., Mallet, L., Bussereau, F., Eckerskorn, C., Lottspeich, F., Dietmeier, K., Jacquet M., & Pfanner, N. (1995). The mitochondrial receptor complex: MOM22 is essential for cell viability and directly interacts with preproteins. Mol. Cell. Biol. 15, 3382–3389.

Horst, M., Oppliger, W., Feifel, B., Schatz, G., & Glick, B. S. (1996). The mitochondrial import motor: Dissociation of mitochondrial hsp70 from its membrane anchor requires ATP binding rather than ATP hydrolysis. Protein Science 5, 759–767.

Horst, M., Jenö, P., Kronidou, N. G., Bolliger, L., Oppliger, W., Scherer, P., Manning-Krieg, U., Jascur, T., & Schatz, G. (1993). Protein import into yeast mitochondria: The inner membrane import site protein ISP45 ist the *MPI1* gene product. EMBO J. 12, 3035–3041.

Hwang, S. T., & Schatz, G. (1989). Translocation of proteins across the mitochondrial inner membrane, but not into the outer membrane, requires nucleoside triphosphates in the matrix. Proc. Natl. Acad. Sci. USA 86, 8432–8436.

Hwang, S., Jascur, T., Vestweber, D., Pon, L., & Schatz, G. (1989). Disrupted yeast mitochondria can import precursor proteins directly through their inner membrane. J. Cell Biol. 109, 487–493.

Hwang, S. T., Wachter, C., & Schatz, G. (1991). Protein import into the yeast mitochondrial matrix: a new translocation intermediate between the two mitochondrial membranes. J. Biol. Chem. 266, 21083–21089.

Jensen, R. E., & Yaffe, M. P. (1988). Import of proteins into yeast mitochondria: The nuclear *MAS2* gene encodes a component of the processing protease that is homologous to the *MAS1*-encoded subunit. EMBO J. 7, 3863–3871.

Kang, P. J., Ostermann, J., Shilling, J., Neupert, W., Craig, E. A., & Pfanner, N. (1990). Requirement for hsp70 in the mitochondrial matrix for translocation and folding of precursor proteins. Nature 348, 137–143.

Kellems, R. E., Allison, V. F., & Butow, R. A. (1975). Cytoplasmic type 80S ribosomes associated with yeast mitochondria. IV. Attachment of ribosomes to the outer membrane of isolated mitochondria. J. Cell Biol. 65, 1–14.

Krämer, R., & Palmieri, F. (1989). Molecular aspects of isolated and reconstituted carrier proteins from animal mitochondria. Biochim. Biophys. Acta 974, 1–23.

Kronidou, N. G., Oppliger, W., Bolliger, L., Hannavy, K., Glick, B. S., Schatz, G., & Horst, M. (1994). Dynamic interaction between Isp45 and mitochondrial hsp70 in the protein import system of the yeast mitochondrial inner membrane. Proc. Natl. Acad. Sci. USA 91, 12818–12822.

Kübrich, M., Keil, P., Rassow, J., Dekker, P. J. T., Blom, J., Meijer, M., & Pfanner, N. (1994). The polytopic mitochondrial inner membrane proteins MIM17 and MIM23 operate at the same preprotein import site. FEBS Lett. 349, 222–228.

Laloraya, S., Gambill, B. D., & Craig, E. A. (1994). A role for a eukaryotic *grpE*-related protein, Mge1p, in protein translocation. Proc. Natl. Acad. Sci. USA 91, 6481–6485

Langer, T., Lu, C., Echols, H., Flanagan, J., Hayer, M. K., & Hartl, F-U. (1992). Successive action of DnaK, DnaJ and GroEL along the pathway of chaperone-mediated protein folding. Nature 356, 683–689.

Lithgow, T., Junne, T., Suda, K., Gratzer, S., & Schatz, G. (1994). The mitochondrial outer membrane protein Mas22p is essential for protein import and viability of yeast. Proc. Natl. Acad. Sci. USA 91, 11973–11977.

Maarse, A. C., Blom, J., Grivell, L. A., & Meijer, M. (1992). *MPI1*, an essential gene encoding a mitochondrial membrane protein, is possibly involved in protein import into yeast mitochondria. EMBO J. 11, 3619–3628.

Maarse, A. C., Blom, J., Keil, P., Pfanner, N., & Meijer, M. (1994). Identification of the essential yeast protein MIM17, an integral mitochondrial inner membrane protein involved in protein import. FEBS Lett. 349, 215–221.

Maduke, M., & Roise, D. (1993). Import of a mitochondrial presequence into protein-free phospholipid vesicles. Science 260, 364–367.

Manning-Krieg, U. C., Scherer, P. E., & Schatz, G. (1991). Sequential action of mitochondrial chaperones in protein import into the matrix. EMBO J. 10, 3273–3280.

Martin, J., Mahlke, K., & Pfanner, N. (1991). Role of an energized inner membrane in mitochondrial protein import: $\Delta\Psi$ drives the movement of presequences. J. Biol. Chem. 266, 18051–18057.

Murakami, H., Pain, D., & Blobel, G. (1988). 70-kD heat shock-related protein is one of at least two distinct cytosolic factors stimulating protein import into mitochondria. J. Cell Biol. 107, 2051–2057.

Nicholls, D.G., & Ferguson, S.J. (1992). *Bioenergetics 2*. Academic Press, London.

Ohba, M., & Schatz, G. (1987). Disruption of the outer membrane restores protein import to trypsin-treated yeast mitochondria. EMBO J. 6, 2117–2122.

Ostermann, J., Horwich, A. L., Neupert, W., & Hartl, F-U. (1989). Protein folding in mitochondria requires complex formation with hsp60 and ATP hydrolysis. Nature 341, 125–130.

Ostermann, J., Voos, W., Kang, P. J., Craig, E. A., Neupert, W., & Pfanner, N. (1990). Precursor proteins in transit through mitochondrial contact sites interact with hsp70 in the matrix. FEBS Lett. 277, 281–284.

Pfanner, N., & Meijer, M. (1995). Protein sorting: Pulling in the proteins. Curr. Biol. 5, 132–135.

Pfanner, N., & Neupert, W. (1985). Transport of proteins into mitochondria: A potassium diffusion potential is able to drive the import of ADP/ATP carrier. EMBO J. 4, 2819–2825.

Pfanner, N., & Neupert, W. (1987). Distinct steps in the import of ADP/ATP carrier into mitochondria. J. Biol. Chem. 262, 7528–7536.

Pfanner, N., Douglas, M. G., Endo, T., Hoogenraad, N. J., Jensen, R. E., Meijer, M., Neupert, W., Schatz, G., Schmitz, U. K., & Shore, G. C. (1996). Uniform nomenclature for the protein transport machinery of the mitochondrial membranes. Trends Biochem. Sci. 21, 51–52.

Pfanner, N., Hoeben, P., Tropschug, M., & Neupert, W. (1987a). The carboxyl-terminal two-thirds of the ADP/ATP carrier polypeptide contains sufficient information to direct translocation into mitochondria. J. Biol. Chem. 262, 14851–14854.

Pfanner, N., Tropschug, M., & Neupert, W. (1987b). Mitochondrial protein import: nucleoside triphosphates are involved in conferring import-competence to precursors. Cell 49, 815–823.

Pfanner, N., Pfaller, R., Kleene, R., Ito, M., Tropschug, M., & Neupert, W. (1988). Role of ATP in mitochondrial protein import: Conformational alteration of a precursor protein can substitute for ATP requirement. J. Biol. Chem. 263, 4049–4051.

Pfanner, N., Rassow, J., Guiard, B., Söllner, T., Hartl, F-U., & Neupert, W. (1990). Energy requirements for unfolding and membrane translocation of precursor proteins during import into mitochondria. J. Biol. Chem. 265, 16324–16329.

Pfanner, N., Rassow, J., van der Klei, I. J., & Neupert, W. (1992). A dynamic model of the mitochondrial protein import machinery. Cell 68, 999–1002.

Pfanner, N., Craig, E. A., & Meijer, M. (1994). The protein import machinery of the mitochondrial inner membrane. Trends Biochem. Sci. 19, 368–372.

Pollock, R. A., Hartl, F-U., Cheng, M. Y., Ostermann, J., Horwich, A., & Neupert, W. (1988). The processing peptidase of yeast mitochondria: The two cooperating components MPP and PEP are structurally related. EMBO J. 7, 3493–3500.

Rapoport, T. A. (1992). Transport of proteins across the endoplasmic reticulum membrane. Science 258, 931–936.

Rassow, J., & Pfanner, N. (1991). Mitochondrial preproteins en route from the outer membrane to the inner membrane are exposed to the intermembrane space. FEBS Lett. 293, 85–88.

Rassow, J., Maarse, A. C., Krainer, E., Kübrich, M., Müller, H., Meijer, M., Craig, E. A., & Pfanner, N. (1994). Mitochondrial protein import: Biochemical and genetic evidence for interaction of matrix hsp70 and the inner membrane protein MIM44. J. Cell Biol.

Rothman, J. E. (1989). Polypeptide chain binding proteins: Catalysts of protein folding and related processes in cells. Cell 59, 591–601.

Rowley, N., Prip-Buus, C., Westermann, B., Brown, C., Schwarz, E., Barrell, B., & Neupert, W. (1994). Mdj1p, a novel chaperone of the DnaJ family, is involved in mitochondrial biogenesis and protein folding. Cell 77, 249–259.

Ryan, K. R., & Jensen, R. E. (1993). Mas6p can be cross-linked to an arrested precursor and interacts with other proteins during mitochondrial protein import. J. Biol. Chem. 268, 23743–23746.

Ryan, K. R., Menold, M. M., Garrett, S., & Jensen, R. E. (1994). *SMS1*, a high-copy suppressor of the yeast *mas6* mutant, encodes an essential inner membrane protein required for mitochondrial protein import. Mol. Biol. Cell 5, 529–538.

Sanders, S. L., & Schekman, R. (1992). Polypeptide translocation across the endoplasmic reticulum membrane. J. Biol. Chem. 267, 13791–13794.

Scherer, P. E., Manning-Krieg, U. C., Jenö, P., Schatz, G., & Horst, M. (1992). Identification of a 45-kDa protein at the protein import site of the yeast mitochondrial inner membrane. Proc. Natl. Acad. Sci. USA 89, 11930–11934.

Schiebel, E., Driessen, A. J. M., Hartl, F-U., & Wickner, W. (1991). $\Delta\mu_{H^+}$ and ATP function at different steps of the catalytic cycle of preprotein translocase. Cell 64, 927–939.

Schleyer, M., & Neupert, W. (1985). Transport of proteins into mitochondria: Translocational intermediates spanning contact sites between outer and inner membranes. Cell 43, 339–350.

Schmidt, B., Wachter, E., Sebald, W., & Neupert, W. (1984). Processing peptidase of *Neurospora* mitochondria: Two-step cleavage of imported ATPase subunit 9. Eur. J. Biochem. 144, 581–588.

Schneider, H-C., Berthold, J., Bauer, M. F., Dietmeier, K., Guiard, B., Brunner, M., & Neupert, W. (1994). Mitochondrial Hsp70/MIM44 complex facilitates protein import. Nature 371, 768–774.

Skerjanc, I. S., Sheffield, W. P., Randall, S. K., Silvius, J. R., & Shore, G. C. (1990). Import of precursor proteins into mitochondria: Site of polypeptide unfolding. J. Biol. Chem. 265, 9444–9451.

Smagula, C. S., & Douglas, M. G. (1988). ADP-ATP carrier of *Saccharomyces cerevisiae* contains a mitochondrial import signal between amino acids 72 and 111. J. Cell. Biochem. 36, 323–328.

Soll, J., & Alefsen, H. (1993). The protein import apparatus of chloroplasts. Physiol. Plant. 87, 433–440.

Söllner, T., Rassow, J., & Pfanner, N. (1991). Analysis of mitochondrial protein import using translocation intermediates and specific antibodies. Meth. Cell Biol. 34, 345–358.

Stuart, R. A., Cyr, D. M., Craig, E. A., & Neupert, W. (1994). Mitochondrial molecular chaperones: Their role in protein translocation. Trends Biochem. Sci. 19, 87–92.

Ungermann, C., Guiard, B., Neupert, W., & Cyrr, D. M. (1996). The $\Delta\Psi$- and Hsp70/MIM44-dependent reaction cycle driving early steps of protein import into mitochondria. EMBO J. 15, 735–744.

Ungermann, C., Neupert, W., & Cyr, D. M. (1994). The role of hsp70 in conferring unidirectionality on protein translocation into mitochondria. Science 266, 1250–1253.

Verner, K., & Schatz, G. (1987). Import of an incompletely folded precursor protein into isolated mitochondria requires an energized inner membrane, but no added ATP. EMBO J. 6, 2449–2456.

Vestweber, D., & Schatz, G. (1988). A chimeric mitochondrial precursor protein with internal disulfide bridges blocks import of authentic precursors into mitochondria and allows quantitation of import sites. J. Cell Biol. 107, 2037–2043.

Von Ahsen, O., Voos, W., Henninger, H., & Pfanner, N. (1995). The mitochondrial protein import machinery: Role of ATP in dissociation of the Hsp70.Mim44 complex. J. Biol. Chem. 270, 29848–29853.

Voos, W., Gambill, D., Guiard, B., Pfanner, N., & Craig, E. A. (1993). Presequence and mature part of preproteins strongly influence the dependence of mitochondrial protein import on heat shock protein 70 in the matrix. J. Cell Biol. 123, 119–126.

Voos, W., Gambill, B. D., Laloraya, S., Ang, D., Craig, E. A., & Pfanner, N. (1994). Mitochondrial GrpE is present in a complex with hsp70 and preproteins in transit across the membranes. Mol. Cell. Biol. 14, 6627–6634.

Voos, W., von Ahsen, O., Müller, H., Guiard, B., Rassow, J., & Pfanner, N. (1996). Differential requirement for the mitochondrial Hsp70-Tim44 complex in unfolding and translocation of preproteins. EMBO J. 15, 2668–2677.

Wienhues, U., Becker, K., Schleyer, M., Guiard, B., Tropschug, M., Horwich, A. L., Pfanner, N., & Neupert, W. (1991). Protein folding causes an arrest of preprotein translocation into mitochondria *in vivo*. J. Cell Biol. 115, 1601–1609.

Witte, C., Jensen, R. E., Yaffe, M. P., & Schatz, G. (1988). *MAS1*, a gene essential for yeast mitochondrial assembly, encodes a subunit of the mitochondrial processing protease. EMBO J. 7, 1439–1447.

UNRAVELING THE PROTEIN TRANSLOCATION MACHINERY IN THE MITOCHONDRIAL INNER MEMBRANE

Nafsika G. Kronidou and Martin Horst

I.	Introduction	149
II.	Two Dynamically Interacting Machineries	151
III.	Dynamic Channel Model and Its Implications for Protein Sorting to the Intermembrane Space	153
IV.	The Inner Mitochondrial Membrane Translocation Machinery	154
V.	Outlook	159
	Acknowledgments	159
	References	159

I. INTRODUCTION

The study of protein import into mitochondria provides a model system for the study of protein translocation into organelles that are enveloped by two membranes. Unlike protein import into the endoplasmic reticulum, import into mitochondria thus possesses an additional complexity: the double membrane. How are cytoplas-

Advances in Molecular and Cell Biology,
Volume 17, pages 149–162.

ISBN: 0-7623-0144-9

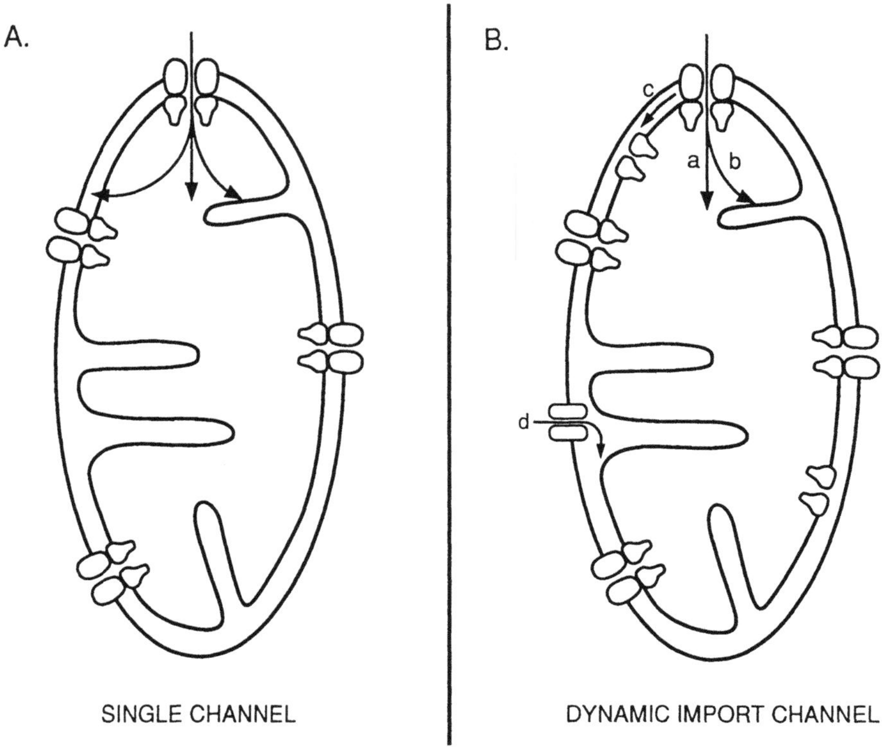

Figure 1. The single channel versus the dynamic model of mitochondrial protein import. According to the single channel model (A), nuclear-encoded mitochondrial proteins are initially imported into the matrix through a continuous, stable channel spanning both membranes. These proteins either remain in the matrix or are sorted to the inner membrane or the intermembrane space. In the dynamic model (B), which is now generally accepted, each of the two mitochondrial membranes has its own distinct import system. When proteins are targeted to (a) the matrix or (b) the inner membrane, the two channels are in close contact. They separate during import of intermembrane space-targeted proteins such as cytochromes c_1 and b_2 (pathway c) or cytochrome c heme lyase (pathway d).

mically synthesized proteins translocated across one or both membranes, and how are they sorted to the various mitochondrial subcompartments? Early electron microscopy studies revealed the presence of "contact sites" between the outer and inner mitochondrial membranes (Hackenbrock, 1968; Kellems et al., 1975). There is now general agreement that these "contact sites" are the ports through which cytoplasmically-made precursor proteins enter the mitochondrial matrix (Ades and Butow, 1980; Schwaiger et al., 1987; Pon et al., 1989; Glick et al., 1991). It was reasonable to speculate that a stable channel (Figure 1a) through which proteins enter the matrix spans both membranes at contact sites. This view has since been disproved and more recent evidence strongly favors a dynamic channel (Figure 1b) where the outer and inner membrane translocation machineries interact reversibly (Glick and Schatz, 1991; Glick et al., 1991; Pfanner et al., 1992; Horst et al., 1993b, 1995). This review summarizes the evidence for this dynamic model and describes the so-far characterized components of the inner membrane translocation machinery. The targeting of proteins to the inner membrane will be discussed in another chapter of this volume.

II. TWO DYNAMICALLY INTERACTING MACHINERIES

The first evidence that the import channel in the mitochondrial inner membrane can operate independently from that in the outer membrane came from experiments in which the outer membrane translocation machinery was inactivated by protease treatment. When these mitochondria were subsequently subjected to osmotic shock, which selectively ruptures the outer membrane to create so-called "mitoplasts," these mitoplasts were now able to import a precursor protein into the mitochondrial matrix (Ohba and Schatz, 1987). Similar results were obtained when protein import into intact mitochondria was first inhibited by adding antibodies against components of the outer membrane translocation machinery, or by jamming the import sites with a chimeric precursor protein containing a tightly-folded C-terminal domain (Hwang et al., 1989). The simplest interpretation of these results is that the inner membrane contains import sites that are unmasked by disrupting the outer membrane barrier. Indeed, purified mitochondrial inner membrane vesicles that have the same "sidedness" as the inner membrane of intact mitochondria import mitochondrial precursor proteins with the same characteristics as intact mitochondria (Hwang et al., 1989). Protein import into these vesicles requires an electrochemical potential across the inner membrane, ATP in the vesicle lumen, and a functional mitochondrial targeting signal on the precursor protein. However, protein import into such vesicles, unlike that into intact mitochondria, is sensitive to antibodies against the inner membrane and insensitive to antibodies against components of the outer membrane translocation machinery (Hwang and Schatz, 1989; Scherer et al., 1992). Furthermore, the import machinery of these vesicles is approximately tenfold more sensitive to protease treatment than the import machinery of intact mitochondria, suggesting the presence of a very protease-sensitive inner membrane import component (Hwang et al. 1989).

If the outer and inner membranes contain two distinct import machineries, a temporary arrest of import of a matrix-targeted protein should provide a "snapshot" of its import pathway. The observation that ATP in the matrix is required for import of precursor proteins across the inner membrane (Hwang and Schatz, 1989) provided an experimental system to obtain such a snapshot. If a matrix-targeted precursor was added to isolated yeast mitochondria whose matrix had been depleted of ATP the precursor accumulated in the mitochondrial inner membrane. This ATP-depletion intermediate was protease-inaccessible in mitochondria but became accessibleupon disruption of the outer membrane (Hwang et al., 1991), suggesting that the protein had completely moved across the outer membrane. Upon addition of ATP, the ATP-depletion intermediate could be chased into the matrix, indicating that it was on the correct import pathway. A similar result was obtained when ATP-depleted mitochondria were allowed to import a chimeric precursor consisting of a mutant form of bovine pancreatic trypsin inhibitor (BPTI) coupled to the C-terminus of a purified precursor protein (Jascur et al., 1992). The BPTI moiety contained three intramolecular disulfide bonds that kept the protein in a tightly folded conformation. In the absence of a reducing agent, the fusion protein remained stuck in the import channel, with the BPTI moiety exposed on the mitochondrial surface and the N-terminal part of the precursor in the matrix. Upon addition of dithiothreitol, the disulfide bonds of the mutant BPTI were cleaved, allowing the BPTI to unfold and the protein to translocate across the outer membrane. Again, this intermediate remained stuck across the inner membrane, but could be chased into the matrix upon addition of ATP. The ability to chase the precursor into the matrix must have reflected the independent operation of the inner membrane translocation system in intact mitochondria.

A similar dissection of the import pathway under low ATP conditions has been reported for the import of two abundant inner mitochondrial membrane proteins: the ADP/ATP carrier and the β-subunit of the F1-ATPase (Rassow and Pfanner, 1991).

Deciphering the import pathway of cytochrome *c* heme lyase (CCHL) has confirmed the previous evidence that the outer and inner membrane translocation machineries can act independently. Cytochrome *c* heme lyase is an intermembrane space enzyme (Nicholson et al., 1988). It catalyzes the attachment of heme to apocytochrome *c* and thereby the import of this protein into the intermembrane space (Nicholson et al., 1988; Dumont et al., 1988). Import of CCHL into the intermembrane space requires neither ATP nor an electrochemical potential across the inner membrane. On the other hand, this precursor seems to employ the generalized import machinery in the outer mitochondrial membrane (Lill et al., 1992). Thus in the case of CCHL, the machinery in the outer membrane can function independently from that in the inner one (Figure 1, panel B, pathway d). These observations have been extended by showing that highly purified "right-side-out" outer membrane vesicles can import and assemble both CCHL and integral outer membrane proteins (Mayer et al., 1993).

When CCHL is modified by the addition of an N-terminal matrix-targeting signal, the resulting fusion protein is imported into the matrix. This process can be experimentally dissected into two steps. If the mitochondria lack a potential across the inner membrane, the fusion protein is transported across the outer membrane into the intermembrane space. Restoration of the membrane potential then allows the protein to be transported into the matrix (Segui-Real et al., 1993). The driving force for the translocation of CCHL across the outer membrane remains to be identified. It is perhaps driven by folding of the protein itself once in the intermembrane space, by attachment to the inner membrane or outer membrane or by binding to an as yet unidentified chaperone in the intermembrane space.

The ability to import precursors into isolated outer (Mayer et al., 1993) and inner membrane vesicles (Hwang et al., 1989) provides additional evidence that the two machineries can function independently, at least *in vitro*. With the identification of the various components of the two systems it should be possible to reconstitute the import machineries into liposomes and learn how they work.

When a matrix-targeted precursor protein attached to an internally cross-linked molecule such as RNAse A or BPTI is imported into isolated mitochondria the construct remains stuck across both membranes. When these mitochondria are solubilized under mild conditions the stuck precursor can be specifically immunoprecipitated with antibodies against either Isp42p (a component of the outer membrane import channel; Vestweber et al., 1989; Baker et al., 1990), or Isp45p (an inner membrane translocation component; see following). Moreover, under these conditions anti-Isp45p antibodies will co-immunoprecipitate Isp42p and vice versa. Thus, in the presence of a stuck precursor spanning both membranes, the outer and inner membrane translocation machineries remain attached to each other. However, in the absence of a stuck precursor, antibodies against Isp42p or against Isp45p immunoprecipitate the respective protein only indicating that the two import machineries are not permanently attached to each other (Horst et al., 1995).

The combined results strongly suggest that mitochondria contain two distinct import systems, one in the outer and one in the inner membrane. These two systems probably operate in tandem during import of a matrix-targeted protein, but can separate under certain conditions. The dynamic nature of the import machinery has implications for import of some proteins to the intermembrane space as will be discussed below.

III. DYNAMIC CHANNEL MODEL AND ITS IMPLICATIONS FOR PROTEIN SORTING TO THE INTERMEMBRANE SPACE

Why are there two dynamically interacting import channels instead of a single stable channel spanning both membranes? For proteins that are targeted to the matrix there is no obvious reason; indeed the two channels most likely remain attached to each other during translocation of a precursor protein to the matrix. However, the functional independence of the two channels is essential for the import and sorting of several proteins to the intermembrane space. These proteins include CCHL, cytochrome c_1, cytochrome b_2, and probably cytochrome c peroxidase and mitochondrial creatine kinase.

As discussed previously, import of CCHL into the intermembrane space requires the generalized import machinery in the mitochondrial outer membrane, but is apparently independent of the inner mitochondrial membrane import machinery. The simplest interpretation of these results is that the two import channels are not permanently coupled and that the outer membrane translocation machinery can function independently from the inner one.

A different import pathway exists for another class of intermembrane space proteins exemplified by cytochromes c_1 and b_2 (Figure 1, panel B, pathway c). The import pathway for these two proteins has been controversial until recently (for reviews see: Glick et al., 1992a; Segui-Real et al., 1992). However, their import pathway is now fairly well characterized and these proteins are sorted to the intermembrane space by a stop-transfer mechanism (van Loon and Schatz, 1987; Glick et al., 1992b; 1993; Wachter et al., 1992; Rospert et al., 1994). Cytochromes c_1 and b_2 are synthesized with bipartite presequences. The amino-terminal portion of each presequence resembles a matrix-targeting signal and the carboxy-terminal portion contains a cluster of positive charges followed by a stretch of hydrophobic residues (Sadler et al., 1984; Guiard, 1985; Jensen et al., 1992; Beasley et al., 1993; Schwarz et al., 1993). The first part of the presequence translocates across the inner membrane in a membrane potential-dependent manner. The general processing protease in the matrix cleaves this part of the presequence (Böhni et al., 1980), but further translocation of the precursor into the matrix is inhibited by the second part of the presequence, which acts as a stop-transfer signal (Hurt and van Loon, 1986; van Loon and Schatz, 1987). The mature portion of the protein then slowly translocates across the outer membrane into the intermembrane space. The resulting import intermediates are all protease accessible in mitoplasts and extractable at alkaline pH, suggesting that they face the intermembrane space and are associated with a proteinaceous channel (Glick et al., 1992b). Although no direct evidence exists, this channel is most likely the protein translocation machinery of the mitochondrial inner membrane. Proteases located on the outer surface of the inner membrane cleave the membrane-anchored intermediates immediately downstream of the stop-transfer domains, releasing the mature proteins (Pratje and Guiard, 1986; Schneider et al., 1991; Nunnari et al., 1993). The current model for import of cytochromes c_1 and b_2 proposes that after the stop-transfer event takes place, the outer and inner membrane translocation machineries separate from one another and move laterally within the plane of the membranes (reviewed in Glick and Schatz, 1991; Figure 1, panel B, pathway c). It is still unknown what force drives movement of the two machineries relative to each other or what is the signal for the two machineries to separate. Simple diffusion of the channels may provide this force. An import intermediate that was anchored to the inner membrane would thus be pulled across the outer membrane by an entropy-driven process. This idea has not yet been directly tested.

IV. THE INNER MITOCHONDRIAL MEMBRANE TRANSLOCATION MACHINERY

The realization that the mitochondrial inner membrane contains it own translocation machinery triggered an active pursuit of the components comprising this

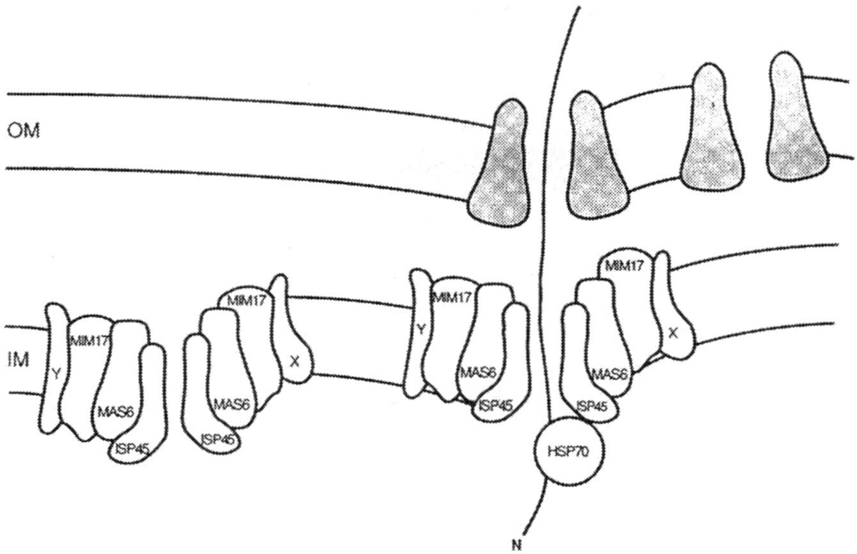

Figure 2. The inner mitochondrial membrane import machinery. The picture of the inner mitochondrial membrane translocation machinery is becoming more comprehensive. The components characterized so far (Isp45p/Mpi1p, Mas6p/Mim23p, Mim17p/Sms1p and mhsp70) may form a stable complex, or may come together during the import of a precursor protein. The arrangement of the various components is of course putative. Possible unidentified components are indicated by X and Y. The outer membrane import machinery is fairly well characterized (Baker and Schatz, 1991, Söllner et al., 1992).

machinery. In the last two years, at least three components have been identified; Isp45p (for 45 kDa import site protein, Scherer et al. 1992; Horst et al. 1993a) and also known as Mpi1p or MIM44, (Maarse et al. 1992; Blom et al., 1993), Mas6p (for mitochondrial assembly; Emtage and Jensen, 1993), and also known as Mim23p (Dekker at al., 1993); and Mim17p or Sms1p (for 17 kDa mitochondrial inner membrane protein; Dekker at al., 1993; Ryan et al., 1994; Figure 2). As suggested by the nonuniformity in the nomenclature, several laboratories employing different approaches have contributed to the rapid progress in the isolation of these components.

Our laboratory identified one of these components using two complementary biochemical approaches (Scherer et al., 1992; Horst et al., 1993a). In one approach, we crosslinked putative subunits of the inner membrane transport system to a precursor protein that had been arrested during translocation through the inner membrane (the "ATP-depletion intermediate" mentioned earlier). Crosslinks were obtained to four mitochondrial polypeptides of apparent molecular weights 55, 45, 20, and 5 K. The crosslinked product containing the 45 kDa protein could be immunoprecipitated with an antiserum raised against a mixture of mitochondrial proteins. We were therefore able to purify the 45 kDa protein on the basis of its ability to quench this immunoprecipitation. The second approach used antisera raised against inner membrane vesicles that inhibited protein import into these vesicles, but not into intact mitochondria. One of the antigens responsible for this inhibition was identified by affinity-purifying the antibodies against different subpopulations of mitochondrial proteins that had been fractionated by anion exchange chromatography. Both biochemical approaches identified the same 45 kDa inner membrane protein, which we termed Isp45p. When we purified the protein to homogeneity and determined its partial amino acid sequence, it proved to match that of the recently described *MPI1* gene product (Maarse et al., 1992). This gene had been discovered by an elegant genetic screen that selected for yeast mutants that were partially defective in the import of a matrix-targeted fusion protein. *MPI1* encodes a peripheral inner mitochondrial membrane protein that is synthesized with a cleavable N-terminal presequence. Import of this protein into mitochondria requires a membrane potential across the inner membrane, an observation that is consistent with a localization of the protein in the inner membrane. The Isp45p/Mpi1p protein is essential for viability, and its depletion leads to the accumulation of uncleaved mitochondrial precursors in the cells (Maarse et al., 1992). The protein is insensitive to protease in either mitochondria or mitoplasts but is degraded when the inner membrane is solubilized by detergents (Blom et al., 1993). Carboxy-terminal epitope tagging of the Isp45/Mpi1 protein indicated that at least part of the C-terminal portion of the protein is exposed to the intermembrane space (Maarse et al., 1992). This observation is in agreement with the finding that antibodies against Isp45p/Mpi1p block import into right-side out inner membrane vesicles (Horst et al., 1993a) and that these antibodies bind to mitoplasts (Kronidou et al., 1994). Exposing mitochondria to alkaline pH releases the protein from the membranes, suggesting that it is in a proteinaceous environment. Isp45p/Mpi1p seems therefore to be embedded in a proteinaceous environment in the inner membrane with at least part of the protein exposed to the intermembrane space. Isp45p/Mpi1p has recently been shown to interact directly with the matrix localized mitochondrial hsp70 (mhsp70; Kronidou et al., 1994; Rassow et al., 1994; Schneider et al., 1994). Therefore, Isp45p most likely spans the inner membrane without actually being integrated into the lipid bilayer and may be attached to other proteins of the import machinery as is depicted in Figure 3.

In an *in vitro* binding assay, precursor proteins were added to solubilized mitochondria followed by immunoprecipitation with antibodies against Isp45p/Mpi1p. Under these conditions the precursors for inner membrane proteins that were tested co-immunoprecipitated with Isp45p/Mpi1p (Blom et al. 1993). On the other hand, the precursor of an outer membrane protein did not show any significant binding to Isp45p/Mpi1p. These results suggest that Isp45p/Mpi1p and therefore the inner membrane translocation machinery may show some selectivity for preproteins.

The same genetic screen used by Maarse and colleagues (1992) to identify the *MPI1* gene also led to the identification of three other complementation groups (Dekker et al., 1993). Two of these complementation groups represent novel genes, *MPI2* and *MPI3*. The *MPI3* gene encodes an integral inner membrane protein with a predicted molecular weight of 23 K (Dekker et al., 1993) and which is identical to the *MAS6* gene product (Emtage and Jensen, 1993). The *MPI3* gene encodes a 17 kDa integral inner membrane protein (Dekker et al., 1993). The third complementation group corresponds to the *SSC1* gene, which encodes mhsp70, (Craig et al., 1989; Kang et al., 1990; Scherer et al., 1990).

The temperature-sensitive *mas6* mutant was first identified in a genetic screen that selected for mutants that accumulated uncleaved mitochondrial precursor proteins at the nonpermissive temperature (Yaffe and Schatz, 1984). The wild-type gene has been cloned and shown to be essential for life (Emtage and Jensen, 1993). The Mas6 protein is located in the mitochondrial inner membrane; antibodies against Mas6p block import of precursor proteins into mitoplasts but not into mitochondria (Emtage and Jensen, 1993). Mas6p is one of the proteins crosslinked to the same ATP depletion intermediate used to identify Isp45p (Scherer et al, 1992; Emtage and Jensen, 1993b).

Genetic and biochemical evidence implicate mhsp70 as a component of the inner membrane import machinery. Mitochondrial hsp70 is an essential protein (Craig et al., 1987). *In vitro* import studies of mitochondria isolated from a temperature sensitive mhsp70 mutant show compromised protein import and an accumulation of precursor proteins at contact sites (Kang et al., 1990). Furthermore, mhsp70 can be crosslinked to or co-immunoprecipitated with precursor proteins on their way to the matrix (Scherer et al., 1990; Kang et al., 1990). Subsequent experiments have shown that mhsp70 forms a transient, ATP-dependent interaction with newly imported precursor proteins (Manning-Krieg et al., 1991). These results have led to the hypothesis that mhsp70 acts as an ATP-dependent motor that pulls precursor proteins into the matrix. This model is consistent with the observation that ATP-depletion in the matrix arrests translocation of precursor proteins across the inner membrane (Hwang and Schatz, 1989; Hwang et al., 1991). In addition, it has recently been shown that a temperature-sensitive mhsp70 allele with a mutated ATPase domain fails to bind and complete import of a partially translocated precursor protein (Gambill et al., 1993). Interestingly, import and assembly of precursors into the yeast endoplasmic reticulum also requires a luminal hsp70 homologue, BiP (Munro and Pelham, 1986; Vogel et al., 1990; Sanders et al., 1992). How hsp70 mediates translocation in either system is not yet understood. In one possible mechanism, mhsp70 exerts force on the translocating polypeptide and pulls the protein into the matrix by virtue of being anchored to the inner membrane translocation machinery (Figure 3, panel A). Consistent with this model is the observation that mhsp70 and its partner protein GrpEp (Bolliger et al., 1994) dynamically interact with Isp45p/Mpi1p (Kronidou et al., 1994; Rassow et al., 1994; Schneider et al., 1994). An alternative mechanism for mhsp70 action is

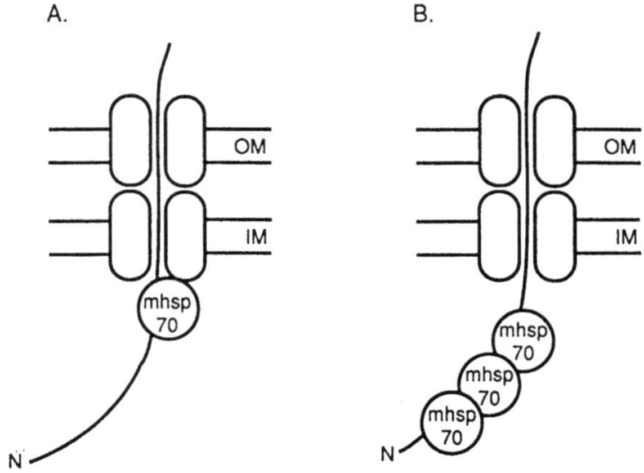

Figure 3. Model for mhsp70 mechanism. Two alternative models have been proposed for the mechanism of mhsp70. In the first model (panel A), mhsp70 is anchored to the inner membrane translocation machinery via Isp45p and exerts force on the translocating polypeptide. In an alternative model (panel B), a mhsp70 molecule binds to the precursor and prevents diffusion of the precursor out of the translocation channel. The binding of additional mhsp70 molecules to the precursor may provide the force to complete translocation of the precursor across the import channel.

illustrated in Figure 3B. According to the "Brownian ratchet" model (Neupert et al., 1990; Simon et al., 1992) the matrix- targeting presequence of a mitochondrial precursor would insert across the inner membrane and the matrix-targeting signal would be cleaved. In the absence of the matrix-targeting signal the precursor could diffuse within the import channel in either direction. However, the binding of a mhsp70 molecule to the precursor would prevent diffusion out of the import channel and ensure that movement is unidirectional. Binding of additional mhsp70 molecules to the precursor could then pull the precursor completely into the matrix. According to this model, no membrane-associated mhsp70 would be required. Obviously, more experiments are needed to determine the mechanism of mhsp70 action.

Have all the components of the inner membrane translocation machinery been identified? The answer is probably no, but we now have powerful tools to identify the remaining components. Co-immunoprecipitation and genetic screens for synthetic lethality or second-site suppressors can now be used to track down proteins that interact with the already known inner membrane components.

According to the endosymbiotic theory of evolution, mitochondria are descendants of gram-negative bacteria. It is intriguing that none of the import channel components identified so far in mitochondria share any detectable similarity with channel proteins found in present-day bacteria. Of course, protein import into mitochondria occurs in the reverse direction to protein export in bacteria, and novel proteins have probably evolved in mitochondria to allow for protein import. How are the few proteins that are synthesized in the mitochondrial matrix inserted into the inner membrane (for a review see Poyton et al., 1992)? Insertion of proteins from the matrix into the inner membrane might resemble the well-characterized export machinery in the bacterial plasma membrane or it might involve mechanisms that have arisen de novo during eukaryotic evolution. These questions are still unanswered.

V. OUTLOOK

The dynamic nature of the import channel has been established, but many intriguing questions remain unanswered. For example, what is the nature of contact sites between the two mitochondrial membranes? It has been proposed that contact sites are held together by as yet unknown adhesion proteins (Glick et al., 1991). The import channels in the two membranes may diffuse freely in and out of contact sites or the channels may interact with one another in a regulated manner.

How do mitochondria distinguish between precursors targeted to the matrix and those targeted to the intermembrane space? The inner membrane import machinery is evidently a complex structure that not only mediates protein import but also recognizes specific stop-transfer signals. It will be interesting to determine whether this translocation arrest is mediated by the same proteins that form the channel. After the stop-transfer event, the two import channels must separate. It is not clear whether this process is mediated by simple diffusion or whether the translocation arrest triggers the separation of the two channels. Finally, we have only partial answers to the two most fundamental questions in the protein translocation field: what is the environment through which a polypeptide traverses a membrane and what force drives the translocation?

ACKNOWLEDGMENTS

We thank Drs. G. Schatz and B. Glick for helpful comments and discussions.

REFERENCES

Ades, I. Z., & Butow, R. A. (1980). The products of mitochondria-bound cytoplasmic polysomes in yeast. J. Biol. Chem. 225, 9918–9924.

Baker, K. P., Schaniel, A., Vestweber, D., & Schatz, G. (1990). A yeast mitochondrial outer membrane protein essential for protein import and cell viability. Nature (London) 348, 605–609.

Baker, K. P., & Schatz, G. (1991). Mitochondrial proteins essential for viability mediate protein import into yeast mitochondria. Nature (London) 349, 205–208.

Beasley, E. M., Müller, S., & Schatz, G. (1993). The signal that sorts yeast cytochrome b_2 to the mitochondrial intermembrane space contains three distinct functional regions. EMBO J. 12, 2303–2311.

Blom, J., Kübrich, M., Rassow, J., Voos, W., Dekker, P. J. T., Maarse, A. C., Meijer, M., & Pfanner, N. (1993). The essential yeast protein MIM44 (encoded by *MPI1*) is involved in an early step of preprotein translocation across the mitochondrial inner membrane. Mol. Cell. Biol. 13,7364–7371.

Böhni, P., Gasser, S., Leaver, C., & Schatz, G. (1980). A matrix-localized mitochondrial protease processing cytoplasmically-made precursors to mitochondrial proteins. In: *The Organization and Expression of the Mitochondrial Genome* (Kroon, A. M., & Saccone, C., Eds.). Elsevier/North Holland, Amsterdam, pp. 423–433.

Bolliger, L., Deloche, O., Glick, B. S., Georgopoulos, C., Jenö, P., Kronidou, N., Horst, M., Morishima, N., & Schatz, G. (1994). A mitochondrial homolog of bacterial GrpE interacts with mitochondrial hsp70 and is essential for viability. EMBO J. 13, 1998–2006.

Craig, E. A., Kramer, J., & Kosic-Smithers, J. (1987). SSC1, a member of the 70-kDa heat shock protein multigene family of *Saccharomyces cerevisiae*, is essential for growth. Proc. Natl. Acad. Sci. USA 84, 4156–4160.

Craig, E. A., Kramer, J., Shilling, J., Werner-Washburne, M., Holmes, S., Kosic-Smithers, J., & Nicolet, C. M. (1989). SSC1, an essential member of the yeast *HSP70* multigene family, encodes a mitochondrial protein. Mol. Cell. Biol. 9, 3000–3008.

Dekker, P. J. T., Keil, P., Rassow, J., Maarse, A., Pfanner, N., & Meijer, M. (1993). Identification of MIM23, a putative component of the protein import machinery of the mitochondrial inner membrane. FEBS Lett. 330, 66–70.

Dumont, M. E., Ernst, J. F., & Sherman, F. (1988). Coupling of heme attachment to import of cytochrome c into yeast mitochondria. Studies with heme lyase deficient mitochondria and altered apocytochromes c. J. Biol. Chem. 263, 15928–15937.

Emtage, J., & Jensen, R. (1993a). MAS6 encodes an essential inner membrane component of the yeast mitochondrial protein import pathway. J. Cell Biol. 122, 1003–1012.

Emtage, J., & Jensen, R. (1993b). MAS6 encodes an essential inner membrane component of the yeast mitochondrial protein import pathway. J. Cell Biol. 122, 1003–1012.

Gambill, B. D., Voos, W., Kang, P. J., Miao, B., Langer, T., Craig, E. A., & Pfanner, N. (1993). A dual role for mitochondrial heat shock protein 70 in membrane translocation of preproteins. J. Cell Biol. 123, 109–117.

Glick, B., Wachter, C., & Schatz, G. (1991). Protein import into mitochondria: Two systems acting in tandem? Trends in Cell Biol. 1, 99–103.

Glick, B., & Schatz, G. (1991). Import of proteins into mitochondria. Annu. Rev. Genet. 25, 21–44.

Glick, B. S., Beasley, E. M., & Schatz, G. (1992a). Protein sorting in mitochondria. TIBS 17, 453–459.

Glick, B. S., Brandt, A., Cunningham, K., Müller, S., Hallberg, R. L., & Schatz, G. (1992b). Cytochromes-c_1 and cytochromes-b_2 are sorted to the intermembrane space of yeast mitochondria by a stop-transfer mechanism. Cell 69, 809–822.

Glick, B., Wachter, C., Reid, G. A., & Schatz, G. (1993). Import of cytochrome b_2 to the mitochondrial intermembrane space: The tightly folded heme-binding domain makes import dependent upon matrix ATP. Protein Science, 2, 1901–1911.

Guiard, B. (1985). Structure, expression and regulation of a nuclear gene encoding a mitochondrial protein: the yeast L(+)-lactate cytochrome c reductase (cytochrome b_2). EMBO J. 4, 3265–3272.

Hackenbrock, C. R. (1968). Chemical and physical fixation of isolated mitochondria in low-energy & high-energy states. Proc. Natl. Acad. Sci. USA 61, 598–605.

Horst, M., Jenö, P., Kronidou, N., Bolliger, L., Oppliger, W., Scherer, P., Manning-Krieg, U., Jascur, T., & Schatz, G. (1993a). Protein import into yeast mitochondria: The inner membrane import site protein Isp45p is the *MPI1* gene product. EMBO J. 12, 3035–3041.

Horst, M., Kronidou, N. G., & Schatz, G. (1993b). Through the mitochondrial inner membrane. Curr. Biol. 3, 175–177.

Hurt, E. C., & van Loon, A. P. M. G. (1986). How proteins find mitochondria and intamitochondrial compartments. Trends Biochem. Sci. 11, 204–206.

Hwang, S., Jascur, T., Vestweber, D., Pon, L., & Schatz, G. (1989). Disrupted yeast mitochondria can import precursor proteins directly through their inner membrane. J. Cell Biol. 109, 487–493.

Hwang, S. T., & Schatz, G. (1989). Translocation of proteins across the mitochondrial inner membrane, but not into the outer membrane, requires nucleoside triphosphates in the matrix. Proc. Natl. Acad. Sci. USA 86, 8432–8436.

Hwang, S. T., Wachter, C., & Schatz, G. (1991). Protein import into the yeast mitochondrial matrix—A new translocation intermediate between the two mitochondrial membranes. J. Biol. Chem. 266, 21083–21089.

Jascur, T., Goldenberg, D. P., Vestweber, D., & Schatz, G. (1992). Sequential translocation of an artificial precursor protein across the two mitochondrial membranes. J. Biol. Chem. 267, 13636–13641.

Jensen, R. E., Schmidt, S., & Mark, R. J. (1992). Mutations in a 19-amino-acid hydrophobic region of the yeast cytochrome c_1 presequence prevent sorting to the mitochondrial intermembrane space. Mol. Cell. Biol. 12, 4677–4686.

Kang, P-J., Ostermann, J., Shilling, J., Neupert, W., Craig, E. A., & Pfanner, N. (1990). Requirement for hsp70 in the mitochondrial matrix for translocation and folding of precursor proteins. Nature (London) 348, 137–143.

Kellems, R. E., Allison, V. F., & Butow, R. A. (1975). Cytoplasmic-type 80S ribosomes associated with yeast mitochondria. J. Cell Biol. 65, 1–14.

Kronidou, N. G., Oppliger, W., Bolliger, L., Hannavy, K., Glick, B. S., Schatz, G., & Horst, M. (1994). Dynamic interaction between Isp45 and mitochondrial hsp70 in the protein import system of the yeast mitochondrial inner membrane. Proc. Natl. Acad. Sci. USA.

Lill, R., Stuart, R. A., Drygas, M. E., Nargang, F. E., & Neupert, W. (1992). Import of cytochrome c heme lyase into mitochondria—A novel pathway into the intermembrane space. EMBO J. 11, 449–456.

Maarse, A. C., Blom, J., Grivell, L. A., & Meijer, M. (1992). *MPI1*, an essential gene encoding a mitochondrial membrane protein, is possibly involved in protein import into yeast mitochondria. EMBO J. 11, 3619–3628.

Manning-Krieg, U. C., Scherer, P. E., & Schatz, G. (1991). Sequential action of mitochondrial chaperones in protein import into the matrix. EMBO J. 10, 3273–3280.

Mayer, A., Lill, R., & Neupert, W. (1993). Translocation and insertion of precursor proteins into isolated outer membranes of mitochondria. J. Cell Biol. 121, 1233–1243.

Munro, S., & Pelham, H. R. B. (1986). An hsp70-like protein in the ER: Identity with the 78 kD glucose-regulated protein and immunoglobulin heavy chain binding protein. Cell 46, 291–300.

Neupert, W., Hartl, F. U., Craig, E. A., & Pfanner, N. (1990). How do polypeptides cross the mitochondrial membranes? Cell 63, 447–450.

Nicholson, D. W., Hergersberg, C., & Neupert, W. (1988). Role of cytochrome c heme lyase in the import of cytochrome c into mitochondria. J. Biol. Chem. 263, 19034–19042.

Nunnari, J., Fox, T., & Walter, P. (1993). A mitochondrial protease with two catalytic subunits of nonoverlapping specificities. Science 262, 1997–2004.

Ohba, M., & Schatz, G. (1987). Disruption of the outer membrane restores protein import to trypsin-treated yeast mitochondria. EMBO J. 6, 2117–2122.

Pfanner, N., Rassow, J., van der Klei, I. J., & Neupert, W. (1992). A dynamic model of the mitochondrial protein import machinery. Cell 68, 999–1002.

Pon, L., Moll, T., Vestweber, D., Marshallsay, B., & Schatz, G. (1989). Protein import into mitochondria: ATP-dependent protein translocation activity in a submitochondrial fraction enriched in membrane contact sites and specific proteins. J. Cell Biol. 109, 2603–2616.

Poyton, R. O., Duhl, D. M. J., & Clarkson, G. H. D. (1992). Protein export from the mitochondrial matrix. Trends in Cell. Biol. 2, 369–375.

Pratje, E., & Guiard, B. (1986). One nuclear gene controls the removal of transient pre-sequences from two yeast proteins: One encoded by the nuclear, the other by the mitochondrial genome. EMBO J. 5, 1313–1317.

Rassow, J., & Pfanner, N. (1991). Mitochondrial preproteins en route from the outer membrane to the inner membrane are exposed to the intermembrane space. FEBS Lett. 293, 85–88.

Rassow, J., Maarse, A. C., Krainer, E., Kübrich, M., Müller, H., Meijer, M., Craig, E. A., & Pfanner, N. (1994). Mitochondrial protein import: Biochemical and genetic evidence for interaction of matrix Hsp70 and the inner membrane protein MIM44. J. Cell Biol. 127, 1547–1556.

Ryan, K. R., Menold, M. M., Garrett, S., & Jensen R. E. (1994). *SMS1*, a high-copy suppressor of the yeast *mas6* mutant, encodes an essential inner membrane protein required for mitochondrial protein import. Mol. Biol. Cell 5, 529–538.

Rospert, S., Müller, S., Schatz, G., & Glick, B. S. (1994). Fusion proteins containing the cytochrome b_2 presequence are sorted to the mitochondrial intermembrane space independently of hsp60. J. Biol. Chem. 269, 17279–17288.

Sadler, I., Suda, K., Schatz, G., Kaudewitz, F., & Haid, A. (1984). Sequencing of the nuclear gene for the yeast cytochrome c1 precursor reveals an un-usually complex amino-terminal presequence. EMBO J. 3, 2137–2143.

Sanders, S. L., Whitfield, K. M., Vogel, J. P., Rose, M. D., & Schekman, R. W. (1992). Sec61p and BiP directly facilitate polypeptide translocation into the endoplasmic reticulum. Cell 69, 353–365.

Scherer, P. E., Krieg, U. C., Hwang, S. T., Vestweber, D., & Schatz G. (1990). A precursor protein partly translocated into yeast mitochondria is bound to a 70 kD mitochondrial protein. EMBO J. 12, 4315–4322.

Scherer, P. E., Manning-Krieg, U. C., Jenö, P., Schatz, G., & Horst, M. (1992). Identification of a 45 kD protein at the protein import site of the yeast mitochondrial inner membrane. Proc. Natl. Acad. Sci. USA 89, 11930–11934.

Schneider, A., Behrens, M., Scherer, P. E., Pratje, E., Michaelis, G., & Schatz, G. (1991). Inner membrane protease I, an enzyme mediating intramitochondrial protein sorting in yeast. EMBO J. 10, 247–254.

Schneider, H-C., Berthold, J., Bauer, M. F., Dietmeier, K., Guiard, B., Brunner, M., & Neupert, W. (1994). Mitochondrial Hsp70/MIM40 complex facilitates protein import. Nature 371, 768–774.

Schwaiger, M., Herzog, V., & Neupert, W. (1987). Characterization of translocation contact sites involved in the import of mitochondrial proteins. J. Cell Biol. 105, 235–246.

Schwarz, E., Seytter, T., Guiard, B., & Neupert, W. (1993). Targeting of cytochrome b_2 into the mitochondrial intermembrane space—Specific recognition of the sorting signal. EMBO J. 12, 2295–2302.

Segui-Real, B., Stuart, R. A., & Neupert, W. (1992). Transport of proteins into the various subcompartments of mitochondria. FEBS Lett. 313, 2–7.

Segui-Real, B., Kispal, G., Lill, R., & Neupert, W. (1993). Functional independence of the protein translocation machineries in mitochondrial outer and inner membranes: Passage of preproteins through the intermembrane space. EMBO J. 12, 2211–2218.

Simon, S. M., Peskin, C. S., & Oster, G. F. (1992). What drives the translocation of proteins? Proc. Natl. Acad. Sci. USA 89, 3770–3774.

Söllner, T., Rassow, J., Wiedmann, M., Schlossmann, J., Keil, P., Neupert, W., & Pfanner, N. (1992). Mapping of the protein import machinery in the mitochondrial outer membrane by crosslinking of translocation intermediates. Nature (London) 355, 84–87.

van Loon, A. P. G. M., & Schatz, G. (1987). Transport of proteins to the mitochondrial intermembrane space: The "sorting" domain of the cytochrome c_1 presequence is a stop-transfer sequence specific for the mitochondrial inner membrane. EMBO J. 6, 2441–2448.

Vestweber, D., Brunner, J., Baker, A., & Schatz, G. (1989). A 42K outer-membrane protein is a component of the yeast mitochondrial protein import site. Nature (London) 341, 205–209.

Vogel, J. P., Misra, L. M., & Rose, M. D. (1990). Loss of Bip/Grp78 function blocks translocation of secretory proteins in yeast. J. Cell Biol. 110, 1885–1895.

Wachter, C., Schatz, G., & Glick, B. S. (1992). Role of ATP in the intramitochondrial sorting of cytochrome c_1 and the adenine nucleotide translocator. EMBO J. 11, 4787–4794.

Yaffe, M. P., & Schatz, G. (1984). Two nuclear mutations that block mitochondrial protein import in yeast. Proc. Natl. Acad. Sci. USA 81, 4819–4823.

PROTEOLYTIC PROCESSING OF MITOCHONDRIAL PRECURSOR PROTEINS

Wayne A. Fenton and Frantisek Kalousek

I. Introduction . . . 164
  A. Substrates . . . 164
  B. Import Overview . . . 164
  C. Nomenclature . . . 165
II. Mitochondrial Processing Peptidase (MPP) . . . 165
  A. Function and Requirements . . . 165
  B. Substrate Structure . . . 166
  C. Purification, Cloning, and Characterization . . . 170
III. Mitochondrial Intermediate Peptidase (MIP) . . . 177
IV. Inner Membrane Peptidase (IMP) . . . 184
V. Other Peptidases . . . 185
VI. Summary . . . 186
  Acknowledgment . . . 186
  References . . . 186

Advances in Molecular and Cell Biology,
Volume 17, pages 163–191.

ISBN: 0-7623-0144-9

I. INTRODUCTION

As soon as initial *in vitro* translations and immunoprecipitations of mitochondrially localized proteins showed that most were synthesized as larger precursors, it was clear that a proteolytic step (or steps) was required to generate many of the active protein species normally present in this organelle. In the subsequent 15 years, a great deal has been learned about this process, its substrates, and its enzymes (for earlier reviews see Hartl et al., 1989; Geli and Glick, 1990; Arretz et al., 1991; and Glick and Schatz, 1991). This chapter will review current information about the known peptidases—their structure, biochemical characteristics, and cleavage-site specificities—and point out areas where questions remain to be answered.

A. Substrates

Most proteins localized to the mitochondria are synthesized in the cytosol as larger precursors by virtue of amino-terminal extensions, which we call "leader peptides," in distinction to the "signal peptides" directing transport across the endoplasmic reticulum. Leader peptides vary in length from 10 to 80 amino acids and, in general, have little sequence homology to each other. Their most notable feature is the usually complete absence of acidic residues, so that, overall, they are positively charged. Structural prediction algorithms suggest that some could form amphiphilic α-helices, with the positively charged and hydrophilic side-chains clustered on one side. In a few cases, physical biochemical measurements have indicated that synthetic peptides based on certain leader peptides have significant α-helical content in solution. Nevertheless, the lack of clear structural or cleavage site motifs initially led to speculation that multiple specific proteases might be involved in the processing of these precursors, but that has since proven to be incorrect (see following).

B. Import Overview

The mechanics of mitochondrial protein transport and the role of the leader peptides in this process are reviewed in detail elsewhere in this volume. All of the precursors subject to proteolytic processing are recognized by the apparatus at the "contact site," and most are translocated fully into the mitochondrial matrix before or concomitant with the proteolysis. Most precursors are processed in a single proteolytic step, although a significant subset is cleaved twice during import (see below). A few species, destined for the intermembrane space, may be only partially exposed to the matrix, so that their leader peptides can be cleaved before they are localized to their final destination. Analysis of both leader peptide and import apparatus mutants (see following) has suggested that translocation into the matrix is not dependent on proteolysis and that cleavage likely occurs subsequent to the initial action of mitochondrial Hsp70 and before folding of mature polypeptide, assisted by Hsp60.

Table 1. Nomenclature for the Mitochondrial Processing Peptidases

Recommended Name	*Other Names*
Mitochondrial Processing Peptidase (MPP)	
α-subunit (α-MPP)	MPP
	MAS2
	MIF2
	P-55
	Cyto *c* reductase, subunit III
β-subunit (β-MPP)	PEP
	MAS1
	MIF1
	P-52
	Cyto *c* reductase, subunits I and II*
Mitochondrial Intermediate Peptidase (MIP)	Protease II
Inner Membrane Peptidase (IMP)	
subunit 1 (IMP1)	
subunit 2 (IMP2)	

Note: *Whether the two β-like subunits in plants should be β^1 and β^2 or β and γ remains to be decided.

C. Nomenclature

Before the mitochondrial proteases are described in detail, a comment about nomenclature is required. As noted in a letter to TIBS (Kalousek et al., 1993), there is now general agreement among workers in this field as to the names by which the individual enzymes and their subunits are called (Table 1). This is particularly important for the general matrix peptidase, which has had a variety of designations in different laboratories. We will use the agreed-upon names and will point out previous usages in our description of individual experiments.

II. MITOCHONDRIAL PROCESSING PEPTIDASE (MPP)

A. Function and Requirements

The function of the mitochondrial processing peptidase (MPP; EC 3.4.99.41) was apparent from the earliest experiments demonstrating that nuclearly encoded mitochondrial proteins, newly synthesized by *in vitro* translation systems, were larger than their "mature" counterparts (Conboy et al., 1979; Côté et al., 1979; Maccecchini et al., 1979) and that isolated mitochondria *in vitro* carried out the conversion from precursor to mature concomitant with translocation (Harmey and Neupert, 1979; Maccecchini et al., 1979; Michel et al., 1979). Shortly thereafter, a soluble extract from yeast mitochondria was shown to contain this activity (Böhni et al., 1980; McAda and Douglas, 1982), with similar demonstrations following for both *Neurospora* and rat liver organelles (Conboy et al., 1982; Miura et al., 1982; Schmidt et al., 1984). Initial experiments with crude extracts of mitochondria from each of these species showed that peptidase activity was inhibited by divalent-cation chelators, such as EDTA, and activated most strongly by Co^{2+}, Mn^{2+}, and Zn^{2+} (Böhni et al., 1980; Conboy et al., 1982; McAda and Douglas, 1982; Schmidt et al., 1984). It was established biochemically that the cleavage occurred near the amino-terminus, with the assumption that it was an endoproteolytic event (Kraus et al., 1981). It was not until the amino acid sequences of precursor and mature proteins could be compared, however, either directly through amino-terminal sequencing or indirectly by cloning and DNA sequencing, that the precise sites of proteolysis could be determined. Confirmation that the cleavage was endoproteolytic has been achieved in a few cases by recovering a released amino-terminal leader peptide (Ou et al., 1989; Arretz et al., 1994).

B. Substrate Structure

The nature of the substrates for MPP has been explored by two approaches: on paper, by comparison of the amino acid sequences of leader peptides, and, *in vitro*, by site-directed mutagenesis of leader peptides and evaluation of their uptake by mitochondria and proteolysis by mitochondrial extracts. As cloned sequences for a number of mitochondrial precursor proteins became available, their general features became apparent: a strongly positive overall charge, a high content of hydrophilic amino acids, and the predicted ability to form amphiphilic α-helices. The extent to which each of these features was required for the individual steps of mitochondrial recognition, translocation, or proteolysis was unclear, however. Extensive surveys using confirmed cleavage sites revealed that many MPP sites had an arginine residue two (sometimes three) amino acids amino-terminal to the cleaved peptide bond (Hendrick et al., 1989; von Heijne et al., 1989; Gavel and von Heijne, 1990). This observation, known as the "Arg at -2" rule, appeared to account for more than 70% of the likely MPP cleavage sites, including those in leader peptides that were known or predicted to be cleaved twice (Hendrick et al., 1989; see following). Gavel and von Heijne (1990) developed a comparison algorithm based on a more complete analysis of these data, but still only achieved 45–75% accuracy in predicting cleavage sites. Sophisticated secondary structural predictions (Gavel and von Heijne, 1990), based on hydrophobic moment analysis, suggested that amphiphilic α-helices were more likely in the amino-terminal parts of the leader sequences. On the other hand, the carboxy-terminal regions, within 10–15 residues of the cleavage sites, were predicted to have a different structure, either an α-helix with a spiraling apolar "ridge" or an amphiphilic β-strand, depending on whether the leader was from a mitochondrial or a chloroplast protein, respectively. These predictions were further complicated by the fact that some precursors were known or suspected to be cleaved twice (see below), making it necessary to attempt to separate the recognition and structural requirements of the second enzymes from those of MPP.

Other researchers have taken an experimental approach, using site-directed and random mutagenesis to produce mutant leader peptides with amino acid substitutions or deletions; both natural substrates and chimeric precursors formed from modified leader peptides and convenient reporter proteins, such as dihydrofolate reductase (DHFR), have been the starting points. Many of these experiments have been carried out with intact mitochondria or crude mitochondrial extracts, rather than purified MPP, and their interpretation is thus subject to some uncertainty because of the need for normal import, in the case of mitochondria, and because of the presence of other protease activities, inhibitors, or effectors in the crude extracts. Nevertheless, some conclusions seem generally valid. Studies substituting the positively charged Arg residues in the leader peptide of the ornithine transcarbamylase precursor (pOTC) with Gly, either multiply or singly, showed that the Arg at position 23, two residues amino-terminal to the MPP cleavage site between 24 and 25, was critical for MPP cleavage, whether determined in whole rat liver mitochondria or in crude soluble extracts of mitochondria (Horwich et al., 1985, 1986). For the human OTC precursor (phOTC), Arg^{23}→Gly was uncleaved, but other substitutions at 23 were either partially cleaved (Asn or Ala) or almost fully cleaved (Lys; Horwich et al., 1986). Interestingly, for the rat OTC precursor (prOTC), which has a somewhat different leader peptide sequence, both Arg^{23}→Gly and Arg^{23}→Asp substitutions were cleaved at a novel site between residues 16 and 17, rather than at the usual 24–25 site (Sztul et al., 1987). Similar experiments replacing basic residues in the leader peptide of mitochondrial aspartate aminotransferase (pmAAT) showed that, when Arg^{28}, two residues from the cleavage site, was replaced with Leu, cleavage was reduced to 17% of that of wild-type (Nishi et al., 1989). Likewise, substitutions for the Arg at -2 in a chimeric precytochrome b_2-DHFR protein reduced processing by purified *N. crassa* MPP significantly (Lys or Ala) or completely (Gly; Arretz et al., 1994). In a study of binding and cleavage of synthetic peptides, based on the malate dehydrogenase (MDH) leader sequence, by purified MPP, Niidome and colleagues (1994) showed that replacement of the -2 Arg by Ala or Lys both increased K_M and decreased V_{max}, such that overall catalytic efficiency dropped to 5% or less relative to the wild-type peptide.

Thus, there seems to be general agreement that the basic residue two (or sometimes three) residues amino-terminal to the cleavage site is an important recognition element for MPP, at least in those precursors where it is present. As noted in the survey data, however, a significant number (20–30%) of leader peptides lack a basic residue in this position (Hendrick et al., 1989). These data might be accounted for by the observations above that some substitutions at this position are tolerated, even in Arg at -2 leaders; alternatively, another peptidase with somewhat different specificity might be responsible for cleaving these precursors.

The importance of other individual amino acids in leader peptides for MPP activity is much less clear. In the most extensive study reported to date, Horwich and colleagues tested a large number of random and directed single substitutions in the OTC leader peptide for processing by MPP and found that all, except the Arg^{23} mutations mentioned above, were substrates (Horwich et al., 1987). Most were cleaved as well as wild-type, and the rest were only slightly reduced (50%). A few single amino acid changes in the leader of alcohol dehydrogenase III have been tested and are essentially without effect on MPP cleavage (Mooney et al., 1990). In synthetic peptides based on the MDH leader, changing the distal Arg at residue 7 (-10 to the cleavage site) to Ala had some effect on cleavage, but did not abolish it (Niidome et al., 1994). Multiple substitutions for basic residues (Arg and His) have been investigated in pOTC (Horwich et al., 1987), pmAAT (Nishi et al., 1989) and pre-adrenodoxin (pAd; Ou et al., 1994), but most showed only moderate effects unless the critical Arg residues were changed as well. In pAd, however, multiple changes of Arg to neutral residues in the middle of the leader peptide did reduce (two changes) or eliminate (four changes) cleavage by MPP (Ou et al., 1994). Because these residues are 21–33 residues away from the cleavage site—farther than the length of many leader peptides—it is difficult to visualize a general role for such distant groups. Another intriguing experiment involved substitution of Gly or Ala for all four of the Leu residues near the amino-terminal end of prOTC (positions 2, 5, 8, and 9; Kraus et al., 1988). In this case, proteolysis was abolished, even though it was only slightly reduced when precursors with three of the four substitutions were tested. The authors suggested that these data could reflect a specificity for a particular structure in the amino-terminal region of the leader or the ability of the amino-terminus to vary its structure from α-helix to β-sheet in response to changing environments encountered in the transport apparatus and processing enzymes.

Deletions in leader peptides and the adjacent mature sequences have also indicated that secondary or even tertiary structural elements may play roles in MPP recognition and cleavage. Experiments with pOTC (Horwich et al., 1986), pmAAT (Nishi et al., 1989), pAd (Ou et al., 1994), and a pCOXIV-DHFR chimera (Hurt et al., 1985, 1987) have all indicated that the amino-terminus is important for MPP recognition and cleavage, although the precise structure involved and its position relative to the cleavage site are unclear. For pmAAT and the chimera, the immediate amino-terminal four to eight residues appear to contain the element, while for pOTC and pAd, the region appears to be internal, between residues 8 and 22 for pOTC and residues 18 and 39 in pAd. Such deletion experiments must, of course, be interpreted with care, given that the deletions juxtapose residues and structures without regard for their normal configurations and, in addition, remove and change the distribution of positive charges.

The possible role of residues and structures on the carboxyl-side of the cleavage site has been addressed extensively only for pOTC, a twice-cleaved percursor. Isaya and her colleagues generated a number of chimeric precursors, using both once-

and twice-cleaved leader peptides and their corresponding mature sequences (also see the MIP discussion below), and tested whether they could be cleaved by either or both purified proteases (Isaya et al., 1991). Interestingly, when the amino-terminal 24 residues of the pOTC leader were juxtaposed to the mature OTC amino-terminus (deleting the 8 residues of the intermediate, iOTC), no cleavage was observed, while a similar construct with the mature sequence of methylmalonyl-CoA mutase (a once-cleaved mitochondrial protein) was processed at its normal site by MPP. Conversely, the mutase leader sequence, joined to mature OTC, was not cleaved, while a chimera between the mutase leader and iOTC was processed normally, at the correct sites, by both proteases. These data and others have led to the conclusion that a structural element on the carboxyl-side of the MPP cleavage site (i.e., the mature amino-terminus of a once-cleaved protein or the amino-terminal octapeptide of the intermediate in twice-cleaved proteins) is critical for recognition and proteolysis of precursors by MPP (Isaya et al., 1991). Only one point mutation has been reported that affects this region. In the cytochrome b_2-DHFR chimera, replacing Tyr at +1 (the carboxyl residue of the cleavage site) with Lys prevented proteolysis by MPP (Arretz et al., 1994). Determining whether this result reflects a specific amino acid side-chain requirement in the context of this precursor or whether the substitution disrupts a structure at the cleavage site will involve additional experiments.

Finally, a few synthetic peptides, corresponding to partial or complete leader peptides, have been tested for their ability to inhibit MPP. In general, only those containing at least the complete amino-terminal side of the cleavage site have been inhibitory, although carboxyl-terminal residues have been required as well for maximum effect (Côté et al., 1988; Chu et al., 1989; Yang et al., 1991). For the COXIV leader, a 25-residue peptide extending 8 residues carboxy-terminal of the MPP cleavage site not only inhibited proteolysis of full-length pCOXIV-DHFR chimera, but was itself a substrate for purified *S. cerevisiae* MPP (Yang et al., 1991). Likewise, a peptide comprising the amino-terminal 30 residues of pMDH inhibited processing of pAd (Ou et al., 1994) and was a substrate for bovine MPP (Niidome et al., 1994).

In sum, then, although a large number of leader peptide variants have been tested for their ability to be substrates for MPP in a few model systems and the wide range of natural substrates has been aligned and compared, the recognition elements and cleavage site specificity of this peptidase remain somewhat unclear. Positive charge certainly plays a role, particularly two residues amino-terminal to the cleavage site, with Arg the preferred residue. Some secondary structural elements seem to be involved as well, both at the amino-terminal end of the leader and at the cleavage site itself, but their nature remains obscure—possibly an amphiphilic α-helix at the amino-terminus and a transition region between a helix with a spiraling apolar ridge and a "loosely structured" region in the mature amino-terminus. Further experiments with purified MPP and additional precursors with site-directed mutations at the putatively critical residues are necessary before these questions can be answered definitively.

One point that is clear from the studies of the various mutant precursors is that import into mitochondria does not require proteolysis. A number of mutations, particularly the amino-terminal deletions and multiple substitutions of arginines, inhibit both processes (Horwich et al., 1985, 1986; Hurt et al., 1987; Kraus et al., 1988; Nishi et al., 1989), but incubations with partially purified MPP have allowed workers to distinguish between these effects. In a few cases such as the $Arg^{23} \rightarrow Gly$ mutant of phOTC, where this distinction was not clear, expression of mutant constructs *in vivo* has demonstrated transport in the absence of proteolysis (Isaya et al., 1988).

C. Purification, Cloning, and Characterization

A major goal for laboratories working with MPP has been its purification so that a defined reaction system could be established to investigate the enzyme's substrate specificity, kinetics, and structure-function relationships. Early work established that MPP was a neutral protease requiring divalent metal ions for activity (Böhni et al., 1980; Conboy et al., 1982; McAda and Douglas, 1982; Miura et al., 1982; Schmidt et al., 1984), and partial purifications of a precursor-cleaving activity were reported by several groups (Miura et al., 1982; Schmidt et al., 1984). It was not until 1988, however, that complete purifications of the enzyme from *N. crassa* (Hawlitschek et al., 1988), *S. cerevisiae* (Yang et al., 1988), and rat (*R. norvegicus*) liver (Ou et al., 1989; Kleiber et al., 1990) mitochondria were reported. MPP has also been reported purified from bovine liver mitochondria by one of the rat liver protocols (Niidome et al., 1994). Recently, purification of the homologous enzyme from potato (*Solanum tuberosum*) tuber mitochondria has been achieved (Braun et al., 1992). MPPs from the first three sources are soluble heterodimers of 100–110 kDa. In *Neurospora*, the subunits separate from each other, at least partially, during many of the purification steps (Hawlitschek et al., 1988); in yeast and rat, the dimer is stable (Yang et al., 1988; Kleiber et al., 1990). Because the larger of the *Neurospora* subunits appeared to retain a small amount of activity that was greatly enhanced by addition of the smaller subunit (Hawlitschek et al., 1988), Neupert and his colleagues initially named the proteins *M*atrix *P*rocessing *P*eptidase (MPP) and *P*rocessing *E*nhancing *P*rotein (PEP), respectively. Experiments with individually expressed subunits (see following) from *Neurospora* have since indicated that the subunits are inactive by themselves *in vitro* and likely function as heterodimers (Arretz et al., 1994). Thus, current nomenclature refers to the two proteins as the α- (larger) and β- (smaller) subunits of MPP (see Table 1). The situation is different for the potato enzyme, however, and probably for other higher plant homologues as well. Here, the activity is membrane-bound in the inner mitochondrial membrane and co-purifies with the cytochrome *c* reductase complex (Braun et al., 1992). Attempts to extract active MPP from the isolated complex have been unsuccessful, suggesting that it may be an integral part of the complex. Sequence analysis indicates that there is one subunit in the complex that is homologous to α-MPP, but two with homology to β-MPP (see following). The stoichiometry of the MPP subunits in the complex and the roles of each in proteolysis and electron transport remain to be established.

The purifications themselves used mostly conventional methodologies, starting with mitochondrial matrix fractions in the case of yeast and rat, whole cell homogenates for *Neurospora*, and whole mitochondria for potato. Chromatography on a Zn-chelate column was an important step for the first three sources, while the potato enzyme was purified by steps routinely used for isolating the cytochrome *c* reductase complex. The final preparation of yeast MPP consisted of subunits of 51 and 48 kDa, for *Neurospora* MPP, 57 and 52 kDa, and for rat, 55 and 52 kDa; resolution of the cytochrome *c* reductase complex from potato and sequence analysis (see following) indicated that peptides of 55, 53, and 51 kDa were likely responsible for its MPP activity.

The cloning and DNA sequencing of the yeast genes encoding the MPP subunits were aided by the early description of two mutant strains defective in mitochondrial protein biogenesis, *mas1* and *mas2* (Yaffe and Schatz, 1984). A temperature-sensitive allele of one of these (*mas1*) was later shown to import precursors normally, but to be defective in their cleavage at nonpermissive temperature (Yaffe et al., 1985). Independent selection by other workers of mutant yeast strains deficient in mitochondrial import function generated two mutants designated *mif1* and *mif2*, which were shown to be allelic to *mas1* and *mas2*, respectively (Pollock et al., 1988). When the corresponding wild-type genes were isolated and sequenced, *MAS1/MIF1* encoded the smaller MPP subunit, while *MAS2/MIF2* encoded the larger (Pollock et al., 1988; Witte et al., 1988; Yang et al., 1988); antibody cross-reactivity and the sequence data confirmed that these subunits comprised the purified enzyme (Pollock et al., 1988; Witte et al., 1988). At about the same time, the corresponding *Neurospora* subunits were cloned by antibody screening of a cDNA library and sequenced, yielding similar results (Hawlitschek et al., 1988; Schneider et al., 1990). Mas1 (β-MPP) of yeast was homologous to the smaller *Neurospora* subunit, previously called PEP; Mas2 (α-MPP) was homologous to the larger one. Rat MPP subunits were cloned by PCR amplification of liver cDNA using degenerate primers based on the amino acid sequences of the purified, separated subunits (Kleiber et al., 1990; Paces et al., 1993). The potato subunits were cloned by library screening with degenerate oligonucleotides (Braun et al., 1992; Emmermann et al., 1993). By comparing the predicted amino acid sequences with the amino-terminal sequences determined from the purified proteins, it was possible to show that all of the subunits are synthesized as larger precursors, with typical leader peptides that are cleaved at typical "Arg at -2" MPP cleavage sites. Thus, MPP is responsible for the post-translocational cleavage of its own newly imported subunits.

Alignments and comparisons of the amino acid sequences predicted from the cloned cDNAs show that yeast and *Neurospora* a-MPP are about 43% identical (Schneider et al., 1990) and that rat α-MPP is about 36% identical to both (Kleiber et al., 1990); similarities, allowing conservative amino acid substitutions, are greater than 60%. Interestingly, it is the smallest subunit from potato that shows the highest homology to α-MPP from the other organisms, ranging from 30–35% identity (Braun et al., 1992). For the β-MPP subunits, identities are somewhat higher, ranging from 45–52% among rat, yeast, and *Neurospora* (Hawlitschek et al., 1988; Paces et al., 1993). Each of the two largest subunits from potato shows good homology with the other β-subunits, ranging from 42–50%; they are 70% identical to each other (Emmermann et al., 1993). It was originally suggested that potato MPP might act as a trimer ($\alpha\beta^1\beta^2$) in the membrane complex, but further experiments have indicated that there are likely two forms of potato cytochrome *c* reductase, one containing an $\alpha\beta^1$ dimer, the other an $\alpha\beta^2$, preserving the quaternary structural homology with other MPPs (Jänsch et al., 1995). The functional differences, if any, between the two forms are unclear at present. Furthermore, a second α-MPP homologue has also been uncovered in potato, but its abundance is much less than that of the other subunits and its function is not known (Emmermann et al., 1994). Interestingly, all of the α- and β-subunits show some homology to each other, about 20–30% within a species, with slightly lower identities across species (Schulte et al., 1989; Paces et al., 1993).

Although initial searches of sequence databases with the *Neurospora* and yeast MPP sequences revealed no other homologies, isolation and cloning of the subunit I (core I) protein of *Neurospora* cytochrome *c* reductase produced a surprising result. This protein, and the gene encoding it, was identical to *Neurospora* β-MPP (Schulte et al., 1989). No other gene encoding either core I or β-MPP could be demonstrated in *Neurospora*, suggesting that this protein is bifunctional in this organism. This finding accounts for the earlier observations that β-MPP is 15-fold more abundant in *Neurospora* mitochondria than α-MPP and that 75% or more of it fractionates as a peripheral inner membrane protein (as does core I), rather than as a soluble matrix protein (Hawlitschek et al., 1988). In contrast, core I in yeast and mammals is a protein distinct from β-MPP, encoded by a different gene. Yeast core I is 24% identical to yeast β-MPP (Schulte et al., 1989), while bovine core I is 52% identical to rat β-MPP (Gencic et al., 1991; Paces et al., 1993; the rat core I sequence has not been reported, but partial amino-terminal protein sequence shows that it is indeed different from rat β-MPP; FK and WAF, unpublished). The more-recent finding that all subunits of potato MPP are part of that organism's cytochrome *c* reductase complex thus becomes more easily understood. Here, MPP itself appears to be bifunctional, forming an integral part of the electron transport chain as well as processing imported mitochondrial precursors. The specific role of the MPP subunits here, and of the core I and core II proteins in other species, remains unclear, although they are required both for oxidoreductase activity and assembly of the complex *in vivo*, based on the phenotypes of yeast strains bearing mutations in *COR1* and *COR2* (Tzagoloff et al., 1986; Oudshoorn et al., 1987).

Not surprisingly, computer program generated alignments of the α- and β-subunits from different species show certain regions where the homologies are very

Table 2. Comparison of Amino Acid Sequences in the Acidic Motif of MPP

Rat MPP α-subunit	**D**	**E**	**E**	I	**E**	M	T	R	M	A	V	Q	F	**E**	L	**E**	**D**	L	N
Neurospora MPP α-subunit	**D**	**E**	**E**	L	**E**	G	Q	I	M	T	A	Q	Y	**E**	V	N	**E**	I	W
Yeast MPP α-subunit	**E**	Q	**E**	L	Q	**E**	Q	K	L	S	A	**E**	Y	**E**	I	**D**	**E**	V	M
Potato MPP α-subunit	**D**	W	**E**	V	K	**E**	Q	L	**E**	K	V	K	**E**	I	S	**E**	Y	S	K
Rat MPP β-subunit	**E**	A	**E**	I	**E**	R	**E**	**R**	**G**	**V**	**I**	**L**	**R**	**E**	M	Q	**E**	V	**E**
Neurospora MPP β-subunit	**E**	S	A	I	**E**	R	**E**	**R**	**D**	**V**	**I**	**L**	**R**	**E**	S	**E**	**E**	V	**E**
Yeast MPP β-subunit	N	S	A	I	**E**	R	**E**	**R**	**D**	**V**	**I**	**I**	**R**	**E**	S	**E**	**E**	V	**D**
Potato subunit I (P55)	**E**	**D**	K	I	I	R	**E**	**R**	**S**	**V**	**I**	**L**	**R**	**E**	M	**E**	**E**	V	**E**
Potato subunit II (P53)	**E**	R	K	I	**E**	R	**E**	**R**	**D**	**V**	**I**	**L**	**R**	**E**	M	**E**	**E**	V	**E**

high. In general, overall homology is greater in the amino-terminal 200 or so amino acids of the mature sequences than in the carboxyl-terminal regions. Within the amino-terminus, two regions stand out. One is a highly negatively charged region between residues 125–145 of the mature sequences of both subunits (all three in potato) in all species (Table 2). Structural predictions suggest that these residues could form an amphiphilic α-helix, with the negatively charged residues clustered on one face (Kleiber et al., 1990). We have speculated that this could be part of the substrate binding site, recognizing a complementary positively charged amino-terminal α-helix in leader peptides (Kleiber et al., 1990).

The second region is about 70 residues amino-terminal to this and includes the motif H-F-L-E-H embedded in a larger homologous area (Table 3). This motif is precisely conserved in all of the β-MPP subunits, including the two larger subunits in potato (Rawlings and Barrett, 1991; Paces et al., 1993). Interestingly, this motif is only somewhat conserved among the α-subunits, varying from H-F-L-E-K in rat to H-I-L-D-R in yeast, with the consensus histidine-hydrophobic-hydrophobic-acidic-basic.

The core I proteins from various species (except the bifunctional core I/β-MPP proteins of *Neurospora* and potato) show much less homology in this region, even to the relaxed α-MPP motif. On the other hand, the H-X-X-E-H motif does identify MPP as a member of a protease family (Rawlings and Barrett, 1991; Becker and Roth, 1992) that includes the insulin-degrading enzyme (IDE) of man (Affholter et al., 1988) and its homologue in *Drosophila* (Kuo et al., 1990) and protease III of *E. coli* (Finch et al., 1986). Overall homology to these proteases is low (20–24%) and is largely confined to the region around this motif (Becker and Roth, 1993; Paces et al., 1993). The motif itself is an inversion of the H-E-X-X-H motif, commonly found in Zn^{2+}-dependent metallopeptidases of the thermolysin family, which is responsible in part for Zn^{2+} binding (Vallee and Auld, 1990). In fact, IDE and protease III are both Zn^{2+}-dependent enzymes (Ebrahim et al., 1991; Ding et al., 1992), and mutations changing either His in the motif in protease III completely

Table 3. Comparison of Sequences Around the Putative Metal-Binding Motif

Rat MPP α-subunit	L	S	G	I	A	**H**	**F**	**L**	**E**	**K**	L	A	F	S	S	T
Neurospora MPP α-subunit	V	R	G	A	S	**H**	**I**	**M**	**D**	**R**	L	A	F	K	S	T
Yeast MPP α-subunit	L	K	G	C	T	**H**	**I**	**L**	**D**	**R**	L	A	F	K	S	T
Potato MPP α-subunit	S	Y	G	A	T	**H**	**L**	**L**	**E**	**R**	M	A	F	K	S	T
Rat MPP β-subunit	N	N	G	T	A	**H**	**F**	**L**	**E**	**H**	M	A	F	K	G	T
Neurospora MPP β-subunit	T	N	G	T	A	**H**	**F**	**L**	**E**	**H**	L	A	F	K	G	T
Yeast MPP β-subunit	N	N	G	T	A	**H**	**F**	**L**	**E**	**H**	L	A	F	K	G	T
Potato subunit I (P55)	N	N	G	V	A	**H**	**F**	**L**	**E**	**H**	M	I	F	K	G	T
Potato subunit II (P53)	T	N	G	T	A	**H**	**F**	**L**	**E**	**H**	M	I	F	K	G	T

abrogates Zn^{2+} binding and eliminates protease activity (Becker and Roth, 1992). The precise roles of these residues in the two MPP subunits is unclear, particularly because MPP appears to prefer divalent cations other than Zn^{2+} for maximum activity (see below).

In order to obtain sufficient quantities of pure MPP for further characterization, workers have employed several different approaches. For yeast, overexpression was achieved by placing the inducible *GAL10* promoter upstream of a chromosomal copy of each gene (Geli et al., 1990). The resulting subunits were properly transported and cleaved, so that MPP represented about 3% of soluble mitochondrial matrix protein. The enzyme was then purified as the active dimer by conventional techniques. More recently, the mature subunits have been overexpressed directly in *E. coli* (Geli, 1993). For *Neurospora*, the subunits were purified separately by standard methods, the α-subunit from an *E. coli* strain overexpressing the mature subunit with Met replacing the first residue, the β-subunit from *Neurospora* mitochondrial membranes, taking advantage of the natural excess of this subunit

Figure 1. Both α- and β-subunits are required for MPP activity, as demonstrated for the enzyme from several sources. **A**, Yeast: Processing of the chimeric precursor, subunit 9-DHFR (pSu9DHFR) by soluble extracts of *E. coli* overexpressing *S. cerevisiae* α- or β-MPP, or both. Radiolabelled pSu9DHFR was incubated with soluble extracts or purified subunit, then analyzed by SDS/PAGE and fluorography. Lane 1, yeast mitochondrial matrix (20 μg protein); lane 2, extract from co-expression of α- and β-subunits (40 μg); lanes 3 and 4, extract from β-MPP overexpression (80 μg), mixed with 80 or 160 ng, respectively, of α-MPP purified from an overexpressing strain; lane 5, β-MPP extract (80 μg) mixed with α-MPP extract (400 ng); lane 6, α-MPP extract alone (80 μg); lane 7, purified α-MPP alone (16 μg); lane 8, β-MPP extract alone (400 μg). pSu9DHFR has two MPP cleavage sites (Schmidt et al., 1984); the markers indicate the position of the precursor (P), the first cleavage product (m1), and the second cleavage product (m2; from Géli, 1993, with permission).

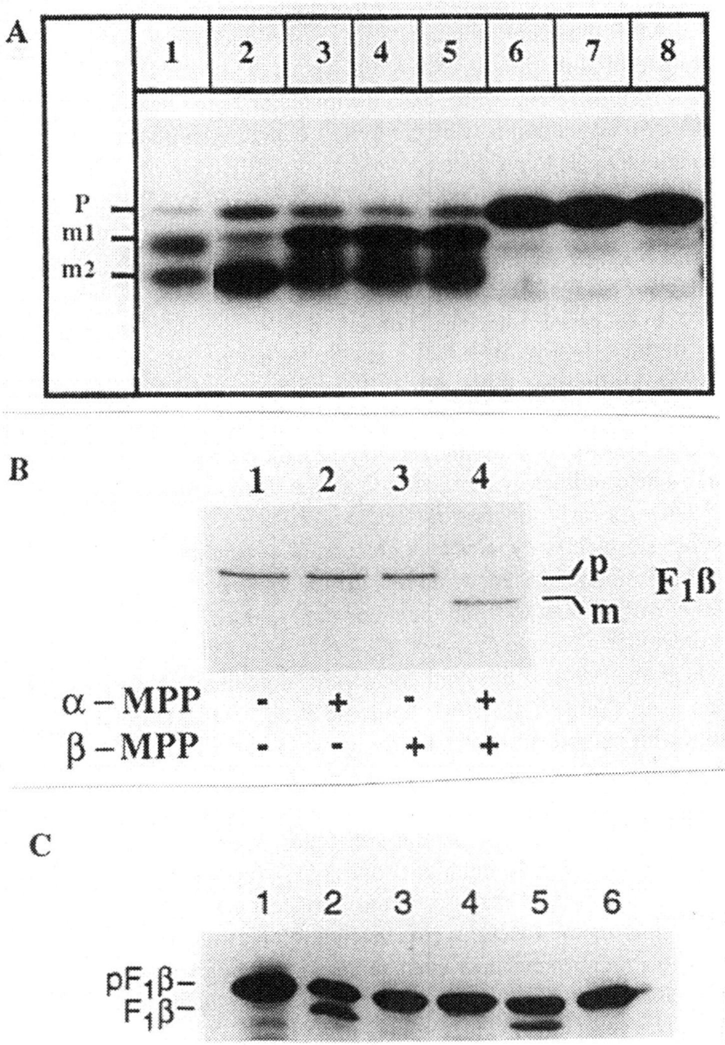

Figure 1. (Continued) **B**, *Neurospora*: Processing by separately produced and purified MPP subunits. Radiolabelled pF_1-β (lane 1) was incubated with 0.04 μM α-MPP (lane 2), 0.04 μM β-MPP (lane 3), and 0.04 μM both (lane 4). Precursor (p) and mature (m) forms of F_1-β are indicated. (From Arretz et al., 1994, with permission.) **C**, Rat: Processing by separately expressed, purified, and renatured MPP subunits. Radiolabelled pF_1-β (lane 1) was the substrate with: lane 2, 100 ng purified rat liver MPP; lane 3, 200 ng α-MPP alone; lane 4, 200 ng β-MPP alone; lane 5, 100 ng of each subunit together; lane 6, 100 ng factor X_a as control (from Saavedra-Alanis et al., 1994, with permission).

(Arretz et al., 1994). Combining the two preparations yielded active enzyme. In the case of rat MPP, the mature subunits were overproduced in *E. coli* as fusion proteins with maltose-binding protein (Saavedra-Alanis et al., 1994). After affinity purification, the fusion proteins were cleaved with factor X_a to release the subunits, which were inactive. The subunits were combined, denatured with urea, and renatured by dilution and dialysis to produce active dimeric MPP.

Investigations into the enzymatic reaction carried out by MPP has been somewhat limited to date. Over-expression and independent purification of each subunit has shown that both are required for activity. Figure 1 shows the combined data from three laboratories for MPP from yeast (panel A), *Neurospora* (panel B), and rat (panel C). In each case, cleavage is observed only when both subunits are present.

Early findings that α-MPP had a small amount of activity by itself probably reflect contamination with sub-stoichiometric amounts of β-subunit. For *Neurospora* MPP, maximum activity is achieved with a 1:1 ratio of the subunits (Arretz et al., 1994); for yeast and rat enzymes, the purified or reconstituted enzyme is clearly a heterodimer by size determinations (Geli et al., 1990; Saavedra-Alanis et al., 1994). As early reported for crude preparations, the purified enzyme has a wide substrate specificity, cleaving precursors of several mitochondrial proteins, although the number tested with the purified enzymes remains small. Synthetic peptides corresponding to leader peptides bind to purified MPP and competitively inhibit precursor cleavage (Yang et al., 1991; Arretz et al., 1994; Niidome et al., 1994; Ou et al., 1994). If the synthetic peptide contains a cleavage site, it can be a substrate itself (Yang et al., 1991; Niidome et al., 1994). Estimates of the kinetic constants with natural substrates for the yeast (Yang et al., 1991) and *Neurospora* (Arretz et al., 1994) enzymes yield k_{cat} values of 0.3–6.75 min^{-1} and K_M values in the 0.1–1.0 μM range, respectively, although these will likely vary with the substrate tested. For synthetic peptide substrates, purified bovine MPP has similar values of K_M (0.7–1.2 μM), but higher ones of k_{cat} (up to 30 min^{-1}) (Niidome et al., 1994). All of the purified enzymes require divalent cations for maximum activity, and EDTA inhibits the reaction. No systematic investigation of the specificity of the metal ion requirement has been reported, although it has been shown that purified yeast MPP does not contain stoichiometric amounts of any metal (Yang et al., 1991). The yeast enzyme is routinely assayed with a mixture of Zn^{2+} and Co^{2+} (Yang et al., 1991), while the *Neurospora* and rat enzymes are assayed with Mn^{2+} (Kleiber et al., 1990; Arretz et al., 1994). Although the potato enzyme complex does not normally require a metal, its MPP activity can be inhibited by EDTA, and this inhibition can be reversed by excess Mn^{2+} or Zn^{2+} (Braun et al., 1992). Other characteristics, such as the sulfhydryl reagent sensitivity of *Neurospora* α-MPP (Schneider et al., 1990) and the insensitivity to high ionic strength of the potato complex (Emmermann et al., 1993), seem to be species-specific and may be unrelated to the reaction mechanism. Initial data relevant to the active site came from site-directed mutagenesis of the related enzyme, *E. coli* protease III (Becker and Roth, 1992, 1993). As mentioned previously, the histidine residues in the

H-X-X-E-H motif are required for Zn^{2+} binding; a glutamate 76 residues downstream may also participate in binding. The glutamate within the motif is required for catalytic activity, suggesting that it may have a role in the reaction analogous to the glutamate in the thermolysin (inverted) motif (Becker and Roth, 1992). More recently, both subunits of rat MPP have been subjected to site-directed mutagenesis in the putative metal-binding site (Table 3) and the acidic motif (Table 2; Kalousek et al., 1994; Kitada et al., 1995; Striebel et al., 1996). As expected, changes in the histidine and glutamate residues in the metal-binding motif in the β-subunit had major effects on activity, although a glutamate to aspartate change was well tolerated. In contrast, similar changes in the α-subunit were generally less disruptive, although here the Glu→Asp mutation strongly inhibited (Striebel et al., 1996). Likewise, in the acidic motif, replacement of the acidic residues or disruptions of their spacing in the β-subunit reduced MPP activity, while most of the analogous changes in the α-subunit had little effect (Striebel et al., 1996). The significance of this apparent functional asymmetry between the subunits in the mechanism or overall function of MPP remains unclear.

Finally, the sum of the genetic and biochemical data on MPP from all sources strongly suggests that this enzyme is alone responsible for the post-translocational cleavage of most mitochondrial precursors, although some are cleaved further by another enzyme (see following). The yeast *mas/mif1-2* strains are defective in proteolysis of a variety of different precursors, both once- and twice-cleaved. Each of these genes is essential in yeast, indicating that no additional cleavage activity with overlapping specificity exists (Yaffe and Schatz, 1984). The purified enzyme from each organism appears to account for all of the mitochondrial processing activity, based on specific activities, and amino acid sequencing of the subunits produces a unique sequence, rather than a mixture. Antibody cross-reactivity and low-stringency DNA hybridization have not identified any other related protein or DNA species. Therefore, it seems likely that the wide range of substrate specificity and selectivity demonstrated in mitochondrial precursor proteolysis are intrinsic to the structure and catalytic mechanism of MPP itself. More extensive site-directed mutagenesis and three-dimensional structure analysis will be required to fully appreciate the complexities of this reaction.

III. MITOCHONDRIAL INTERMEDIATE PEPTIDASE (MIP)

Initial observations of the processing of pOTC by rat liver mitochondria or crude mitochondrial extracts indicated that an additional protein species, intermediate in migration on SDS/PAGE between pOTC and mature OTC, was produced during the reaction, although its relevancy to the overall processing pathway was unclear at that time (Mori et al., 1980; Conboy and Rosenberg, 1981). Other precursors were apparently cleaved twice as well, including those of yeast cytochrome b_2 (Guiard, 1985), cytochrome c_1 (Sadler et al., 1984), cytochrome *c* peroxidase (Kaput et al., 1982), and the Fe/S protein of ubiquinol:cytochrome *c* reductase of

Neurospora and yeast (Hartl et al., 1986; Fu et al., 1990), although the first three have since been shown to follow a more specialized route (see following). The demonstrations that the Fe/S and OTC intermediate proteins (iFe/S and iOTC) could be chased into the corresponding mature subunits with a precursor/product relationship were accompanied by the determination of their amino acid sequences (Hartl et al., 1986; Sztul et al., 1987). Each was shown to contain the final (carboxyl-terminal) eight residues of the respective precursors attached to the mature sequences. The chase experiments, the precise nature of the cleavages, and the fact that the intermediate forms could be observed *in vivo* under certain conditions (Hartl et al., 1986; Isaya et al., 1988) were strong indications that these species were true intermediates in the maturation pathway. The activity carrying out the proteolytic removal of the final eight residues of the leader peptides was designated *M*itochondrial *I*ntermediate *P*eptidase (MIP; EC 3.4.24.93D).

The surveys of cleavage sites in mitochondrial precursors mentioned in the MPP section above (Hendrick et al., 1989; von Heijne et al., 1989; Gavel and von Heijne, 1990) revealed that there was a motif in the carboxyl-terminus of a number of precursors, including those such as pOTC already shown to be twice-cleaved, that identified them as putative substrates of MIP. The motif consisted of an Arg at -10, a bulky hydrophobic side-chain at -8, and a small hydrophilic side-chain at -5, all relative to the mature amino-terminus. MPP cleavage occurred (or was predicted to occur) between -9 and -8, obeying the "Arg at -2" rule (see preceding), and leaving an amino-terminal octapeptide to be removed by MIP. In both surveys, almost 40% of leader peptides displayed this motif, although the number of these clearly proven to be cleaved twice was small. In addition, there was some potential duplication because homologous precursors from different species were included. On the other hand, the fact that the motif was conserved between species as divergent as yeast and man (in cytochrome oxidase subunit IV, for example), even though precise amino acid conservation was not maintained, suggested that the requirement for two cleavages might reflect a more general functional importance.

Definitive evidence for MIP function has come from its complete purification, cloning, and sequencing. Only the rat liver enzyme has been purified to homogeneity so far, starting from freshly prepared mitochondrial matrix (Kalousek et al., 1992). MIP is labile in crude extracts, likely accounting for the difficulties investigators have had recovering this activity in other organisms. MIP co-purified with MPP through much of the procedure, but was separated completely in a final step (Kalousek et al., 1992). The final preparation contained three proteins of 75kDa, 47kDa, and 28kDa on SDS/PAGE, in a ratio of approximately 5:1:1. Amino acid sequencing of each showed that the two smaller ones were proteolytic fragments of the largest (Kalousek et al., 1992). On gel filtration chromatography, MIP migrates with an apparent size of 75kDa, indicating that the enzyme functions as a monomer. Whether the proteolysis has any functional significance remains to be determined.

The enzymatic characteristics of MIP have been defined largely with the rat liver enzyme. The enzyme carries out the second cleavage of pFe/S (*Neurospora*), pMDH (rat), and pOTC (human), but only after the corresponding intermediate forms have been generated by MPP (Isaya et al., 1992a; Kalousek et al., 1992). It is inactive against once-cleaved precursors. An artificial intermediate, generated by removing most of the leader coding sequence and placing a methionine at position -8 to generate Met-iOTC, was cleaved specifically by MIP, without requiring the presence of MPP, establishing MIP as an independent protease (Isaya et al., 1991). The divalent cation dependence of MIP was studied using Met-iOTC translation mixtures, dialyzed to remove Mg^{2+} and other potentially interfering cations (Kalousek et al., 1992). Surprisingly, Mg^{2+} was as effective as Mn^{2+} in stimulating MIP activity; Ca^{2+} also was stimulatory, but to a lesser extent. Other cations (Co^{2+}, Fe^{2+}, and Zn^{2+}) did not support activity and even appeared to inhibit the minimal activity present without added metal ion. This specificity had been obscured previously by the use of undialyzed translation mixtures as a source of precursor substrates and by the need for MPP activity, which may have a different metal ion requirement (see preceding), to generate the intermediate species. As expected from earlier studies, the purified enzyme was inhibited reversibly by EDTA and completely by *N*-ethyl maleimide (NEM); other sulfhydryl group inhibitors, such as *p*-hydroxymercuribenzoate and iodoacetamide, also inhibited, although less strongly (Kalousek et al., 1992). Certain class-specific protease inhibitors, including pepstatin, N^{α}–tosyllysyl chloromethyl ketone, and bestatin, also inhibited MIP activity. This pattern suggested that MIP might belong to the family of cysteine proteases, although these enzymes generally do not require divalent cations for activity (see following).

The substrate specificity of rat MIP has been studied in attempts to better define its biologic function. The experiments of Isaya and her colleagues, referred to in the MPP section, using chimeric and artificial leader peptides and precursors, showed that MIP recognized specific amino-terminal octapeptides and cleaved them, while not acting on others with similar amino acid compositions (Isaya et al., 1991). Surprisingly, one artificial octapeptide, corresponding to the eight amino-terminal residues of mature methylmalonyl-CoA mutase, a once-cleaved enzyme, was cleaved when substituted for the OTC octapeptide (Isaya et al., 1991). In addition, an intermediate with a nonamer at its amino-terminus, formed by placing an initiator Met before the OTC octapeptide, was not a substrate for MIP despite the presence of the entire octapeptide and the OTC mature sequences (Isaya et al., 1992a). These results, and others with partial deletions of the OTC octapeptide (Isaya et al., 1991), suggest that MIP recognizes the context of the octapeptide, both its position directly at the amino-terminus and the residues immediately carboxyl-terminal to its cleavage site. This hypothesis was strengthened by experiments placing the octapeptide from Met-iOTC in front of mature sequences of rat MDH and *Neurospora* Fe/S, both twice-cleaved, and yeast F_1-β, a once-cleaved protein (Isaya et al., 1992a). The first two were cleaved to their normal amino-termini by

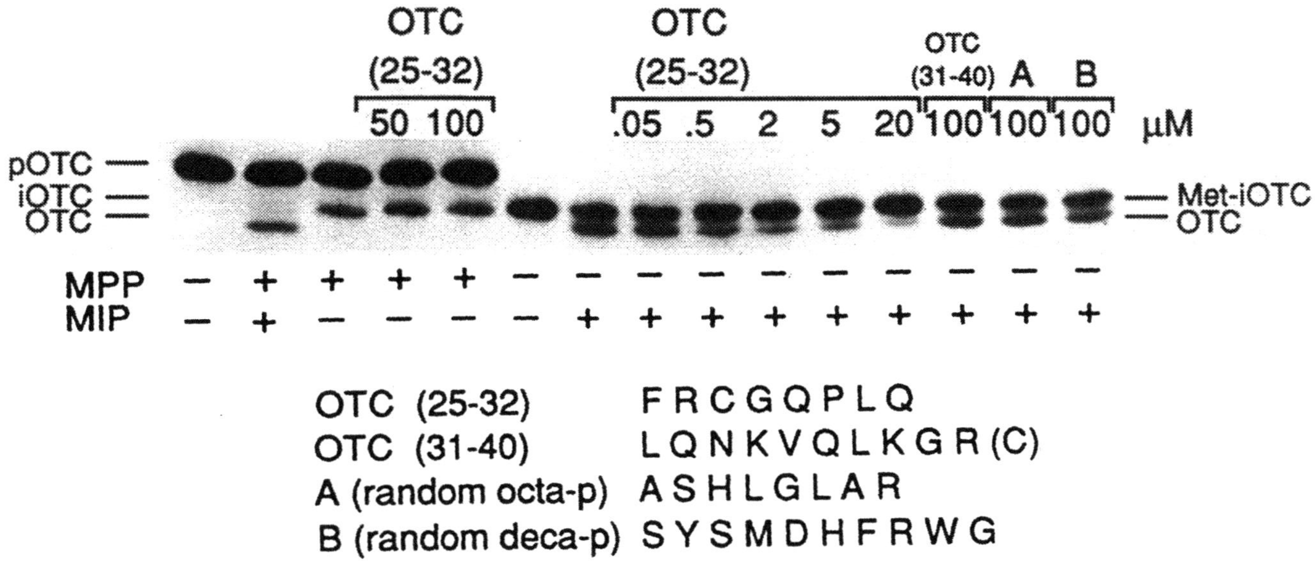

Figure 2. The OTC octapeptide inhibits MIP without affecting MPP activity. Radiolabelled pOTC was incubated with purified MPP plus MIP, or MPP alone; radiolabelled Met-iOTC was incubated with purified MIP; differing concentrations of several peptides were added at the start of the processing reactions, as indicated; products were analyzed by SDS/PAGE. OTC(25–32) corresponds to the OTC octapeptide specifically cleaved by MIP; OTC(31–40) is an adjacent decamer; two random peptides, A and B, were also used; the amino acid sequences of these peptides are indicated in the Figure. Markers indicate the positions of pOTC, iOTC, OTC, and Met-iOTC. (Reproduced from Isaya et al., 1992a, by permission.)

Table 4. Inhibition of MIP by Synthetic Peptides

	Peptide Concentration (μM)							
Peptide	*0*	*0.05*	*0.5*	*2*	*5*	*20*	*50*	*100*
OTC(25–32)	100	93 ± 15	61 ± 12	47 ± 1	36 ± 1	3 ± 2		
OTC(25^Y–32)	100				73 ± 15	71 ± 15	47 ± 22	12 ± 18
OTC(26–32)	100				89 ± 4	95 ± 7	94 ± 12	86 ± 9
OTC(25–31)	100				92 ± 4	74 ± 18	43 ± 26	19 ± 8
OTC(31–40)	100							77 ± 10
Random octapep	100							95 ± 10
Random decapep	100							86 ± 14

Notes: Labeled Met-iOTC was incubated with purified MIP in the presence of the indicated peptides at different concentrations. Conversion to mature OTC was determined by densitometry of fluorograms after SDS/PAGE. The amount of mature OTC formed in the presence of the peptides is expressed as the percent of that formed in their absence. The numbering of the peptides reflects the residues of the OTC precursor included; OTC(25^Y–32) designates an octamer with tyrosine (Y) substituted for phenylalanine at the amino-terminus of the octapeptide, position 25 in pOTC.

Source: Adapted from Isaya et al. (1992a) with permission.

MIP (without MPP), while the last one was not digested by MIP. Thus, the Rx↓ΦxxSxxxx↓ pattern (Hendrick et al., 1989), while likely forming an optimized MPP site (Isaya et al., 1991), also defines in its last eight residues part of the MIP site, the rest of which extends for a presently unknown distance into the mature sequence.

Inhibition studies with synthetic peptides derived from these sequences have provided additional information. An undecamer, consisting of the OTC octapeptide, the first two amino-terminal residues of mature OTC, and a carboxyl-terminal cysteine was an effective inhibitor of MIP (Isaya et al., 1992a). A second undecamer, containing the last two residues of the octapeptide, the eight amino-terminal residues of mature OTC, and a cysteine, gave little inhibition, even at 100 μM, although it contains both sides of the site normally cleaved by MIP. An octamer identical to the eight amino-terminal residues of iOTC also was an effective inhibitor of MIP cleavage of both iOTC and iMDH, produced by an initial MPP cleavage of the respective precursors, and of Met-iOTC, synthesized directly by *in vitro* translation, requiring a concentration of ~2 μM to give 50% inhibition (Figure 2). A heptapeptide consisting of residues 25–31 of pOTC (the seven amino-terminal residues of the octapeptide) was less effective than the octamer, requiring 20–50 μM concentration for 50% inhibition, while a heptamer of residues 26–32 did not inhibit at 100 μM (Table 4). The difference between these two is that the amino-terminal residue of the first is Phe, the conserved hydrophobic residue in the recognition site motif, while the second has an Arg at that position. The role of this residue was further emphasized by testing an octamer with Tyr substituted for Phe at the amino-terminus. This peptide was about 10-fold less effective in inhibiting MIP, requiring 20 μM for 50% inhibition (Table 4).

Interestingly, Tyr is not found at the amino-terminus of the octapeptide in known or predicted twice-cleaved precursors. Whether the change in hydrophobicity or the change in size of the side-chain, both resulting from the additional hydroxyl group, is responsible for this effect is not known. Another natural octapeptide, that from *Neurospora* Fe/S, was also tested and found to be less effective than the OTC octapeptide in inhibiting both Met-iOTC cleavage and iFe/S cleavage (Isaya et al., 1992a). The amino-terminus of this peptide is Leu, which is found at this position in many twice-cleaved precursors. All of these data suggest that MIP recognizes amino-terminal octapeptides that conform to the motif indicated previously, but only cleaves them within the structural context of a twice-cleaved mitochondrial protein precursor.

Rat MIP has been cloned by degenerate oligonucleotide-directed PCR amplification of mRNA and library screening (Isaya et al., 1992b). The predicted protein sequence includes a 33-residue leader peptide and has a typical "Arg at -2" cleavage site, suggesting that it is processed by MPP (Isaya et al., 1992b). Indeed, purified MPP carries out this cleavage on *in vitro* translated pMIP. The mature sequence is characterized by the presence of 16 cysteine residues and a typical, thermolysin-like, Zn^{2+}-binding motif (H-E-X-X-H) toward the carboxy-terminus (residues 492–496; Isaya et al., 1992b). Note that this is the inverse of the MPP metal ion-binding motif (see MPP section). Database searches with this sequence revealed two similarities. The first was to the predicted protein sequence from an unassigned open reading frame on yeast chromosome III, known as YCL57w (Oliver and et al., 1992). YCL57w was predicted to be about the same size (78 kDa) as MIP, was 24% identical (46% similar) to MIP, and appeared to have an amino-terminal mitochondrial leader peptide with a consensus MPP cleavage site. A second similarity was with a subclass of zinc metallopeptidases including bacterial dipeptidyl carboxypeptidases (Hamilton and Miller, 1992), oligopeptidase A (*S. typhimurium*; Conlin and Miller, 1992), and metalloendopeptidase EC 3.4.24.15 (EP24.15) from rat (Pierotti et al., 1990), with 20–25% identity and 44–47% similarity overall; YCL57w also fit into this group. The region with the greatest similarity is toward the carboxy-terminus in about 200 amino acids surrounding the H-E-X-X-H motif (Isaya et al., 1992b). The significance of this homology is not clear, given the diversity of substrates recognized by these enzymes. In addition, the other members of the family are clearly Zn^{2+}-dependent peptidases, while MIP is inhibited by this cation and requires Mn^{2+} or Mg^{2+} instead (Kalousek et al., 1992). The third metal ion-binding residue (in addition to the two His residues in the motif), usually present in metallopeptidases, has not been identified in any members of this family; its position and nature may account for the differences in metal ion requirements. EP24.15 is inhibited by sulhydryl reagents (Pierotti et al., 1990), as is MIP, but no information is available on the others, however, no conserved cysteine residues are apparent in sequence alignments (Isaya et al., 1995).

The yeast (*S. cerevisiae*) MIP gene (*MIP1*) has been cloned by reduced stringency library screening using the rat cDNA and a degenerate oligonucleotide corresponding to the conserved metal-binding motif (Isaya et al., 1994). The predicted amino acid sequence of 772 residues (85 kDa) contains a potential leader peptide of 37 amino acids with a typical MPP cleavage site. Overall, the rat and yeast proteins are 31% identical, 54% similar; a region of 16 residues around the putative metal binding site is absolutely conserved. As expected, YCL57w also shows homology to MIP1, but the conservation around the binding site is less extensive. Chromosomal disruption mutants of *YCL57w* showed no apparent biochemical phenotype, while disruptions of *MIP1* generated strains unable to grow on nonfermentable substrates (Isaya et al., 1994). This respiration-deficient phenotype was rescued by a plasmid encoding MIP1. Notably, *MIP1* is not essential in yeast, in contrast to the MPP subunit genes (see above). The biochemical effects in *mip1* mutants included complete absence of cytochrome oxidase and succinate-cytochrome *c* reductase activities, 90% reduction of succinate dehydrogenase, 72% reduction of NADH-cytochrome *c* reductase, and even a 66% reduction of NADH dehydrogenase (Isaya et al., 1994). These deficiencies reflect the inability of *mip1* mitochondria to fully process certain proteins of the respiratory chain to their mature forms. Two of these, the Fe/S protein of cytochrome *c* reductase and subunit IV of cytochrome oxidase (COXIV) accumulated in *mip1* mitochondria as intermediates, iFe/S and iCOXIV, while maturation of cytochrome c_1, carried out by MPP and a different second peptidase (see below) proceeded normally. Extracts of *mip1* mitochondria were unable to cleave iOTC or *Neurospora* iFe/S (produced by MPP cleavage) to their mature forms; likewise, these extracts were unable to cleave Met-iOTC. On the other hand, the MPP-directed cleavages of pOTC and pFe/S to intermediates was unaffected, as was the ability of the extracts to process pF_1-β, a once-cleaved protein. As expected from the studies of mutants in MPP (above), MIP1 was not required for import of its potential substrates. MIP1 precursor itself was imported and processed in the usual manner, and the mature protein appeared to localize largely to the soluble (matrix) fraction (Isaya et al., 1994).

The pleiotropic effects of the *mip1* mutants on respiration can be explained based on the functions of known twice-cleaved proteins. The Fe/S protein is an integral part of complex III of the respiratory chain, ubiquinol:cytochrome *c* reductase, required for oxidation of both FADH- and NADH-linked substrates, including succinate. Likewise, COXIV is an important subunit of complex IV, cytochrome oxidase, so that its incomplete processing impairs the activity of this part of the respiratory chain. Interestingly, the leader peptide of the flavoprotein of complex II, succinate dehydrogenase, has a typical MIP cleavage site, suggesting that it is normally twice-cleaved (Schulke et al., 1992); a defect here would thus account for the reduced succinate dehydrogenase activity of the *mip1* extracts. Other secondary effects on the physical structure of the respiratory chain might also occur in *mip1* mutants, as suggested by the apparent inability of iCOXIV to assemble into the inner membrane location normally occupied by mature COXIV (Isaya et al., 1994).

A further explanation for the pleiotropic effects of *mip1* can be found in the observation by Branda and Isaya (1995) that a number of proteins involved in the function of the mitochondrial genome, including its replication, recombination, and repair, are known or likely to be substrates for MIP1. Thus, mutations in this nuclear gene may lead to instability and dysfunction of the mitochondrial genome, resulting in turn in defects in respiratory function. On the other hand, it is clear that none of the essential functions in yeast mitochondria depend on MIP1 for processing. This agrees with observations that all of the essential mitochondrial proteins examined to date have precursors known or predicted to be cleaved once by MPP. Nevertheless, the respiratory deficiency produced by MIP1 defects shows clearly the important role played by this protease in normal mitochondrial metabolism.

Recently, MIP has been cloned from a third organism, the fungus *Schizophyllum commune*, by reduced-stringency library screening with rat MIP sequences (Isaya et al., 1995). Pairwise comparisons of the three predicted amino acid sequences show 31–37% identity, 54–57% similarity. Other members of the putative family described previously are likewise similar to the new sequence, as expected, and database searches have uncovered two new members, porcine soluble angiotensin-binding protein (Sugiura et al., 1992) and microsomal endopeptidase from rabbit liver (Kawabata et al., 1993). A region of 40–45 amino acids toward the carboxyl-terminus is particularly well-conserved, generating the consensus: F-H-E-X-G-H-X-X-H-X_{12}-G-X_5-D-X-X-E-X-P-S-X-X-X-E. This includes the zinc-binding motif described earlier and likely constitutes at least part of the active site of these enzymes. Details of the roles of any of these residues in metal binding and the catalytic mechanism must await mutagenesis and further biochemical studies.

IV. INNER MEMBRANE PEPTIDASE (IMP)

As noted in the previous section several proteins (cytochrome b_2, cytochrome c_1, and cytochrome *c* peroxidase) that underwent two cleavages during mitochondrial processing were not substrates for MIP, as first suggested by the observation that they were cleaved by a peptidase that localized to the inner membrane (Daum et al., 1982; Gasser et al., 1982). When amino acid sequences of their leader peptides became available, they appeared to contain a bipartite targeting signal (Kaput et al., 1982; Sadler et al., 1984; Guiard, 1985). The amino-terminus was a typical mitochondrial targeting sequence, with an identifiable MPP cleavage site; this was followed by a hydrophobic sequence reminiscent of a bacterial signal peptide and a cleavage site similar to that recognized by *E. coli* signal peptidase (Guiard, 1985). Thus, these proteins seem to follow a specialized import pathway—a mitochondrial targeting sequence brings them into the organelle and is then removed by MPP to reveal a membrane-insertion sequence, which in turn targets the proteins to or across the inner membrane and is itself cleaved (van Loon et al., 1986, 1987). At present, these are the only three proteins known to follow this pathway, although there may be others (Haas and Strauss, 1990; see the chapters by Stuart et al. and

Rospert for more details). A fourth protein in yeast, subunit II of cytochrome oxidase (COXII), uses part of this system. It is encoded by the mitochondrial genome and is synthesized within the mitochondria with an amino-terminal signal peptide that is cleaved following insertion into the inner membrane (Pratje et al., 1983).

The genes in yeast responsible for this activity, called *I*nner *M*embrane *P*rotease (IMP), have been isolated by taking advantage of two mutant strains. One, designated *pet ts2858*, accumulated both iCytb$_2$ and pCOXII, suggesting a deficiency of IMP activity (Behrens et al., 1991). The wild-type gene (*IMP1*) was cloned by complementation and shown to encode a 21.4 kDa protein with 31% homology in its amino-terminal region to *E. coli* signal peptidase. A second strain, *JN109*, carried a mutation that resulted in accumulation of an intermediate form of cytochrome c_1 (iCytc$_1$), as well as defective maturation of pCOXII and iCytb$_2$ (Nunnari et al., 1993). The gene (designated *IMP2*), cloned by complementation, predicted a protein of 177 amino acid residues (19 kDa), different from IMP1. Overall, IMP2 shows 25% amino acid identity (44% similarity) with IMP1. Several regions of highest homology also show homology to bacterial signal peptidases. Because both accumulated intermediate species were properly localized to the inner membrane and exposed to the intermembrane space in *imp2* mutants, it was concluded that IMP activity was required for maturation, not membrane transport or localization, analogous to the function of MPP and MIP (see preceding). Both IMP1 and IMP2 were localized to the outer face of the inner membrane (Schneider et al., 1991; Nunnari et al., 1993); IMP1 was not cleaved during import (Schneider et al., 1994). Interestingly, in *imp2* disruption mutants, IMP1 appeared to be unstable, suggesting an interaction. Furthermore, velocity gradient centrifugation of extracts and co-immunoprecipitation experiments showed that the two proteins formed a complex (Nunnari et al., 1993). Recently, partial purification of IMP (based on conversion of iCytb$_2$) has resulted in the recovery of both proteins, confirming that they form a heterodimer with nonoverlapping substrate specificities (Schneider et al., 1994). In contrast to disruption mutants, when the active site Ser in IMP2 was changed to Ala, only iCytc$_1$ accumulated, and IMP1 was present in normal amounts, indicating that the role of IMP2 in proteolysis of pCOXII and iCytb$_2$ is to stabilize IMP1 and accounting for the pleiotropic effects of disruption of *IMP2* (Nunnari et al., 1993). Data are not yet available to indicate which of these proteins is specifically responsible for the final step of cytochrome *c* peroxidase maturation, or if another IMP activity is required.

V. OTHER PEPTIDASES

Although the three peptidases described previously appear to account for almost all of the proteolytic processing of imported mitochondrial proteins, it seems possible that other specialized proteases may exist, particularly for unusual substrates or different routes of import. For example, the so-called hinge protein of ubiquinol:cytochrome *c* reductase (subunit VI in yeast) is extremely acidic in overall composition and has a predicted amino-terminus that is likewise acidic, in both yeast and human cDNAs (Van Loon et al., 1984; Ohta et al., 1987). The details of its import and cleavage are unknown at this time, but it seems unlikely that one of the enzymes described earlier is responsible for its proteolysis. As sequences of other unusual mitochondrial proteins become available, other exceptional cleavage sites and specificities may be recognized.

VI. SUMMARY

Thus, the pattern of proteolytic maturation of imported mitochondrial proteins involves a hierarchy of cleavages by a limited number of mitochondrial peptidases. Most cleaved precursors, whether ultimately destined for the matrix, the inner membrane, or the intermembrane space, are acted on by MPP in its role as the general mitochondrial peptidase. A major subset of these, eventually localizing to either the matrix or the inner membrane, is cleaved specifically by MIP only after MPP has exposed a suitable octapeptide at the amino-terminus of the intermediate. A few others, targeted to the intermembrane space (or that face of the inner membrane) by exposed sequences reminiscent of bacterial signal peptides, are cleaved by a localized protease, IMP, after the second targeting step is complete. In all cases examined, the proteolytic steps are not required for transport, but serve to generate mature amino-termini that permit protein folding, membrane insertion, and/or macromolecular complex assembly in order to produce the active enzymes or functional structures of mitochondria.

ACKNOWLEDGMENT

This work was supported by a grant (DK09527) from the NIH.

REFERENCES

Affholter, J. A., Fried, V. A., & Roth, R. A. (1988). Human insulin-degrading enzyme shares structural and functional homologies with *E. coli* protease III. Science 242, 1415–1418.

Arretz, M., Schneider, H., Wienhues, U., & Neupert, W. (1991). Processing of mitochondrial precursor proteins. Biomed. Biochim. Acta 50, 403–412.

Arretz, M., Schneider, H., Guiard, B., Brunner, M., & Neupert, W. (1994). Characterization of the mitochondrial processing peptidase of *Neurospora crassa*. J. Biol. Chem. 269, 4959–4967.

Becker, A. B., & Roth, R. A. (1992). An unusual active site identified in a family of zinc metalloendopeptidases. Proc. Natl. Acad. Sci. USA 89, 3835–3839.

Becker, A. B., & Roth, R. A. (1993). Identification of glutamate-169 as the third zinc-binding residue in proteinase III, a member of the family of insulin-degrading enzymes. Biochem. J. 292, 137–142.

Behrens, M., Michaelis, G., & Pratje, E. (1991). Mitochondrial inner membrane protease I of *Saccharomyces cerevisiae* shows sequence similarity to the *Escherichia coli* leader peptidase. Mol. Gen. Genet. 228, 167–176.

Böhni, P., Gasser, S., Leaver, C., & Schatz, G. (1980). A matrix-localized mitochondrial protease processing cytoplasmically-made precursors to mitochondrial proteins. In: *The Organization and Expression of the Mitochondrial Genome* (Kroon, A. M., & Saccone, C., Eds.). Elsevier/North-Holland, Amsterdam, pp. 423–433.

Branda, S. S., & Isaya, G. (1995). Prediction and identification of new natural substrates of the yeast mitochondrial intermediate peptidase. J. Biol. Chem. 270, 27366–27373.

Braun, H.-P., Emmermann, M., Kruft, V., & Schmitz, U. K. (1992). The general mitochondrial processing peptidase from potato is an integral part of cytochrome c reductase of the respiratory chain. EMBO J. 11, 3219–3227.

Chu, T. W., Eftime, R., Sztul, E., & Strauss, A. W. (1989). Synthetic transit peptides inhibit import and processing of mitochondrial precursor proteins. J. Biol. Chem. 264, 9552–9558.

Conboy, J. G., & Rosenberg, L. E. (1981). Posttranslational uptake and processing of *in vitro* synthesized ornithine transcarbamoylase precursor by isolated rat liver mitochondria. Proc. Natl. Acad. Sci. USA 78, 3073–3077.

Conboy, J. G., Kalousek, F., & Rosenberg, L. E. (1979). *In vitro* synthesis of a putative precursor of mitochondrial ornithine transcarbamoylase. Proc. Natl. Acad. Sci. USA 76, 5724–5727.

Conboy, J. G., Fenton, W. A., & Rosenberg, L. E. (1982). Processing of pre-ornithine transcarbamylase requires a zinc-dependent protease localized to the mitochondrial matrix. Biochem. Biophys. Res. Commun. 105, 1–7.

Conlin, C. A., & Miller, C. G. (1992). Cloning and nucleotide sequence of *opdA*, the gene encoding oligopeptidase A in *Salmonella typhimurium*. J. Bacteriol. 174, 1631–1640.

Côté, C., Solioz, M., & Schatz, G. (1979). Biogenesis of the cytochrome bc_1 complex of yeast mitochondria. J. Biol. Chem. 254, 1437–1439.

Côté, C., Poirier, J., Boulet, D., Dionne, G., & Lacroix, M. (1988). Structural identity between the NH_2-terminal domain of the rat and human ornithine carbamyltransferase "targeting" sequences. J. Biol. Chem. 263, 5752–5756.

Daum, G., Gasser, S. M., & Schatz, G. (1982). Import of proteins into mitochondria: Energy-dependent, two-step processing of the intermembrane space enzyme cytochrome b_2 by isolated yeast mitochondria. J. Biol. Chem. 257, 13075–13080.

Ding, L., Becker, A. B., Suzuki, A., & Roth, R. A. (1992). Comparison of the enzymatic and biochemical properties of human insulin-degrading enzyme and *Escherichia coli* protease III. J. Biol. Chem. 267, 2414–2420.

Ebrahim, A., Hamel, F. G., Bennett, R. G., & Duckworth, W. C. (1991). Identification of the metal associated with the insulin degrading enzyme. Biochem. Biophys. Res. Commun. 181, 1398–1406.

Emmermann, M., Braun, H.-P., Arretz, M., & Schmitz, U. K. (1993). Characterization of the bifunctional cytochrome *c* reductase-processing peptidase complex from potato mitochondria. J. Biol. Chem. 268, 18936–18942.

Emmermann, M., Braun, H. P., & Schmitz, U. K. (1994). The mitochondrial processing peptidase from potato: A self-processing enzyme encoded by two differentially expressed genes. Mol. Gen. Genet. 245, 237–245.

Finch, P. W., Wilson, R. E., Brown, K., Hickson, I. D., & Emmerson, P. T. (1986). Complete nucleotide sequence of the *Escherichia coli ptr* gene encoding protease III. Nucleic Acids Res. 14, 7695–7703.

Fu, W., Japa, S., & Beattie, D. S. (1990). Import of the iron-sulfur protein of the cytochrome bc_1 complex into yeast mitochondria. J. Biol. Chem. 265, 16541–16547.

Gasser, S. M., Ohashi, A., Daum, G., Böhni, P. C., Gibson, J., Reid, G. A., Yonetani, T., & Schatz, G. (1982). Imported mitochondrial proteins cytochrome b_2 and cytochrome c_1 are processed in two steps. Proc. Natl. Acad. Sci. USA 79, 267–271.

Gavel, Y., & von Heijne, G. (1990). Cleavage-site motifs in mitochondrial targeting peptides. Prot. Eng. 4, 33–37.

Geli, V. (1993). Functional reconstitution in *Escherichia coli* of the yeast mitochondrial matrix peptidase from its two inactive subunits. Proc. Natl. Acad. Sci. USA 90, 6247–6251.

Geli, V., & Glick, B. (1990). Mitochondrial protein import. J. Bioenerg. Biomemb. 22, 725–751.

Geli, V., Yang, M., Suda, K., Lustig, A., & Schatz, G. (1990). The *MAS*-encoded processing protease of yeast mitochondria. Overproduction and characterization of its two nonidentical subunits. J. Biol. Chem. 265, 19216–19222.

Gencic, S., Schagger, H., & von Jagow, G. (1991). Core I protein of bovine ubiquinol-cytochrome-*c* reductase; an additional member of the mitochondrial-protein-processing family. Cloning of bovine core I and core II cDNAs and primary structure of the proteins. Eur. J. Biochem. 199, 123–131.

Glick, B., & Schatz, G. (1991). Import of proteins into mitochondria. Ann. Rev. Genet. 25, 21–44.

Guiard, B. (1985). Structure, expression and regulation of a nuclear gene encoding a mitochondrial protein: The yeast L(+)-actate cytochrome *c* oxidoreductase (cytochrome b_2). EMBO J. 4, 3265–3272.

Haas, R. C., & Strauss, A. W. (1990). Separate nuclear genes encode sarcomere-specific and ubiquitous human mitochondrial creatine kinase isozymes. J. Biol. Chem. 265, 6921–6927.

Hamilton, S., & Miller, C. G. (1992). Cloning and nucleotide sequence of the *Salmonella typhimurium dcp* gene encoding dipeptidyl carboxypeptidase. J. Bacteriol. 174, 1626–1630.

Harmey, M. A., & Neupert, W. (1979). Biosynthesis of mitochondrial citrate synthase in *Neurospora crassa*. FEBS Lett. 108, 385–389.

Hartl, F.-U., Schmidt, B., Wachter, E., Weiss, H., & Neupert, W. (1986). Transport into mitochondria and intramitochondrial sorting of the Fe/S protein of ubiquinol-cytochrome *c* reductase. Cell 47, 939–951.

Hartl, F.-U., Pfanner, N., Nicholson, D. W., & Neupert, W. (1989). Mitochondrial protein import. Biochim. Biophys. Acta 988, 1–45.

Hawlitschek, G., Schneider, H., Schmidt, B., Tropschug, M., Hartl, F-U., & Neupert, W. (1988). Mitochondrial protein import: Identification of processing peptidase and of PEP, a processing enhancing protein. Cell 53, 795–806.

Hendrick, J. P., Hodges, P. E., & Rosenberg, L. E. (1989). Survey of amino-terminal proteolytic cleavage sites in mitochondrial precursor proteins: Leader peptides cleaved by two matrix proteases share a three-amino acid motif. Proc. Natl. Acad. Sci. USA 86, 4056–4060.

Horwich, A. L., Kalousek, F., & Rosenberg, L. E. (1985). Arginine in the leader peptide is required for both import and proteolytic cleavage of a mitochondrial precursor. Proc. Natl. Acad. Sci. USA 82, 4930–4933.

Horwich, A. L., Kalousek, F., Fenton, W. A., Pollock, R. A., & Rosenberg, L. E. (1986). Targeting of pre-ornithine transcarbamylase to mitochondria: Definition of critical regions and residues in the leader peptide. Cell 44, 451–459.

Horwich, A. L., Kalousek, F., Fenton, W. A., Furtak, K., Pollock, R. A., & Rosenberg, L. E. (1987). The ornithine transcarbamylase leader peptide directs mitochondrial import through both its midportion structure and net positive charge. J. Cell Biol. 105, 669–677.

Hurt, E. C., Pesold-Hurt, B., Suda, K., Oppliger, W., & Schatz, G. (1985). The first twelve amino acids (less than half of the pre-sequence) of an imported mitochondrial protein can direct mouse cytosolic dihydrofolate reductase into the yeast mitochondrial matrix. EMBO J. 4, 2061–2068.

Hurt, E. C., Allison, D. S., Müller, U., & Schatz, G. (1987). Amino-terminal deletions in the presequence of an imported mitochondrial protein block the targeting function and proteolytic cleavage of the presequence at the carboxy terminus. J. Biol. Chem. 262, 1420–1424.

Isaya, G., Fenton, W. A., Hendrick, J. P., Furtak, K., Kalousek, F., & Rosenberg, L. E. (1988). Mitochondrial import and processing of mutant human ornithine transcarbamylase precursors in cultured cells. Mol. Cell. Biol. 8, 5150–5158.

Isaya, G., Kalousek, F., Fenton, W. A., & Rosenberg, L. E. (1991). Cleavage of precursors by the mitochondrial processing peptidase requires a compatible mature protein or an intermediate octapeptide. J. Cell Biol. 113, 65–76.

Isaya, G., Kalousek, F., & Rosenberg, L. E. (1992a). Amino-terminal octapeptides function as recognition signals for the mitochondrial intermediate peptidase. J. Biol. Chem. 267, 7904–7910.

Isaya, G., Kalousek, F., & Rosenberg, L. E. (1992b). Sequence analysis of rat mitochondrial intermediate peptidase: Similarity to zinc metallopeptidases and to a putative yeast homologue. Proc. Natl. Acad. Sci. USA 89, 8317–8321.

Isaya, G., Miklos, D., & Rollins, R. A. (1994). *MIP1*, a new yeast gene homologous to the rat mitochondrial intermediate peptidase gene, is required for oxidative metabolism in *Saccharomyces cerevisiae*. Mol. Cell. Biol. 14, 5603–5616.

Isaya, G., Rollins, R. A., Shen, G. P., Hanson, L. C., Ullrich, R. C., & Novotny, C. P. (1995). Mammalian mitochondrial intermediate peptidase: Gene structure of a putative homologue from the fungus *Schizophyllum commune* and relationship to bacterial and eukaryotic thiol-dependent metallo-peptidases. Genomics 28, 450–461.

Jänsch, L., Kruft, V., Schmitz, U. K., & Braun, H-P. (1995). Cytochrome *c* reductase from potato does not comprise three core proteins but contains an additional low-molecular-mass subunit. Eur. J. Biochem. 228, 878–885.

Kalousek, F., Isaya, G., & Rosenberg, L. E. (1992). Rat liver mitochondrial intermediate peptidase (MIP): Purification and initial characterization. EMBO J. 11, 2803–2809.

Kalousek, F., Neupert, W., Omura, T., Schatz, G., & Schmitz, U. K. (1993). Uniform nomenclature for the mitochondrial peptidases cleaving precursors of mitochondrial proteins. Trends Biochem. Sci. 18, 249.

Kalousek, F., Rysavy, P., & Striebel, H-M. (1994). Rat liver mitochondrial processing peptidase: site-directed mutagenesis of zinc binding motif. Mol. Biol. Cell 5, 475a.

Kaput, J., Goltz, S., & Blobel, G. (1982). Nucleotide sequence of the yeast nuclear gene for cytochrome *c* peroxidase precursor. Functional implications of the pre-sequence for protein transport into mitochondria. J. Biol. Chem. 257, 15054–15058.

Kawabata, S.-I., Nakagawa, K., Muta, T., Iwanaga, S., & Davie, E. W. (1993). Rabbit liver microsomal endopeptidase with substrate specificity for processing proproteins is structurally related to rat testes metalloendopeptidase 24.15. J. Biol. Chem. 268, 12498–12503.

Kitada, S., Shimokata, K., Niidome, T., Ogishima, T., & Ito, A. (1995). A putative metal-binding site in the β subunit of rat mitochondrial processing peptidase is essential for its catalytic activity. J. Biochem. 117, 1148–1150.

Kleiber, J., Kalousek, F., Swaroop, M., & Rosenberg, L. E. (1990). The general mitochondrial matrix processing protease from rat liver: Structural characterization of the catalytic subunit. Proc. Natl. Acad. Sci. USA 87, 7978–7982.

Kraus, J. P., Conboy, J. G., & Rosenberg, L. E. (1981). Pre-ornithine transcarbamylse: Properties of the cytoplasmic precursor of a mitochondrial matrix enzyme. J. Biol. Chem. 256, 10739–10742.

Kraus, J. P., Novotny, J., Kalousek, F., Swaroop, M., & Rosenberg, L. E. (1988). Different structures in the amino-terminal domain of the ornithine trascarbamylase leader peptide are involved in mitochondrial import and carboxyl-terminal cleavage. Proc. Natl. Acad. Sci. USA 85, 8905–8909.

Kuo, W.-L., Gehm, B. D., & Rosner, M. R. (1990). Cloning an expression of the cDNA for a *Drosophila* insulin-degrading enzyme. Mol. Endo. 90, 1580–1591.

Maccecchini, M-L., Rudin, Y., Blobel, G., & Schatz, G. (1979). Import of proteins into mitochondria: Precursor forms of the extramitochondrially made F_1-ATPase subunits in yeast. Proc. Natl. Acad. Sci. USA 76, 343–347.

McAda, P. C., & Douglas, M. G. (1982). A neutral metalloendoprotease involved in the processing of an F_1-ATPase subunit precursor in mitochondria. J. Biol. Chem. 257, 3177–3182.

Michel, R., Wachter, E., & Sebald, W. (1979). Synthesis of a larger precursor for the proteolipid subunit of the mitochondrial ATPase complex of Neurospora crassa in a cell-free wheat germ system. FEBS Lett. 101, 373–376.

Miura, S., Mori, M., Amaya, Y., & Tatibana, M. (1982). A mitochondrial protease that cleaves the precursor of ornithine carbamoyl transferase. Eur. J. Biochem. 122, 641–647.

Mooney, D. T., Pilgrim, D. B., & Young, E. T. (1990). Mutant alcohol dehydrogenase (ADH III) presequences that affect both *in vitro* mitochondrial import and *in vitro* processing by the matrix protease. Mol. Cell. Biol. 10, 2801–2808.

Mori, M., Miura, S., Tatibana, M., & Cohen, P. P. (1980). Characterization of a protease apparently involved in processing of pre-ornithine transcarbamylase of rat liver. Proc. Natl. Acad. Sci. USA 77, 7044–7048.

Niidome, T., Kitada, S., Shimokata, K., Ogishima, T., & Ito, A. (1994). Arginine residues in the extension peptide are required for cleavage of a precursor by mitochondrial processing peptidase: Demonstration using synthetic peptide as a substrate. J. Biol. Chem. 269, 24719–24722.

Nishi, T., Nagashima, F., Tanase, S., Fukumoto, Y., Joh, T., Shimada, K., Matsukado, Y., Ushio, Y., & Morino, Y. (1989). Import and processing of precursor to mitochondrial aspartate aminotransferase. Structure-function relationship of the presequence. J. Biol. Chem. 264, 6044–6051.

Nunnari, J., Fox, T. D., & Walter, P. (1993). A mitochondrial protease with two catalytic subunits of nonoverlapping specificities. Science 262, 1997–2003.

Ohta, S., Goto, K., Arai, H., & Kagawa, Y. (1987). An extremely acidic amino-terminal presequence of the precursor for the human mitochondrial hinge protein. FEBS Lett. 226, 171–175.

Oliver, S. G. et al. (1992). The complete DNA sequence of yeast chromosome III. Nature 357, 38–46.

Ou, W.-J., Ito, A., Okazaki, H., & Omura, T. (1989). Purification and characterization of a processing protease from rat liver mitochondria. EMBO J. 8, 2605–2612.

Ou, W.-J., Kumamoto, T., Mihara, K., Kitada, S., Niidome, T., Ito, A., & Omura, T. (1994). Structural requirement for recognition of the precursor proteins by the mitochondrial processing peptidase. J. Biol. Chem. 269, 24673–24678.

Oudshoorn, P., van Steeg, H., Swinkels, B. W., Schoppink, P., & Grivell, L. A. (1987). Subunit II of yeast QH_2:cytochrome-*c* oxidoreductase. Nucleotide sequence of the gene and features of the protein. Eur. J. Biochem. 163, 97–103.

Paces, V., Rosenberg, L. E., Fenton, W. A., & Kalousek, F. (1993). The β subunit of the mitochondrial processing peptidase from rat liver: Cloning and sequencing of a cDNA and comparison with a proposed family of metallopeptidases. Proc. Natl. Acad. Sci. USA 90, 5355–5358.

Pierotti, A., Dong, K.-W., Glucksman, M. J., Orlowski, M., & Roberts, J. L. (1990). Molecular cloning and primary structure of rat testes metalloendopeptidase EC 3.4.24.15. Biochemistry 29, 10323–10329.

Pollock, R. A., Hartl, F.-U., Cheng, M. Y., Ostermann, J., Horwich, A., & Neupert, W. (1988). The processing peptidase of yeast mitochondria: The two cooperating components MPP and PEP are structurally related. EMBO J. 7, 3493–3500.

Pratje, E., Mannhaupt, G., Michaelis, G., & Beyreuther, K. (1983). A nuclear mutation prevents processing of a mitochondrially encoded membrane protein in *Saccharomyces cerevisiae*. EMBO J. 2, 1049–1054.

Rawlings, N. D., & Barrett, A. J. (1991). Homologues of insulinase, a new superfamily of metalloendopeptidases. Biochem. J. 275, 389–391.

Saavedra-Alanis, V. M., Rysavy, P., Rosenberg, L. E., & Kalousek, F. (1994). Rat liver mitochondrial processing peptidase (MPP): Both α-and β- subunits are required for activity. J. Biol. Chem. 269, 9284–9288.

Sadler, I., Suda, K., Schatz, G., Kaudewitz, F., & Haid, A. (1984). Sequencing of the nuclear gene for the yeast cytochrome c_1 precursor reveals an unusually complex amino-terminal presequence. EMBO J. 3, 2137–2143.

Schmidt, B., Wachter, E., Sebald, W., & Neupert, W. (1984). Processing peptidase of *Neurospora* mitochondria. Two-step cleavage of imported ATPase subunit 9. Eur. J. Biochem. 144, 581–588.

Schneider, A., Behrens, M., Scherer, P., Pratje, E., Michaelis, G., & Schatz, G. (1991). Inner membrane protease I, an enzyme mediating intramitochondrial protein sorting in yeast. EMBO J. 10, 247–254.

Schneider, A., Oppliger, W., & Jenö, P. (1994). Purified inner membrane protease I of yeast mitochondria is a heterodimer. J. Biol. Chem. 269, 8635–8638.

Schneider, H., Arretz, M., Wachter, E., & Neupert, W. (1990). Matrix processing peptidase of mitochondria: Structure-function relationship. J. Biol. Chem. 265, 9881–9887.

Schulke, N., Blobel, G., & Pain, D. (1992). Primary structure, import and assembly of the yeast homologue of succinate dehydrogenase flavoprotein. Proc. Natl. Acad. Sci. USA 89, 8011–8015.

Schulte, U., Arretz, M., Schneier, H., Tropschug, M., Wachter, E., Neupert, W., & Weiss, H. (1989). A family of mitochondrial proteins involved in bioenergetics and biogenesis. Nature 339, 147–149.

Striebel, H.-M., Rysavy, P., Adamec, J., Spizek, J., & Kalousek, F. (1996). Mutational analysis of both subunits from rat mitochondrial processing peptidase (MPP). Submitted.

Sugiura, N., Hagiwara, H., & Hirose, S. (1992). Molecular cloning of porcine soluble angiotensin-binding protein. J. Biol. Chem. 267, 18067–18072.

Sztul, E. S., Hendrick, J. P., Kraus, J. P., Wall, D., Kalousek, F., & Rosenberg, L. E. (1987). Import of rat ornithine transcarbamylase precursor into mitochondria: Two-step processing of the leader peptide. J. Cell Biol. 105, 2631–2639.

Tzagoloff, A., Wu, M. A., & Crivellone, M. (1986). Assembly of the mitochondrial membrane system. Characterization of *COR1*, the structural gene for the 44-kilodalton core protein of yeast coenzyme QH_2-cytochrome *c* reductase. J. Biol. Chem. 261, 17163–17169.

Vallee, B. L., & Auld, D. S. (1990). Zinc coordination, function, and structure of zinc enzymes and other proteins. Biochemistry 29, 5647–5659.

van Loon, A. P. G. M., De Groot, R. J., De Haan, M., Dekker, A., & Grivell, L. A. (1984). The DNA sequence of the nuclear gene coding for the 17-kd subunit VI of the yeast ubiquinol-cytochrome *c* reductase: A protein with an extremely high content of acidic amino acids. EMBO J. 3, 1039–1043.

van Loon, A. P. G. M., Brändli, A. W., & Schatz, G. (1986). The presequences of two imported mitochondrial proteins contain information for intracellular and intramitochondrial sorting. Cell 44, 801–812.

van Loon, A. P. G. M., Brändli, A. W., Pesold-Hurt, B., Blank, D., & Schatz, G. (1987). Transport of proteins to the mitochondrial intermembrane space: The "matrix-targeting" and the "sorting" domains in the cytochrome c_1 presequence. EMBO J. 6, 2433–2439.

von Heijne, G., Steppuhn, J., & Herrmann, R. G. (1989). Domain structure of mitochondrial and chloroplast targeting peptides. Eur. J. Biochem. 180, 535–545.

Witte, C., Jensen, R. E., Yaffe, M. P., & Schatz.G (1988). *MAS1*, a gene essential for yeast mitochondrial assembly, encodes a subunit of the mitochondrial processing protease. EMBO J. 7, 1439–1447.

Yaffe, M. P., & Schatz, G. (1984). Two nuclear mutations that block mitochondrial protein import in yeast. Proc. Natl. Acad. Sci. USA 81, 4819–4823.

Yaffe, M. P., Ohta, S., & Schatz, G. (1985). A yeast mutant temperature-sensitive for mitochondrial assembly is deficient in a mitochondrial protease activity that cleaves imported precursor polypeptides. EMBO J. 4, 2069–2074.

Yang, M., Jensen, R. E., Yaffe, M. P., Oppliger, W., & Schatz, G. (1988). Import of proteins into yeast mitochondria: The purified matrix processing protease contains two subunits which are encoded by the nuclear *MAS1* and *MAS2* genes. EMBO J. 7, 3857–3862.

Yang, M., Geli, V., Oppliger, W., Suda, K., James, P., & Schatz, G. (1991). The *MAS*-encoded processing protease of yeast mitochondria. Interaction of the purified enzyme with signal peptides and purified precursor protein. J. Biol. Chem. 266, 6416–6423.

SORTING OF PROTEINS TO THE MITOCHONDRIAL INTERMEMBRANE SPACE

Rosemary A. Stuart, Heike Fölsch, Albrecht Gruhler, and Walter Neupert

I. Introduction 194
II. Submitochondrial Sorting—Which Pathway? 195
III. Import of Cytochrome *c* 196
IV. Import of Other IMS Proteins without Cleavable Presequences 197
V. Sorting of Presequence Containing Precursors to the IMS: Alternative Views . 197
VI. The Nature of IMS Sorting Signals in Bipartite Presequences 199
A. Defining the IMS Sorting Signal Requirements 199
B. Reconciling Newly Defined Sorting Signals in Terms of the Stop-Transfer Model 201
C. IMS Sorting Signals: Functional Conservation of Prokaryotic Features? . 202
VII. IMP, The Inner Membrane Protease Has Prokaryotic Origins 203
VIII. Identification of a Novel Cytochrome b_2 Sorting Intermediate 204
IX. Role of Molecular Chaperones in the Import of IMS Proteins 207
A. Requirement for mt-Hsp70/Matrix ATP for Import Across the IM 207

Advances in Molecular and Cell Biology,
Volume 17, pages 193–220.

ISBN: 0-7623-0144-9

B. IMS-targeted Preproteins Depend on Matrix ATP/mt-Hsp70 to Promote Their Unfolding Outside Mitochondria 208
C. Can the IMS Sorting Machinery Render mt-Hsp70 Function Dispensable? . . 209
D. Can Apparent Lack of mt-Hsp70 Dependence Be Used as Evidence For or Against Any One Sorting Model? 210
E. Role of Other Molecular Chaperones in the IMS Sorting Process 210
X. The Rieske FeS Protein is Sorted in a Conservative Manner 212
XI. Summary and Future Perspectives 214
References 216

I. INTRODUCTION

Mitochondria can be divided into four major subcompartments, the outer membrane (OM), the inner membrane (IM), the intermembrane space (IMS) and finally the matrix. Each of these subcompartments is characterized by its own unique repertoire of enzymes. The vast majority of these proteins are nuclear encoded and are synthesized as precursor proteins on cytoplasmic ribosomes. Maintenance of specific mitochondrial functions is dependent on the correct mitochondrial targeting and furthermore on the accurate submitochondrial sorting of these precursor proteins. Consequently precursor proteins must contain the necessary information to target them to the mitochondria and to then sort them to their correct submitochondrial location. For the majority of precursors, these signals are present in the form of amino-terminal cleavable presequences, which undergo processing during import and sorting within the organelle.

Thus, precursor proteins are targeted to mitochondria where they are specifically recognized by receptor proteins on the mitochondrial surface. Following binding to the mitochondria, they become inserted into the OM. These initial steps of import are mediated by a number of proteinaceous components of the OM, termed the mitochondrial outer membrane machinery (MOM). Further details on the composition and mechanism of operation of this outer membrane translocation machinery is reviewed in the chapter by Lill and colleagues.

Following insertion into the OM, the various submitochondrial sorting pathways begin to diverge. For instance, proteins destined for the outer membrane, for example, porin, are thought to assemble directly from the OM import site into the outer membrane. Some IMS-targeted proteins are also thought to be directly sorted to this location from the OM (Pfaller et al., 1988; Lill et al., 1992b). Hence, the import and assembly of such proteins does not require the action of downstream components of the import machinery in the IM and matrix. Most other proteins must be either inserted into or translocated across the next lipid barrier, the inner membrane. Translocation of proteins across the IM occurs at areas where this membrane is adjacent to the outer membrane. Indeed, *in vivo* almost all of the OM is in close proximity to part of the inner membrane known as the inner boundary membrane (Pfanner et al., 1992). The translocation machineries of the OM and IM can operate independently of each other; however, abundant evidence suggests that normal import across these two membranes occurs as a coupled reaction (Schleyer and Neupert, 1985; Segui-Real et al., 1993; Mayer et al., 1993). The components of the inner membrane import machinery have recently begun to be unraveled. At least three inner membrane proteins have been identified and implicated to be involved in the translocation process: MPI1 (also termed ISP45 and MIM44; Maarse et al., 1992; Scherer et al., 1992; Blom et al., 1993), MAS6 (also termed MIM23 and MPI2; Emtage and Jensen, 1993; Dekker et al., 1993), and MPI3 (also termed MIM17; Dekker et al., 1993). Whether all of these proteins form a functional complex and how they mediate translocation is to date unclear. Increasing evidence, however, indicates that the function of at least MPI1 is tightly coupled to that of mt-Hsp70 (see following; Schneider et al., 1994).

The passage of the targeting presequence across the IM in response to $\Delta\Psi$ represents the first step of translocation across the IM. Completion of translocation of the remaining portion of the precursor, in contrast, does not require $\Delta\Psi$ but does require matrix ATP hydrolysis. Hydrolysis of matrix ATP thus constitutes a part of the driving force for the vectorial movement of preproteins into the matrix (Glick et al., 1993; Stuart et al., 1994a). A decisive role for this matrix ATP in protein import appears to be through its facilitating the reversible binding of mt-Hsp70 to incoming polypeptide chains (Kang et al., 1990; Gambill et al., 1993; Stuart et al., 1994b). Mt-Hsp70 action, as will be discussed ahead, is vital for the import of proteins across the IM into the matrix. Newly imported polypeptides have to refold to attain their native conformation in order to function. In many cases, this appears to be an assisted process involving the transfer of the imported protein from mt-Hsp70 to another matrix-localized molecular chaperone, Hsp60. Folding mediated by Hsp60 is ATP-dependent and requires the action of another chaperone, Hsp10 (Rospert et al., 1993; Höhfeld and Hartl, 1994).

II. SUBMITOCHONDRIAL SORTING—WHICH PATHWAY?

Precursor proteins destined for the mitochondrial matrix are imported along the general import pathway across both the OM and IM as described previously. Those precursors destined for other mitochondrial subcompartments initiate import along this general pathway, however, become deviated off at different points depending on their final destination. This review will focus on protein sorting to the IMS compartment and will summarize the current knowledge about the pathways to the IMS, the IMS-sorting signals, and energetic requirements involved.

The sorting of precursors to the IMS probably represents the most complex situation for submitochondrial sorting and, in addition, a number of different sorting pathways have been described (Glick et al., 1992a; Stuart et al., 1993). For the purpose of simplification, we initially wish to define the term "intermembrane space localization" using a classification based on the final functional location of the protein rather than the sorting pathway used by the protein to reach this location. We use the term "intermembrane space localization" here to describe all proteins that are either soluble in this compartment or that are associated with the inner membrane but have large domains protruding into the intermembrane space, which usually contain their catalytic sites. Examples of soluble IMS proteins that will be dealt with here are cytochrome *c*, cytochrome b_2, and cytochrome *c* peroxidase (CCPO). The inner membrane associated proteins fall into three subcategories. First, there are those proteins that are peripherally associated with the IM such as cytochrome *c* heme lyase. Second, there are those that are anchored to the IM by amino-terminal sequences, hence have carboxy-terminal domains in the IMS, an example of which is the Rieske FeS protein of the cytochrome bc_1 complex. Finally, there are those proteins that are anchored to the IM via carboxy-terminal anchors, for example cytochrome c_1 of the same complex with an amino-terminal domain in the IMS.

At least three different sorting pathways to the IMS exist. The simplest is that taken by cytochrome *c*, which is imported directly across the outer membrane without displaying a dependence on the previously mentioned components of the outer membrane receptor/GIP machinery. A second pathway is used by cytochrome *c* heme lyase which, like outer membrane proteins, requires the outer membrane translocation machinery but not that of the inner membrane. These two pathways will be dealt with only briefly, as they are more extensively reviewed in the chapters by Dumont and Lill and colleagues. The third route to the IMS is a multistep pathway that involves the participation of translocation machineries of both the OM and IM and additional components and co-factors in the matrix. It will form the main focus of this review.

III. IMPORT OF CYTOCHROME *C*

The import pathway taken by cytochrome *c* is relatively simple and differs drastically from that taken by all other mitochondrial proteins (Stuart and Neupert, 1990). Cytochrome *c* is synthesized as a precursor protein, apocytochrome *c*, which does not contain a cleavable targeting sequence and differs from its mature counterpart in that it lacks the covalently bound heme group (Stuart et al., 1987). Apocytochrome *c* is imported into mitochondria in a manner that does not use surface receptors, the OM import machinery, and energy sources used by other precursors, namely ATP hydrolysis (both cytosolic and matrix) and a membrane potential (Nicholson et al., 1988; Stuart et al., 1989). Apocytochrome *c* appears to spontaneously insert into the outer mitochondrial membrane, where it is recognized and bound by cytochrome *c* heme lyase (CCHL; Dumont et al., 1988; Drygas et al., 1989). The driving force for the import is thought to be provided by the folding of cytochrome *c* following covalent heme addition, a reaction catalyzed by CCHL and which requires NADH and flavins. Apocytochrome *c* interacts with CCHL following partial or complete translocation across the outer membrane (Nicholson et al.,

1988; Dumont et al., 1991). Heme addition and subsequent folding finally traps cytochrome *c* in the IMS.

IV. IMPORT OF OTHER IMS PROTEINS WITHOUT CLEAVABLE PRESEQUENCES

This category includes those IMS proteins that are peripherally associated with the IM, the best characterized example being cytochrome *c* heme lyase (CCHL). Like cytochrome *c*, CCHL is synthesized as a precursor form that does not contain an amino-terminal cleavable extension sequence. Furthermore, the import of CCHL across the outer membrane does not display apparent energetic requirements such as ATP hydrolysis or a membrane potential (Lill et al., 1992a, 1992b). In contrast to apocytochrome *c* and in common with the majority of mitochondrial precursors, CCHL uses components of the outer membrane import machinery for mitochondrial binding and OM translocation. Current results give no indication that components of the inner membrane are required for translocation into the IMS. The nature of the driving force for import is unknown and again it is speculated to be connected with folding of the polypeptide chain upon emergence into the IMS (Lill et al., 1992b).

An increasing number of IMS proteins whose precursors do not have cleavable presequences are currently being identified. All these proteins are peripherally associated with the IM. In addition to CCHL, other members of this category most likely include cytochrome c_1 heme lyase (CC_1HL) (A. Haid pers. commun.) and the two presequence processing enzymes of the inner membrane discussed ahead, inner membrane protease 1 and 2 (Imp1p and Imp2p; Schneider et al., 1991; Nunnari et al., 1993). Whether all these precursors follow the same import pathway as CCHL awaits further characterization of their import. Preliminary evidence, however, suggests that this is indeed the case for CC_1HL (Steiner et al., 1995).

V. SORTING OF PRESEQUENCE CONTAINING PRECURSORS TO THE IMS: ALTERNATIVE VIEWS

The third category of IMS includes all those proteins synthesized with cleavable amino-terminal presequences and includes a vast array of precursor proteins encompassing soluble N-terminal and C-terminal anchored intermembrane space proteins (Table 1). Whether all of these proteins are imported by a common mechanism awaits further investigation.

The presequences of most of these precursors can be subdivided into two distinct segments. The amino-terminal portion of the presequence is a typical mitochondrial-targeting sequence found in most preproteins to be transported into the mitochondrial matrix. This is followed by the IMS sorting signal at the carboxy-terminal end. This sorting signal spans approximately 50 amino acids and consists of a hydrophilic sequence that precedes a relatively hydrophobic segment (Sadler

Table 1. Some IMS Localized Proteins

Protein	IMS topology	IMS targeting signal	MPP	IMS signal processing
cyt. c	Soluble	Internal	No	No
cyt. b_2	Soluble	Bipartite	Yes	Yes, Imp1p
CCPO	Soluble	Bipartite?	n.d.	Yes, n.d.
CCHL	IM peripheral	Internal	No	No
CC_1HL	IM peripheral	Internal	No	No
Cyt. c_1	IM C-anchored	Bipartite	Yes	Yes, Imp2p
FeS	IM N-anchored	Internal	Yes	Yes, MIP
Sco1p	IM N-anchored	Internal	Yes	Yes, MIP
COXII	IM N-anchored	Presequence	No	Yes, Imp1p

et al., 1984; Guiard, 1985; van Loon et al., 1986). Cleavage of this bipartite presequence usually occurs in two distinct steps, the first being catalyzed by the matrix processing peptidase (MPP), the second one by a protease on the outer surface of the IM, of which two have been described, namely Imp1p or Imp2p (**i**nner **m**embrane **p**rotease; see Table 1), which remove the sorting signal. In some precursors, for example, the Rieske FeS protein (Hartl et al., 1986) and the Sco1p (Schulze and Rödel, 1989), the IMS sorting information, however, forms an integral part of the mature polypeptide and hence is not cleaved off by Imp1p or Imp2p. Interestingly, both of these proteins are substrates for the mitochondrial intermediate peptidase (MIP) in the matrix (Kalousek et al., 1992; Isaya et al., 1992), which specifically cleaves eight amino acid residues immediately following the MPP cleavage site. The significance of this processing event is not known to date.

The manner in which the IMS sorting signal operates is unclear and remains a controversial issue with two distinct pathways being discussed, namely, the "stop-transfer" pathway and the "conservative sorting" pathway. Recent years have seen modifications to the original versions of these models. However, little, if any, consensus has been reached.

The term "stop transfer" was originally used by Blobel to describe how integral membrane proteins become anchored in the lipid bilayer (Blobel, 1980). The interaction of hydrophobic segments of a preprotein with the hydrophobic phase of the lipid bilayer was proposed to cause an arrest of complete translocation of the polypeptide chain across the membrane. The term "stop transfer" was subsequently used for mitochondrial proteins, initially to describe the sorting of cytochrome *c* peroxidase (Kaput et al., 1982), and was later adopted by Schatz and colleagues to describe the sorting of other IMS proteins, such as cytochrome b_2 and cytochrome c_1 (van Loon and Schatz, 1987; van Loon et al., 1987; Glick et al., 1992a, 1992b). The "stop transfer" model for the sorting of mitochondrial proteins proposed that such precursors initiate import along the general import pathway used by matrix targeted proteins, whereby the import machineries of the OM and IM are in close contact. In contrast to the matrix targeted proteins, complete translocation across the IM of the IMS precursors is prevented, due to an interaction of the hydrophobic domain in their sorting signal with the lipid bilayer of IM import machinery. Thus, these preproteins become arrested in the import channel during their import process. As a result of this, only the extreme amino-terminal matrix targeting signal of this precursor would become exposed to the matrix, where it would be proteolytically cleaved by MPP to generate an intermediate-size species. The mature portion of the precursor would be selectively translocated across the OM only, a process proposed to be supported in some way by the dissociation of the import channel of the two mitochondrial membranes. The precursor arrested with its sorting signal in the IM is thought to leave the import site by a lateral diffusion process in the IM, freeing this site for further import. Anchorage to the IM via the sorting signal would be severed following its proteolytic processing by one of the Imp proteases, thus resulting in the release of a soluble, mature-sized protein in the IMS, which then could fold and assemble into a functional enzyme.

Hartl and Neupert proposed an alternate view of IMS sorting that takes into account evolutionary aspects of organelles such as mitochondria and chloroplasts (Hartl et al., 1986, 1987; Hartl and Neupert, 1990). It is generally accepted that these organelles have originated from prokaryotic ancestors, the closest descendants of which are probably represented by photosynthetic bacteria of the Rhodobacter species. Throughout evolution many prokaryotic features have been conserved in mitochondria, that is, translation system including its sensitivity with respect to antibiotics, many aspects of its electron-transport system, ATP synthesis and the heat shock/chaperone system, to name just a few aspects. The conservative sorting hypothesis stresses that along with these prokaryotic attributes of mitochondria, the mechanism/process of prokaryotic protein secretion has also been maintained. Analogous to how bacterial proteins are secreted across the periplasmic membrane, the sorting signals and their mechanism of operation in mitochondrial IMS proteins are proposed to have been conserved from the prokaryotic secretory pathway. Consequently, such IMS proteins would be imported into the matrix (a compartment corresponding to bacterial cytoplasm) from where they embark on an export pathway (analogous to the bacterial secretory pathway) that is directed by the signal sequence and that uses features/components conserved from their ancestral secretory pathway.

VI. THE NATURE OF IMS SORTING SIGNALS IN BIPARTITE PRESEQUENCES

A. Defining the IMS Sorting Signal Requirements

Although there is little agreement as to how the IMS sorting signal operates, a number of recent attempts to define the essential features of an IMS sorting signal have been unanimous in their findings (Koll et al., 1992; Jensen et al., 1992;

Beasley, 1993; Beasley et al., 1993; Schwarz et al., 1993). From early experiments it was clear that the hydrophobic domain in the sorting signal is essential for correct sorting to the IMS as deletion of 19 amino acid residues encompassing this region in the cytochrome b_2 presequence resulted in the mistargeting of a fusion protein to the matrix (Koll et al., 1992). This observation was also shown to be true for a similar deletion in the cytochrome c_1 sorting signal (Jensen et al., 1992). These findings could initially be interpreted in favor of the stop-transfer model, which emphasized the importance of the hydrophobic interactions of the sorting sequence and the lipid bilayer.

A number of other observations, however, would clearly indicate hydrophobicity alone is insufficient to ensure correct sorting. First, not all proteins with such a hydrophobic domain become arrested in the IM import channel, the best characterized examples being subunit 9 from the F_0 part of the ATP synthase in *Neurospora crassa* (Mahlke et al., 1990; Rojo et al., 1995) and the Rieske FeS protein (Hartl et al., 1986; van Loon and Schatz, 1987; Fu et al., 1990), both of which are targeted to the matrix prior to subsequent sorting to the inner membrane. This observation in itself is hard to reconcile with the stop-transfer hypothesis. Second, recent detailed *in vivo* and *in vitro* studies on sorting signals from a number of laboratories are united in agreement that the sorting signal is indeed very complex in nature and all results emphasize the importance of the hydrophilic sequences flanking the hydrophobic segment.

In a genetic approach aimed at defining the sorting signal in the cytochrome b_2 presequence, the first 167 residues of cytochrome b_2 precursor were fused to a carrier protein, cytochrome oxidase subunit IV, a peripheral inner membrane protein of the matrix. Mutations that caused missorting to the matrix *in vivo* and hence complementation of a coxIV gene deletion were selected, by the ability to restore growth on nonfermentable growth media (Beasley, 1993). This procedure enabled identification of point mutations in three distinct domains of the IMS sorting signal spanning more than 50 amino acids, namely, the cluster of positively charged amino acids immediately preceding the hydrophobic core, the hydrophobic core itself, and unexpectedly the first amino acid residue of the mature sequence of the protein. Surprisingly, the level of missorting caused by mutations within the hydrophobic segment did not always correlate with the effects on the hydrophobicity of this region, but appeared to be related to changes in the conformation of this region, for example upon introduction of proline residues (Beasley 1993; Beasley et al., 1993). This conclusion supported one previously made with the homologous domain in the cytochrome c_1 precursor (Jensen et al., 1992). As pointed out by Beasley and colleagues (1993), however, a role of protein conformation would not have been predicted by a simple structural analysis of these domains.

In a more directed approach *in vitro*, Schwarz and colleagues (1993), used site-directed mutagenesis to address the consequences of serially mutating the cluster of positive amino acid residues Arg-Lys-Arg (RKR located to the amino-terminal side of the hydrophobic core of cytochrome b_2 sorting sequence). Removal of one or more of these positive charges had a deleterious effect on IMS sorting, even when they were replaced by uncharged or hydrophobic ones to extend the hydrophobic region. These mutant proteins accumulated in the matrix in an ATP-dependent manner. Interestingly, if matrix ATP was reduced to interfere with mt-Hsp70 action (see below), these proteins became correctly sorted to the IMS like wild-type precursors. How the missorting is governed by mt-Hsp70 action, or the lack thereof, is an important question that remains unresolved.

In addition to the positive cluster and hydrophobic domain, the sequences downstream of these motifs, closer to the mature protein part, are also essential for sorting. As mentioned previously, mutations in the first amino acid residue of the mature sequence were observed to cause missorting to the matrix (Beasley et al., 1993). Furthermore, the cytochrome b_2-DHFR fusion protein pb_2(1–76)-DHFR in which the final four amino acid residues of the carboxy-terminal domain of the presequence had been removed, is mistargeted to the matrix (Stuart and Neupert, unpublished results). Removal of the hydrophilic flanking region between the MPP cleavage site and the positive cluster RKR significantly affected the efficiency of sorting (Ono et al., 1995).

B. Reconciling Newly Defined Sorting Signals in Terms of the Stop-Transfer Model

In summary, a detailed analysis of the cytochrome b_2 sorting signal demonstrated that efficient sorting to the IMS involves almost the entire second domain of the presequence, which, in addition to the hydrophobic core, includes hydrophilic flanking sequences. This signal therefore encompasses almost 50 amino acid residues. How can one fit these findings into the two models of IMS sorting pathways? As a consequence of these findings the original view of "stop-transfer" has to be modified, whereby it seems clear that if the sorting signal would operate as a stop-transfer signal, it would do so by way of specific hydrophilic interactions with most probably proteinaceous components of the IM and not by pure virtue of its hydrophobicity. A second aspect of these findings that must be borne in mind here is the observation that the newly defined sorting signal extends from the MPP cleavage site to the IMP cleavage site, some 50 amino acid residues. This length represents more than twice the amount of amino acid residues required to span a lipid bilayer, and furthermore is comparable to that shown to be sufficient to concurrently span both the import sites of the OM and IM (Rassow et al., 1990; Ungermann, Cyr, and Neupert, unpublished results). In order to explain how such a long stretch could possibly operate as a stop-transfer signal in the IM, one must envisage the signal not as an extended sequence of amino-acids, but rather as a specific secondary structure element. Attainment of this specific structure in the context of the IM would therefore be a prerequisite for efficient stopping, a conclusion that would be in agreement with that previously made by Beasley and colleagues (1993). Thus, it would be suggested that the flanking sequences fold during transit through the IMS and IM and interact in a conformation dependent manner with proteinaceous components on the periphery of both sides of the IM channel.

Consequently, arresting of preproteins in this fashion would require that such an IM import channel could operate both in a selective and nonselective manner, allowing the complete passage of matrix-targeted proteins by a default mechanism while specifically retaining the IMS-targeted ones. Interestingly, the *in vitro* kinetics of import of matrix targeted precursors through this channel are in fact very rapid. For an IMS-targeted protein to be selectively stopped in such a channel, such a sequence would have to pass slower through the channel in such a fashion that the flanking sequences could fold and attain the structure required for recognition by their binding partners before the precursor is completely drawn into the matrix along the default pathway.

C. IMS Sorting Signals: Functional Conservation of Prokaryotic Features?

On the other hand, such findings regarding the sequence requirements of sorting signals could be more easily explained by the conservative sorting model where no size or folding constraints would operate. The IMS sorting signals bear remarkable similarity, and in some instances identity, to bacterial secretory leader sequences (Hartl and Neupert, 1990). Indeed, the processing enzymes Imp1p and Imp2p are related to bacterial leader peptidase (see following). The importance of the hydrophobic core, and in particular the positively charged motif preceding this domain, for efficient secretion across the bacterial periplasmic membrane has been previously reported (Puziss et al., 1989). The essential role of the corresponding positive charges in the mitochondrial IMS sorting signal therefore reflects a functional conservation. Hence, these findings strongly support the suggestion that IMS targeting signals function as export signals. Furthermore, it was proposed that a protein component exists in the mitochondria that specifically recognizes this sorting signal (Schwarz et al., 1993). In eukaryotes, similar sequence requirements have already been described for interaction with the signal recognition particle (SRP) and subsequent components of the endoplasmic reticulum translocation machinery. Hence, in line with the endosymbiotic origin of mitochondria, the basis for the conservative sorting pathway, it was proposed that the IMS sorting signal is recognized also by a (possibly homologous) mitochondrial signal sequence binding protein (mt-SBP; Schwarz et al., 1993). Such a putative component was postulated to be located either in the matrix or at the inner face of the mitochondrial inner membrane. The IMS sorting signal would therefore be completely translocated into the matrix where it could adopt its necessary secondary conformation required for recognition by an interaction with mt-SBP, and then embark on the export event. It was also speculated that the function of such an SBP component could also be conserved in yet another aspect, namely, that it may assist in the co-translocational transfer of the polypeptide to the IMS, while completing translocation across the general import pathway into the matrix. Indeed, evidence in support of simultaneous import and export reactions has recently been provided (Koll et al., 1992) with independent evidence accumulating to support this (see following).

VII. IMP, THE INNER MEMBRANE PROTEASE HAS PROKARYOTIC ORIGINS

The IMS-targeted proteins that contain cleavable bipartite signal sequences undergo a maturation event in the IMS, which is catalyzed by the inner membrane protease complex (Imp). This protease has recently been shown to contain at least two catalytic subunits, the previously identified Imp1p (Schneider et al., 1991; Behrens et al., 1991) together with a new protein, the Imp2p (Nunnari et al., 1993). Although both Imp1p and Imp2p are believed to be subunits of the same complex, these proteins differ in their substrate specificities (Nunnari et al., 1993).

The Imp1p protease was previously shown to be associated with the IM and facing the IMS. This protease is responsible for the maturation of not only cytochrome b_2 but also interestingly of the precursor of cytochrome oxidase subunit II (COXII). COXII, a component of the inner membrane cytochrome oxidase complex, and like two other subunits of this complex, COXI and COXIII, is mitochondrially encoded. These proteins are synthesized on mitochondrial ribosomes and are thought to be co-translationally inserted into or translocated across the IM. Pre-COXII, unlike the other mitochondrial encoded proteins, contains a cleavable presequence postulated to direct its integration into the IM, which results in the translocation of a large carboxy-terminal part of pre-COXII across the IM to the IMS. Cleavage of pre-COXII by Imp1p indicates that this protease is involved in the maturation of preproteins originating in two distinct cellular subcompartments, the mitochondria and the cytoplasm. Thus, the sorting pathway of an imported nuclear encoded protein converges with that of an exported mitochondrially encoded gene product, whereby at least one common component used by these proteins during their sorting is the Imp1p.

Furthermore, on the basis of sequence comparisons, Imp1p was shown to be a member of the signal peptidase family, a group of both prokaryotic and eukaryotic membrane-bound endopeptidases that function to process secretory signal sequences. The similarity between Imp1p and other signal peptidases, which include *E. coli* leader peptidase and *S. cerevisiae* endoplasmic reticulum signal peptidase, Sec11p, was shown to span all three distinct regions that are strongly conserved throughout members of this family and are believed to contain essential amino acid residues for their catalytic activity.

Temperature-sensitive yeast mutants of the IMP1 gene product showed therefore defective processing in both cytochrome b_2 and COXII. The processing of cytochrome c_1 and CCPO, for example, appeared unaffected, thus suggesting the presence of a least one other protease. A second protease, Imp2p, was recently identified in a genetic screen designed to identify novel gene products that function at a point after convergence of the biosynthetic pathways of mitochondrial-encoded and nuclear-encoded IMS proteins (Nunnari et al., 1993). Imp2p was shown to be the protease required for the maturation of cytochrome c_1. It remains unclear so far as to how or where the CCPO precursor is matured (Kaput et al., 1989; Nunnari et al., 1993).

Sequence comparison of the Imp2p protease with the Imp1p revealed a close resemblance between these two subunits, 25% amino acid identity and 44% amino acid similarity. Furthermore, like Imp1p, Imp2p has distinct sequence similarities to other prokaryotic and eukaryotic signal peptidases. Deletion of the IMP2 gene resulted in the accumulation of not only i-cytochrome c_1, but also i-cytochrome b_2 and pre-COXII, all of which were located in the IMS but remained bound to the IM. Further investigations revealed that Imp1p and Imp2p do in fact have separate nonoverlapping substrate specificities, but in addition to catalyzing the cleavage of IMS sorting signals, Imp2p is required for the stable expression of a functional Imp1p.

In summary, the mitochondrial inner membrane protease is composed of at least two catalytic subunits, Imp1p and Imp2p. Both subunits are related to each other and to a family of eubacterial and eukaryotic signal peptidases, where both Imp1p and Imp2p show closer primary structural similarity to the eubacterial enzymes. In addition to amino acid sequence, both subunits have a similar topology to other members of this family, namely they are membrane associated enzymes with their active site facing the compartment to which the substrate molecules become localized during their biogenesis. We believe the similarity of both Imp1p and Imp2p to the bacterial leader peptidase argues in favor of the conservative sorting hypothesis and is a clear indication that at least one component of the bacterial secretory machinery has been conserved during the evolution of mitochondria from their prokaryotic ancestors. In further support of this, as discussed previously, IMS sorting signals closely resemble the secretion signals of bacterial proteins, indicating that not only the processing enzyme but also the structural features required for the recognition of such signals, possibly by these proteases, have also been conserved.

VIII. IDENTIFICATION OF A NOVEL CYTOCHROME b_2 SORTING INTERMEDIATE

In a previous study, the complete import of cytochrome b_2-DHFR fusion proteins was shown to be blocked if the DHFR moiety was tightly folded due to the presence of bound methotrexate (MTX; Koll et al., 1992). If the cytochrome b_2 portion of these fusion proteins was of sufficient length (220 amino acid residues or more), both cleavage reactions of the presequence were observed, despite inhibition of complete translocation. According to the conservative sorting model, such arrested cytochrome b_2-DHFR proteins were thought to be spanning the mitochondrial membrane system from the site of import to that of export, with small segments looping through the matrix. The observed length requirement thus reflected the distance from the import to the export machineries. To substantiate this claim, such arrested mature-sized proteins were demonstrated to be concomitantly associated with Hsp60, the matrix-localized chaperone (Koll et al., 1992).

This claim was strongly disputed by Glick and colleagues (1992b) who argued that according to the stop-transfer model, such MTX-blocked fusion proteins following Imp1p cleavage were arrested with their amino-terminus and cytochrome b_2 moiety free in the IMS, while the folded DHFR domain remained outside (Glick et al., 1992a, 1992b). Thus, this species traversed only the OM and was claimed to be on a "dead-end" pathway, unable to complete translocation into the IMS (Glick et al., 1992a).

We have readdressed this problem and can demonstrate that such MTX-arrested, Imp1p processed, cytochrome b_2 are clearly not "dead-end" species as predicted by the stop-transfer model. Furthermore, we can provide independent evidence for the looping of segments of arrested species through the matrix (Gruhler et al., 1995). We consistently find that upon removal of the MTX, up to 60–70% of such arrested mature-sized proteins can be chased into a protease protected location, exposed to the IMS. In the course of the studies another important observation was made. Following chase across the OM after MTX removal, only a certain percentage of the mature-sized cytochrome b_2-DHFR species was soluble in the IMS; the rest remained tightly bound to the IM (Figure 1). The chase across the OM was very efficient, resulting in the protease protected location of the protein; however, this was in contrast to subsequent translocation across the IM, a step proposed to be catalyzed by the export process. It thus appears that once stalled, as was the case for the MTX-arrested DHFR proteins, the export apparatus is either very labile or very inefficiently re-inactivated. Alternatively, one could speculate the absence of some co-factors or energetic requirements needed to optimize such an export event. Thus, the chased protease-protected species is thought to largely accumulate exposed to the IMS but at the same time spanning the IM, looping through the matrix. Such membrane-bound mature-sized cytochrome b_2 species could be found in a complex with the mt-Hsp70 and MIM44, as an independent confirmation for looping through the matrix. Such a topology of a mature-size cytochrome b_2 species is totally inconsistent with the stop-transfer model, but can be simply perceived by the conservative sorting one (Gruhler et al., 1995).

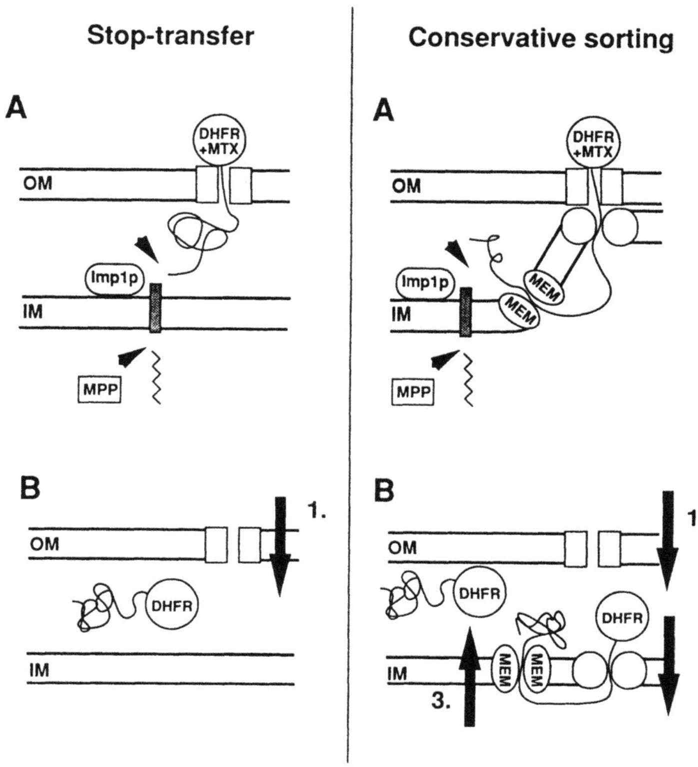

Figure 1. Chase of methotrexate-arrested mature-size cytochrome b_2-DHFR fusion proteins. See text for detailed explanation. *Stop-transfer model*: (**A**) Cytochrome b_2-DHFR fusion protein arrested with methotrexate spans only the OM at the site of import. Processing of the amino-terminal IMS sorting signal by Imp1p results in the release of a free amino-terminus in the IMS. (**B**) Removal of methotrexate and subsequent chase (represented by an arrow) into the mitochondria (although not predicted by Glick et al., 1992a) would involve translocation across the OM only (1) and would result in accumulation of a soluble mature-size cytochrome b_2-DHFR species in the IMS.

Conservative sorting model: (**A**) Cytochrome b_2-DHFR fusion protein arrested with methotrexate spans the OM and IM at the sites of import and once again the IM at the site of export where it undergoes processing to its mature-size species by Imp1p. (**B**) Removal of methotrexate and subsequent chase can result in either soluble or inner membrane bound mature-size b_2-DHFR species in the IMS. Chase across the OM alone (1) or combined with a partial chase across the IM import machinery (1+2) would result in the accumulation of a mature-sized cytochrome b_2-DHFR species protected against exogenously added protease which is spanning the IM and is partially exposed to the matrix. (continued)

Figure 1. (Continued) Complete chase, that is, through the OM and IM import machinery together with chase through the export machinery (1+2+3) would result in the accumulation of soluble mature-sized cytochrome b_2-DHFR species localized in the IMS. MTX, methotrexate; OM, outer membrane; IM, inner membrane; IMS, intermembrane space; MEM, putative mitochondrial export machinery; MPP, matrix processing peptidase; Imp1p, inner membrane protease 1. The zig-zag with the positive charges represent the matrix targeting domain of the precursor presequence, while the shaded box represents the IMS sorting signal. The open squares represents the OM import machinery, open circles represent the inner membrane import machinery. Arrows indicate the three possible membrane translocation events during the chase period.

IX. ROLE OF MOLECULAR CHAPERONES IN THE IMPORT OF IMS PROTEINS

A. Requirement for mt-Hsp70/Matrix ATP for Import Across the IM

Hydrolysis of matrix ATP has been proposed to constitute (at least part of) the driving force for the vectorial movement of precursors into the mitochondrial matrix. One decisive role of matrix ATP in protein import appears to be through its facilitating the action of mt-Hsp70, a molecular chaperone (Neupert et al., 1990; Manning-Krieg et al., 1991; Glick et al., 1993; Stuart et al., 1994a,b). The effects of stringent matrix-ATP depletion on the import of various preproteins are directly paralleled by the effects of specifically inactivating mt-Hsp70 in a temperature-sensitive yeast mutant carrying a mutation in the SSC1 gene (Gambill et al., 1993; Voos et al., 1993; Stuart and Neupert, unpublished results). Recent reports indicate that matrix proteins require matrix ATP/mt-Hsp70 for at least three distinct steps of import which together serve to convey unidirectionality and to provide a driving force for the import process (Cyr et al., 1993; Glick et al., 1993; Stuart et al., 1994a, 1994b). First, matrix ATP and mt-Hsp70 are involved in the translocation of mitochondrial targeting signals across the IM in conjunction with the $\Delta\Psi$. Such an initial passage appears to be a reversible event. Binding of mt-Hsp70 to the incoming polypeptide upon exposure to the matrix, however, prevents the escape of the precursor back to the cis side of the IM. The translocation of such a precursor is thus committed (Cyr et al., 1993). Spontaneous relaxing of the remaining mature parts of the polypeptide chain outside the mitochondria would render segments of the protein free to pass across the membrane. The protein becomes translocated in a step-wise manner where further binding of mt-Hsp70 commits increasing amounts of the protein to the matrix and thus shifts the equilibrium of the protein structure outside the mitochondria towards the unfolded state. This in turn ultimately draws the unfolded segments of the polypeptide across the IM to the mitochondrial matrix. Thus, mt-Hsp70 plays essential roles at the very early stages by securing the presequence in the matrix, unfolding folded domains outside the mitochondria, and completing translocation across the IM into the matrix (Neupert et al., 1990; Stuart et al., 1994b). Thus, under conditions of stringent matrix ATP depletion, the action of mt-Hsp70 is drastically reduced and preproteins targeted to the matrix are neither imported nor processed. The situation with proteins targeted to the IMS in contrast appears to be more complicated.

B. IMS-targeted Preproteins Depend on Matrix ATP/mt-Hsp70 to Promote Their Unfolding Outside Mitochondria

Following a study of the energetic requirements of the import of cytochrome c_1, it was recently suggested that matrix ATP and hence mt-Hsp70 action was not required for correct sorting to the IMS (Wachter et al., 1992). Hence, it was inferred that this protein does not become completely translocated across the IM during its sorting process as it was apparently mt-Hsp70 independent. It was implied that cytochrome c_1 would not be sorted by a conservative sorting mechanism but rather by a stop transfer event. In contrast, however, preliminary results suggested that the import of cytochrome b_2 was dependent on matrix ATP (Hwang and Schatz, 1989) and furthermore, analysis of a conditional mutation in the SSC1 gene indicated that mt-Hsp70 could play an important role in the import of both cytochromes b_2 and c_1 (Kang et al., 1990; Ostermann et al., 1991). Thus, these two lines of evidence appear to be contradictory as both cytochrome c_1 and b_2 are thought to be sorted by a similar mechanism and thus is difficult to explain at first glance why they should differ in their energetic requirements. This issue has recently been partially resolved following detailed analysis of the energetic requirements of the import of these precursor and cytochrome b_2-derived fusion proteins (Glick et al., 1993; Stuart et al., 1994b).

These recent studies as outlined ahead have indicated that the role of matrix ATP/mt-Hsp70 during the import of cytochrome b_2 is to mediate the unfolding of tightly folded segments in mature regions of preproteins outside the mitochondria. Both the processing and import of cytochrome b_2 displayed a very strong requirement for matrix ATP/mt-Hsp70 function. Cytochrome b_2-derived preproteins consisting of 185 or more residues of pb_2 fused to DHFR showed the same energetic/mt-Hsp70 requirements, whereas the import of shorter fusion proteins (up to 167 residues of pb_2) was largely independent of matrix ATP (Glick et al., 1993; Stuart et al., 1994b). Those pb_2 fusion proteins that displayed strong matrix ATP/mt-Hsp70 requirements could, however, be imported into ATP-depleted and mt-Hsp70-mutant mitochondria if they were denatured in 8 M urea prior to import (Glick et al., 1993; Voos et al., 1993; Stuart et al., 1994b).

Cytochrome b_2 (L-lactate dehydrogenase) contains both flavin and heme as prosthetic groups. The approximately 100 initial residues of the mature polypeptide chain form a tightly folded structure, the cytochrome b_5 or heme binding domain (Xia and Matthews, 1990). This folded domain, which has been shown to form also in the precursor protein (Glick et al., 1993), prevents the stable insertion of the presequence into the matrix, unless it becomes unfolded, a process requiring mt-Hsp70 function. Hence all precursors that contain this domain require ATP-dependent binding of mt-Hsp70 to matrix exposed parts of the protein in order to mediate the stabilization of unfolded states of this domain outside the mitochondria. By contrast, cytochrome b_2 constructs where this domain has been deleted (p$b_2$167-DHFR and smaller) or disrupted (p$b_2\Delta$135–226 in which 91 residues spanning part of the heme domain have been deleted) can be efficiently processed and imported in the apparent absence of mt-Hsp70 activity (Stuart et al., 1994b).

C. Can the IMS Sorting Machinery Render mt-Hsp70 Function Dispensable?

At first glance it appears difficult to see why IMS-targeted preproteins that do not have stably folded domains display an apparent independence mt-Hsp70 if they are sorted in a conservative manner. Their sorting would require passage across the IM and therefore could show the same requirements as matrix-targeted proteins since they share the same import pathway. We propose, however, that the apparent independence of mt-Hsp70 for both the stabilization of the presequence in the matrix and for completion of translocation across the IM can be explained by the interaction with specific component(s) of the sorting pathway. Most probably, the mitochondrial signal sequence binding protein (mt-SBP) binds to the sorting sequence upon its exposure to the matrix (Schwarz et al., 1993). We propose, thus, in the absence of a tightly folded domain to physically restrain the preprotein outside, the precursor inserts partially through the mitochondrial import machineries into the matrix in response to $\Delta\Psi$. Stabilization of the presequence in the matrix, a task usually performed by mt-Hsp70 (at least for matrix targeted proteins) could be achieved by binding of mt-SBP to the sorting signal. Naturally if the sorting signal was unable to spontaneously penetrate deeply enough into the matrix, that is, due to physical restriction by a folded domain in the mature sequence, the function of mt-Hsp70 would be indispensable. Second, we would envisage that further completion of translocation into the IMS, at least in the apparent absence of mt-Hsp70 action, could be driven by the export system, which would then transfer the preprotein into the IMS preferentially in a co-translocational manner.

It should be pointed out that the relative independence of certain IMS-targeted proteins on matrix ATP/mt-Hsp70 could also be very easily explained by the stop-transfer model. According to this model, it is suggested that matrix ATP/mt-Hsp70 is needed to "pull" the presequence far enough into the mitochondria for the stop-transfer signal to be recognized and become engaged in the IM (Glick et al., 1993). Such an ATP requirement is particularly obvious when a tightly folded domain, such as the cytochrome b_5-like domain, inhibits the sorting signal from reaching the recognition machinery in the IM. Furthermore, in the absence of such a folded domain, the ATP independent translocation/stabilization of the targeting signal is promoted by binding of the sorting signal to the putative stop-transfer machinery, rather than to mt-Hsp70 or mt-SBP. The passage of the cytochrome b_5 domain across the OM is the step in the import of cytochrome b_2 that shows a distinct requirement for matrix ATP. Further import would involve passage of the polypeptide across the OM only and hence would be independent of mt-Hsp70/matrix ATP. As soon as the stop-transfer signal would be engaged, mt-Hsp70 could no longer exert a "pulling effect" and hence could not influence completion of translocation (Glick et al., 1993).

D. Can Apparent Lack of mt-Hsp70 Dependence Be Used as Evidence For or Against Any One Sorting Model?

In summary, it appears that both sorting models can account for the relative independence on mt-Hsp70 for the import of certain IMS-targeted proteins. Thus, the present data on the energetic requirements unfortunately do not allow a discrimination between these alternative views of sorting. These results imply that matrix ATP/mt-Hsp70 dependence is not a function of the targeting information of the precursor and hence of the selected sorting pathway, but rather is a characteristic of the folded state of the mature segments of the precursor itself. However, the lack of a requirement for ATP in the matrix cannot be interpreted in favor of the stop-transfer model, since it is clear that the conformational properties of the precursor to be sorted and not its IMS sorting signal dictates such an energetic requirement.

E. Role of Other Molecular Chaperones in the IMS Sorting Process

In bacterial systems, it is clear that the function of DnaK, the homologue of mt-Hsp70, is tightly regulated by two other chaperones, DnaJ and GrpE (Langer et al., 1992). Recently, mitochondrial homologues of these two proteins have been found in *Saccharomyces cerevisiae*, a DnaJ homologue termed Mdj1p (mitochondrial DnaJ; Rowley et al., 1994) and Mge1p (mitochondrial GrpE; Ikeda et al., 1994). Mdj1p, a matrix-localized protein associated with the inner face of the IM, is not an essential protein; however, deletion of the MDJ1 gene (Δmdj1 yeast) results in a ts phenotype for growth and a loss of mitochondrial DNA, hence a petite phenotype. Although import of preproteins into mitochondria prepared from the deletion strain was observed not to be affected, folding of newly imported proteins and their stability against heat denaturation was reduced. Whether the Mdj1p is somehow required for the sorting process of proteins to the IMS is unclear to date, but an interesting observation with the Rieske FeS protein suggests that this could indeed be the case, at least for certain preproteins. Imported Rieske FeS protein accumulated as an intermediate-sized species in this mutant and the second processing event, catalyzed by the mitochondrial intermediate peptidase (MIP), was inhibited (Rowley et al., 1994). The processing of other substrate preproteins by MIP was also inhibited (Langer, Stuart, and Neupert, unpublished results). Further recent investigations have revealed that the imported Rieske FeS protein displays a defective sorting in the mdj1 null mutant mitochondria, due to a prolonged interaction with mt-Hsp70 and furthermore appears to aggregate in a time-dependent manner (Stuart, Fölsch, and Neupert, unpublished results). The initial import of this IMS targeted protein is strongly dependent on matrix ATP/mt-Hsp70 action and other evidence exists that its subsequent sorting process involves the participation of the Hsp60/Hsp10 chaperone system (Cheng et al., 1989; Ostermann et al., 1989; Höhfeld and Hartl, 1994). One could speculate that the efficient and correct transfer from mt-Hsp70 to Hsp60 and/or the export system, to prevent misfolding/aggregation, is dependent on Mdj1p action. Whether Mge1p is also involved in this process is not known to date. MGE1, like SSC1, is an essential gene and preliminary evidence indicates an involvement of this protein in the import process (Craig, personal communication). The exact nature of this involvement will only be clearly dissected when a ts mutant is available.

In summary, Mdj1p is involved in the sorting of the Rieske FeS protein through the matrix space to the IMS, but the possible involvement in sorting of other preproteins to this submitochondrial location is not yet known.

Following membrane translocation of newly imported polypeptides, those destined to stay in the matrix have to fold to attain their native conformation in order to be enzymatically active. In many cases this appears to be an assisted process involving transfer from mt-Hsp70 to Hsp60. Folding mediated by Hsp60 is an ATP-mediated process and requires the action of another chaperone, Hsp10 (Rospert et al., 1993; Höhfeld and Hartl, 1994). Evidence that Hsp60 could be involved in protein sorting to the IMS came from analyzing a ts mutant of Hsp60, termed mif4 (Cheng et al., 1989). This mutant was reported to accumulate incompletely processed forms of cytochrome b_2 and of the Rieske FeS protein, suggesting a defective sorting of these proteins. Other evidence in support of the involvement of Hsp60 came from a study by Koll and colleagues (1992) where the import of a number of methotrexate-arrested cytochrome b_2-DHFR fusion proteins was analyzed. Such arrested translocation intermediates, which were processed to their mature size form, were suggested to span the matrix as they could be shown to be interacting with Hsp60. Such evidence would favor the conservative sorting process for these preproteins. The involvement of Hsp60 in cytochrome b_2 sorting to the IMS has, however, been disputed since by Glick and colleagues (1992) and Hallberg and colleagues (1993). The cause for this discrepancy is unclear but indicates that the requirement for Hsp60 is not absolute and can be circumvented. On the other hand, recent results from analysis of a ts mutant of the Hsp10 demonstrated the tendency of FeS to aggregate in the matrix in the absence of a functional Hsp60/Hsp10 system (Höhfeld and Hartl., 1994). These results clearly support the original proposal for the involvement of this chaperone system in the sorting of some IMS proteins in a conservative manner.

X. THE RIESKE FeS PROTEIN IS SORTED IN A CONSERVATIVE MANNER

As mentioned previously, the Rieske FeS protein undergoes two proteolytic processing events upon import into mitochondria. The second processing step, in contrast to that of cytochromes b_2 and c_1, occurs like the first one in the mitochondrial matrix. This event serves to remove an N-terminal octapeptide following the MPP cleavage site. The IMS sorting signal therefore forms an integral part of the mature FeS polypeptide sequence. Although very little is known about the events leading to the assembly of the imported FeS species into the cytochrome bc_1 complex, there appears to be general agreement that the conservative sorting hypothesis holds true for at least this IMS sorted protein (Hartl et al., 1986; van Loon and Schatz, 1987; Fu et al., 1990; Glick et al., 1992a).

The import and export steps in the assembly pathway of the Rieske FeS protein can be experimentally dissected *in vitro* into a number of distinct steps (Fölsch, Stuart, and Neupert, in preparation) and our results are summarized in the model presented in Figure 2. The Rieske FeS precursor protein is initially imported completely into the matrix prior to its export to the IMS. This import event requires energy in the form of a membrane potential ($\Delta\Psi$) and matrix ATP hydrolysis. The import of Rieske FeS is dependent on mt-Hsp70 action as import is abolished if mt-Hsp70 is impaired either in ts ssc1 mutants or in matrix ATP-depleted mitochondria. During or following import the precursor is processed by MPP to generate an intermediate-sized form. Efficient processing by MIP to the mature-sized species requires the function of Mdj1p, probably to release the Fe/S protein from mt-Hsp70. This results in the accumulation of a mature-sized Rieske FeS soluble in the matrix. This step appears to be prevented in mitochondria from the mdj1 deletion strain. Kinetic data suggest that matrix soluble species give rise to a membrane bound form of the Rieske FeS that is localized to the matrix side of the IM. In the absence of a functional Hsp60/Hsp10 system, the Rieske FeS displays a tendency to aggregate in the matrix, thus suggesting that these chaperones play an important role in the maintenance of an assembly competent form of the FeS (Cheng et al., 1989; Höhfeld and Hartl, 1994). The Rieske FeS then becomes assembled, presumably from the membrane bound form, into the cytochrome bc_1 complex and is thought to be the last component to assemble into this multimeric complex (Grivell, 1989).

A number of open questions exist, especially with regard to the final steps of the assembly process. For example, it is unclear whether the membrane bound form becomes directly assembled into the cytochrome bc_1 complex from the matrix, or if it is initially translocated across the IM to the IMS using a putative export machinery and from there becomes assembled into its final location. Interestingly in this respect, it has been show in *E. coli*, which does not have a cytochrome bc_1 complex (unlike Rhodobacter, for example), that *Rhodobacter* Rieske FeS can be expressed and correctly exported across the periplasmic membrane (Van Doren et

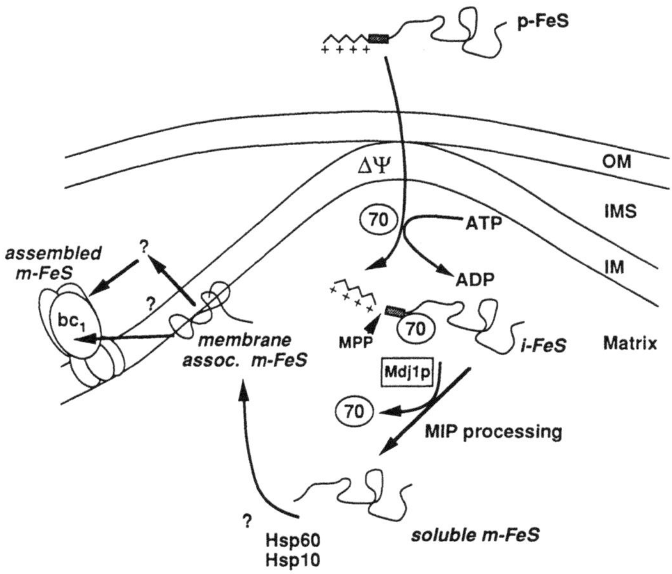

Figure 2. Hypothetical model of the import and sorting of the Rieske FeS protein to the mitochondrial intermembrane space. See text for details. The zig-zag portion with the positive charges denotes the part of the presequence which serves to target the precursor protein of FeS (pFeS) into the matrix. It undergoes cleavage by the matrix processing peptidase (MPP) to result in an intermediate-size FeS species (iFeS). The shaded box represents the octapeptide, which is proteolytically removed by the mitochondrial intermediate peptidase (MIP) to give rise to the mature-sized FeS species (mFeS). mt-Hsp70, mitochondrial Hsp70; Mdj1p, mitochondrial DnaJ homologue; OM, outer membrane; IM, inner membrane; IMS intermembrane space; bc_1, cytochrome bc_1 complex.

al., 1993). These results indicate that export of a Rieske FeS can be facilitated by bacterial secretory machinery and thus can occur independent of direct assembly into a cytochrome bc_1 complex. Whether an export machinery exists in the inner mitochondrial membrane to facilitate such a translocation process to the IMS is, however, unknown. An inner membrane protein termed BCS1 has been identified by genetic means and has been reported to be specifically involved in the assembly process of the Rieske FeS protein (Nobrega et al., 1992). The nature of the involvement of this interesting protein is unfortunately unclear to date, but one could speculate its participation in the translocation process of the FeS from the matrix across the IM.

In addition, the information to target the export of the Rieske FeS from the matrix to the IMS, residing within the mature sequence of the protein, has not been identified to date. Furthermore, open questions exist with regard to the formation and addition of the FeS cluster to the apo-FeS protein. It is not known which enzymes catalyze these events nor which mitochondrial subcompartment they occur in. It is most likely that they occur in the IMS, as it appears unlikely that a folded protein containing an FeS cluster would be competent for membrane translocation.

XI. SUMMARY AND FUTURE PERSPECTIVES

In order to complete this review we wish to take the opportunity to summarize the reasons, both experimental and conceptual, which we believe argue strongly in favor of the conservative sorting of IMS targeted proteins. As mentioned throughout this review, a large amount of experimental evidence exists to support the conservative sorting hypothesis. This model proposes that IMS-targeted proteins embark on the general mitochondrial import pathway crossing the OM and IM, where they emerge into the matrix. Upon exposure to the matrix, the sorting signal is recognized by a putative signal binding protein (mt-SBP). The preprotein then embarks on a translocation event back across the IM, a process postulated to resemble the export of proteins in bacteria across the periplasmic membrane. Initiation of this export step is thought to occur prior to completion of the import of carboxy-terminal portions of the preproteins through the import machinery. Thus, it is highly plausible that, if of sufficient length, such IMS-targeted proteins can span both the OM and IM import machineries while being translocated across the IM during its export step. As mentioned previously, such IM spanning intermediates have been identified and have been shown to be looping through the matrix by two independent approaches.

Detailed characterization of IMS-targeted proteins surprisingly indicated that the sorting information encompassed basically the entire second domain of the bipartite presequence. We believe these sequence requirements are difficult to reconcile with the stop-transfer model of sorting for two basic reasons. First, the results stressed the importance of hydrophilic sequences, presumably to mediate protein-protein interactions in a model that is based on hydrophobic interactions between preprotein and lipid bilayer. Second, the newly defined sorting signal is long enough to span both OM and IM. To act as a stop-transfer signal, therefore, it would have to operate by adopting a specific conformation. Although, as discussed before, experimental evidence may support this, conceptually this is very difficult to accept for the following reasons: Passage of preproteins through the import machinery occurs as an extended polypeptide chain, hence the extreme amino-terminal part of the sorting signal would be exposed to the matrix while carboxy-terminal portions would still be in the OM. Thus, the initial hydrophilic region alone would have to be successful in retarding the translocation in an extended form prior to the engaging of carboxy-terminal regions of the signal. This is difficult to perceive when one bears in mind the rapid kinetics of the mt-Hsp70 mediated import of matrix-targeted proteins across this import channel. The three essential domains of the sorting signal, the initial hydrophilic segment, the hydrophobic core, then the second hydrophilic segment, according to the stop transfer model, would undergo temporally successive recognition events by IM components to ensure the overall efficiency of stopping. Thus, it is also very difficult to explain why mutations in the mature part of the protein should have such adverse consequences for the ability of these upstream domains to function as a stop-transfer signal. In contrast, according to the conservative sorting hypothesis, no size or folding restrictions would operate in the matrix. In addition, the close functional resemblance of such sorting signals with bacterial secretion presequences strongly support the conservative sorting hypothesis.

According to the conservative sorting hypothesis, IMS-targeted proteins pass completely through the matrix as they undergo sorting to the IMS. Although present evidence suggests that import and export events occur co-translocationally, under certain experimental conditions one can detect the accumulation of sorting intermediates in the matrix. These conditions, which we believe to retard the export step, are difficult to reconcile with the stop-transfer model (see below). It should be pointed out here that matrix-localized forms can also be explained by the stop-transfer model under the assumption that the preprotein failed to become arrested in the IM and would then enter the matrix along a default pathway. Such forms would be dead-end intermediates, however. Detailed kinetic studies of the import of cytochrome b_2 and derived fusion proteins followed by localization studies have shown a transient, time-dependent accumulation of intermediate-sized species in the matrix (Ono et al., 1995). The levels accumulating can be elevated if precautions are taken to hinder a potential export step. For example, if one follows the import of a cytochrome b_2 precursor in which the MPP cleavage site has been mutated so that it can no longer be cleaved, efficient but delayed kinetics of maturation are observed. The efficiency of accumulating transient matrix-localized species (precursor in this case) is significantly higher under these conditions (Ono et al., 1995). It is difficult to see why lack of cleavage of the matrix-targeting signal in this manner should influence sorting according to the stop-transfer model, unless it physically restrains it in adopting its required secondary conformation. On the other hand, the fact that such matrix localized precursors display a prolonged but not a permanent matrix-localization clearly identifies them as intermediates on a slowed conservative sorting pathway, rather than dead-end products that failed to be arrested in the IM. Finally, matrix-localized intermediate and mature forms of the Rieske FeS protein have been reported under a number of experimental circumstances, for example upon import at low temperature (to block export), at low levels of matrix ATP (to block mt-Hsp70 to Hsp60 transfer), and in yeast mutant strains temperature-sensitive defective in HSP60 and HSP10.

Another protein whose assembly pathway is best explained by a conservative sorting pathway is cytochrome c_1, a carboxy-terminally anchored IM protein. According to the stop-transfer model, this protein would be sorted to the IMS in a manner similar to cytochrome b_2. However, following maturation, cytochrome c_1

would have to undergo an independent membrane insertion step of its carboxy-terminus in order to achieve correct anchorage. The development of such a mechanism of insertion, together with the necessary targeting and energetic requirements, would be unique for such an IMS targeted proteins. On the other hand, a simpler explanation is provided by the conservative sorting pathway, where this protein, like other IMS proteins, would be exported from the matrix, amino terminus first, but would remain anchored to the IM due to the presence of an uncleaved carboxy-terminal hydrophobic segment.

Finally, we believe, from a purely conceptual point of view, that conservative sorting represents a simpler process than the stop transfer mechanism of sorting for IMS targeted proteins. At first glance, the conservative sorting pathway appears rather complex, as it entails a protein to be translocated across three membranes in order to reach its functional location (two at the site of import into the matrix, and the IM again for export). However, one has to bear in mind that the import step into the matrix involves the general import default-pathway, which had to evolve to allow the import of the large number of matrix-targeted proteins. Then, once exposed to the matrix, proteins destined to the IMS would use an export machinery that evolved already in the prokaryotic ancestors of mitochondria. As a result, one does not have to envisage the evolution of new recognition or translocation machineries specific for this subset of mitochondrial proteins, as would be the case with the "stop-transfer" model of IMS sorting.

To date, this putative export machinery has not been identified. It may bear homology to the previously identified Sec machinery of *E.coli*. Interestingly, homologues of SecY and SecE have been identified in eukaryotic ER membranes, suggesting that these proteins have been functionally conserved throughout evolution. With the exception of the Imp proteases, no mitochondrial homologues of the Sec components have been identified. The search for such proteins is presently one of the focal points of our ongoing research.

REFERENCES

Beasley, E. M. (1993). Genetic characterization of the intermembrane space sorting domains of yeast cytochrome b_2. NATO ASI Series 71, 223–233.

Beasley, E. M., Müller, S., & Schatz, G. (1993). The signal that sorts cytochrome b_2 to the mitochondrial intermembrane space contains three distinct functional regions. EMBO J. 12, 2303–2311.

Behrens, M., Michaelis, G., & Pratje, E. (1991). Mitochondrial inner membrane protease 1 of *Saccharomyces cerevisiae* shows sequence similarity to the *Escherichia coli* leader peptidase. Mol. Gen. Genet. 228, 167–176.

Blobel, G. (1980). Intracellular protein topogenesis. Proc. Natl. Acad. Sci. USA 77, 1496–1500.

Blom, J., Kübrich, M., Rassow, J., Voos, W., Dekker, P. J. T., Maarse, A. C., Meijer, M., & Pfanner, N. (1993). The essential yeast protein MIM44 (encoded by MPI1) is involved in an early step of preprotein translocation across the mitochondrial inner membrane. Mol. Cell. Biol. 13, 7364–7371.

Cheng, M. Y., Hartl., F-U., Martin, J., Pollock, R. A., Kalousek, F., Neupert, W., Hallberg, E. M., Hallberg, R. L., & Horwich, A. L. (1989). Mitochondrial heat shock protein hsp60 is essential for assembly of proteins imported into yeast mitochondria. Nature 337, 620–625.

Cyr, D. M., Stuart, R. A., & Neupert, W. (1993). A matrix ATP requirement for presequence translocation across the inner membrane of mitochondria. J. Biol. Chem. 268, 23751–23754.

Dekker, P. J. T., Keil, P., Rassow, J., Maarse, A. C., Pfanner, N., & Meijer, M. (1993). Identification of MIM23, a putative component of the protein import machinery of the mitochondrial inner membrane. FEBS Lett. 330, 66–70.

Drygas, M. E., Labowitz, A. M., & Nargang, F. E. (1989). Cloning and analysis of the *Neurospora crassa* gene for cytochrome *c* heme lysase. J. Biol. Chem. 264, 17897–17907.

Dumont, M. E., Ernst, J. F., & Sherman, F. (1988). Coupling of heme attachment to import of cytochrome *c* into yeast mitochondria. J. Biol. Chem. 263, 15928–1593.

Dumont, M. E., Cardillo, T. S., Hayes, M. K., & Sherman, F. (1991). Role of cytochrome *c* heme lyase in mitochondrial import and accumulation of cytochrome *c* in *S. cerevisiae*. Mol. Cell. Biol. 11, 5487–5496.

Emtage, J. L. T., & Jensen, R. E. (1993). MAS 6 encodes an essential inner membrane component of the yeast mitochondrial protein import pathway. J. Cell Biol. 122, 1003–1012.

Fu, W., Japa, S., & Beattie, D. S. (1990). Import of the iron-sulfur protein of the cytochrome bc_1 complex into yeast mitochondria. J. Biol. Chem. 265, 16541–16547.

Gambill, B. D., Voos, W., Kang, P. J., Miao, B., Langer, T., Craig, E. A., & Pfanner, N. (1993). A dual role for miochondrial heat shock protein 70 in membrane translocation of preproteins. J. Cell. Biol. 123, 109–118.

Glick, B. S., Beasley, E. M., & Schatz, G. (1992a). Protein sorting in mitochondria. Trends Biochem. Sci. 17, 453–459.

Glick, B. S., Brandt, A., Cunningham, K., Müller, S., & Schatz, G. (1992b). Cytochromes c_1 and b_2 are sorted to the intermembrane space of yeast mitochondria by a stop-transfer mechanism. Cell 69, 809–822.

Glick, B. S., Wachter, C., Reid, G. A., & Schatz, G. (1993). Import of cytochrome b_2 to the mitochondrial intermembrane space: The tightly folded heme-binding domain makes import dependent upon matrix ATP. Protein Sci. 2, 1901–1917.

Grivell, L. A. (1989). Nucleo-mitochondrial interactions in yeast mitochondrial biogenesis. Eur. J. Biochem. 182, 477–493.

Gruhler, A., Ono, H., Guiard, B., Neupert, W., & Stuart, R. A. (1995). A novel intermediate on the import pathway of cytochrome b_2 into mitochondria: evidence for conservative sorting. EMBO J. 14, 1349–1359.

Guiard, B. (1985). Structure, expression and regulation of a nuclear gene encoding a mitochondrial protein: The yeast L(+)-lactate cytochrome c oxidoreductase (cytochrome b_2). EMBO J. 4, 3265–3272.

Hallberg, E. M., Shu, Y., & Hallberg, R. L. (1993). Loss of mitochondrial hsp60 function: Nonequivalent effects on matrix-targeted and intermembrane psace targeted proteins. Mol. Cell. Biol. 13, 3050–3057.

Hartl, F-U., Schmidt, B., Wachter, E., Weiss, H., & Neupert, W. (1986). Transport into mitochondria and intramitochondrial sorting of the Fe/S protein of the ubiquinol-cytochrome *c* reductase. Cell 47, 939–951.

Hartl, F-U., Ostermann, J., Guiard, B., & Neupert, W. (1987). Successive translocation into and out of the mitochondrial matrix: Targeting of proteins to the intermembrane space by a bipartite signal peptide. Cell 51, 1027–1037.

Hartl, F-U., & Neupert, W. (1990). Protein sorting to mitochondria: Evolutionary conservations of folding and assembly. Science 247, 930–938.

Höhfeld, J., & Hartl, F-U. (1994). Requirement of the chaperonin cofactor Hsp10 for protein folding and sorting in yeast mitochondria. J. Cell. Biol. 126, 305–315.

Hwang, S. T., & Schatz, G. (1989). Translocation of proteins across the mitochondrial inner membrane, but not the outer membrane, requires nucleoside triphosphates in the matrix. Proc. Natl. Acad. Sci. USA 86, 8432–8436.

Ikeda, E., Yoshida, S., Mitsuzawa, H., Uno, I., & Toh-e, A. (1994). YGE is a yeast homologue of *Escherichia coli grpE* and is required for maintenance of mitochondrial functions. FEBS Lett. 39, 265–268.

Isaya, G., Kalousek, F., & Rosenberg, L. E. (1992). Sequence analysis of rat mitochondrial intermediate peptidase: Similarity to zinc metallopeptidases and a putative yeast homologue. Proc. Natl. Acad. Sci. USA 89, 8317–8321.

Jensen, R. E., Schmidt, S., & Mark, R. J. (1992). Mutations in a 19-amino acid hydrophobic region of the cytochrome c_1 presequence prevent sorting to the mitochondrial intermembrane space. Mol. Cell. Biol. 12, 4677–4686.

Kalousek, F., Isaya, G., & Rosenberg, L. E. (1992). Rat liver mitochondrial intermediate peptidase (MIP): Purification and initial characterization. EMBO J. 11, 2803–2809.

Kang, P. J., Ostermann, J., Shilling, J., Neupert, W., Craig, E. A., & Pfanner, N. (1990). Requirement for Hsp70 in the mitochondrial matrix for translocation and folding of precursor proteins. Nature 348, 137–142.

Kaput, J., Goltz, S., & Blobel, G. (1982). Nucleotide sequence of the yeast nuclear gene for cytochrome *c* peroxidase precursor: Functional implications of the presequence for protein transport into mitochondria. J. Biol. Chem. 257, 15054–15058.

Kaput, J., Brandriss, M. C., & Prussak-Wieckowska, T. (1989). *In vitro* import of cytochrome *c* peroxidase into the intermembrane space: Release of the processed form by intact mitochondria. J. Cell. Biol. 109, 101–112.

Koll, H., Guiard, B., Rassow, J., Ostermann, J., Horwich, A. L., Neupert, W., & Hartl, F-U. (1992). Antifolding activity of hsp60 couples protein import into the mitochondrial matrix with export to the intermembrane space. Cell 68, 1163–1175.

Langer, T., Lu, C., Echols, H., Flanagan, J., Hayer-Hartl, M., & Hartl, F-U. (1992). Successive action of DnaK, DnaJ and GroEL along the pathway of chaperone mediated protein folding. Nature 356, 683–689.

Lill, R., Hergersberg, C., Schneider, H., Söllner, T., Stuart, R. A., & Neupert, W. (1992a). General and exceptional pathways of protein import into submitochondrial compartments. In: Membrane Biogenesis and Protein Targeting (Neupert, W., & Lill, R., Eds.). Elsevier, Amsterdam.

Lill, R., Stuart, R. A., Drygas, M., Nargang, F. E., & Neupert, W. (1992b). Import of cytochrome *c* heme lyase into mitochondria: A novel pathway into the intermembrane space. EMBO J. 11, 449–456.

Maarse, A. C., Blom, J., Grivell, L. A., & Meijer, M. (1992). MPI1, an essential gene encoding a mitochondrial membrane protein, is possibly involved in protein import into yeast mitochondria. EMBO J. 11, 3619–3628.

Mahlke, K., Pfanner, N., Martin, J., Horwich, A. L., Hartl, F-U., & Neupert, W. (1990). Sorting pathways of mitochondrial inner membrane proteins. Eur. J. Biochem. 192, 551–555.

Manning-Krieg, U. C., Scherer, P. E., & Schatz, G. (1991). Sequential action of mitochondrial chaperones in protein import into the matrix. EMBO J. 10, 3273–3280.

Mayer, A., Lill, R., & Neupert, W. (1993). Insertion and translocation of precursor proteins into isolated outer membranes of mitochondria. J. Cell. Biol. 121, 1233–1243.

Neupert, W., Hartl., F-U., Craig, E. A., & Pfanner, N. (1990). How do polypeptides cross the mitochondrial membranes? Cell 63, 447–450.

Nicholson, D. W., Hergersberg, C., & Neupert, W. (1988). Role of cytochrome *c* heme lyase in the import of cytochrome *c* into mitochondria. J. Biol. Chem. 263, 19034–19042.

Nobrega, F. G., Nobrega, M. P., & Tzagoloff, A. (1992). BCS1, a novel gene required for the expression of functional Rieske iron-sulfur protein in *Saccharomyces cerevisiae*. EMBO J. 11, 3821–3829.

Nunnari, J., Fox, T. D., & Walter, P. (1993). A mitochondrial protease with two catalytic subunits of nonoverlapping specificities. Science 262, 1997–2003.

Ono, H., Gruhler, A., Stuart, R. A., Guiard, B., Schwarz, E., & Neupert, W. (1995). Sorting of cytochrome b_2 to the intermembrane space of mitochondria: kinetic analysis of intermediates demonstrates passage through the matrix. J. Biol. Chem. 270, 16932–16938.

Ostermann, J., Horwich, A. L., Neupert, W., & Hartl, F-U. (1989). Protein folding in mitochondria requires complex formation with hsp60 and ATP hydrolysis. Nature 341, 125–130.

Ostermann, J., Voos, W., Kang, P. J., Craig, E. A., Neupert, W., & Pfanner, N. (1991). Precursor proteins in transit through mitochondrial contact sites interact with hsp70 in the matrix. FEBS Lett. 277, 281–284.

Pfaller, R., Steger, H., Rassow, J., Pfanner, N., & Neupert, W. (1988). Import pathways or precursors into mitochondria: Multiple receptor sites are followed by a common insertion site. J. Cell. Biol. 107, 2483–2490.

Pfanner, N., Rassow, J., van der Klei, I., & Neupert, W. (1992). A dynamic model of the mitochondrial protein import machinery. Cell 68, 999–1002.

Puziss, J. W., Fikes, J. D., & Bassford, P. J. (1989). Analysis of mutational alterations in the hydrophilic segment of the maltose-binding protein signal peptide. J. Bacteriol. 171, 2303–2311.

Rassow, J., Hartl, F-U., Guiard, B., Pfanner, N., & Neupert, W. (1990). Polypeptides traverse the mitochondrial envelope in an extended state. FEBS Lett. 275, 190–194.

Rojo, E. E., Stuart, R. A., & Neupert, W. (1995). Conservative sorting of the F_1F_0-ATPase subunit 9: export from matrix requires ΔpH across inner membrane and matrix ATP. EMBO J. 14, 3445–3451.

Rospert, S., Glick, B. S., Jenö, P., Schatz, G., Todd, M. T., Lorimer, G. H., & Viitanen, P. V. (1993). Identification and functional analysis of chaperone 10, the groES homolog from yeast mitochondria. Proc. Natl. Acad. Sci. USA 90, 10967–10971.

Rowley, N., Prip-Buus, C., Westermann, B., Brown, C., Schwarz, E., Barrell, B., & Neupert, W. (1994). Mdj1p, a novel chaperone of the DnaJ family, is involved in mitochondrial biogenesis and protein folding. Cell 77, 249–259.

Sadler, I., Suda, K., Schatz, G., Kaudewitz, F., & Haid, A. (1984). Sequencing of the nuclear gene for the yeast cytochrome c_1 precursor reveals an unusually complex amino-terminal presequence. EMBO J. 3, 2137–2143.

Scherer, P. E., Manning-Krieg, U. C., Jenö, P., Schatz, G., & Horst, M. (1992). Identification of a 45-kDa protein at the import site of the yeast mitochondrial inner membrane. Proc. Natl. Acad. Sci. USA 89, 11930–11934.

Schleyer, M., & Neupert. (1985). Transport of proteins into mitochondria: Translocation intermediates spanning contact sites between outer and inner membranes. Cell 43, 339–350.

Schneider, A., Behrens, M., Scherer, P. E., Pratje, E., Michaelis, G., & Schatz, G. (1991). Inner membrane protease I, an enzyme mediating intramitochondrial sorting in yeast. EMBO J. 10, 247–254.

Schneider, H. C., Berthold, J., Bauer, M. F., Dietmeier, K., Guiard, B., Brunner, M., & Neupert, W. (1994). Mitochondrial Hsp70/MIM44 complex facilitates protein import. Nature 371, 768–774.

Schulze, M., & Rödel, G. (1989). Accumulation of the cytochrome *c* oxidase subunits I and II in yeast requires a mitochondrial membrane associated protein, encoded by the nuclear SCO1 gene. Mol. Gen. Genet. 216, 37–43.

Schwarz, E., Seytter, T., Guiard, B., & Neupert, W. (1993). Targeting of cytochrome b_2 into the mitochondrial intermembrane space: Specific recognition of the sorting signal. EMBO J. 10, 2295–2302

Segui-Real, B., Kispal, G., Lill, R., & Neupert, W. (1993). Functional independence of the protein translocation machineries in mitochondrial outer and inner membranes: Passage of preproteins through the intermembrane space. EMBO J. 12, 2211–2218.

Steiner, H., Zollner, A., Haid, A., Neupert, W., & Lill, R. (1995). Biogenesis of mitochondrial heme lyases in yeast. Import and folding in the intermembrane space. J. Biol. Chem. 270, 22842–22849.

Stuart, R. A., Neupert, W., & Tropschug, M. (1987). Deficiency in mRNA splicing in a cytochrome *c* mutant of Neurospora crassa; Importance of carboxy-terminus for import of apocytochrome *c* into mitochondria. EMBO J. 6, 2131–2137.

Stuart, R. A., Nicholson D. W., & Neupert, W. (1989). Early steps in mitochondrial protein import: Receptor unctions can be substituted by the membrane insertion activity of apocytochrome *c*. Cell 60, 31–43.

Stuart, R. A., & Neupert, W. (1990). Apocytochrome *c*: An exceptional mitochondrial precursor using an exceptional import pathway. Biochimie 72, 115–121.

Stuart, R. A., Koll, H., & Neupert, W. (1993). Intra-mitochondrial sorting of precursor proteins. NATO ASI Series 71, 213–223.

Stuart, R. A., Cyr, D. M., Craig, E. A., & Neupert, W. (1994a). Mitochondrial molecular chaperones: Their role in protein translocation. Trends. Biochem. Sci. 19, 87–92.

Stuart, R. A., Gruhler, A., van der Klei, I. J., Guiard, B., Koll, H., & Neupert, W. (1994b). The requirement of matrix ATP for the import of precursor proteins into the mitochondrial matrix and intermembrane space. Eur. J. Biochem. 220, 9–18.

Ungermann, C., Neupert, W., & Cyv, D. M. (1994). The role of Hsp70 in conferring unidirectionality on protein translocation into mitochondria. Science 266, 1250–1253.

Van Doren, S. R., Yun, C.-H., Crofts, A. R., & Gennis, R. B. (1993). Assembly of the Rieske iron-sulfur subunit of the cytochrome bc_1 complex in the *Escherichia coli* and *Rhodobacter spaeroides* membranes independent of the cytochrome b and c_1 subunits. Biochem. 32, 628–636.

Van Loon, A. P. G. M., Brändli, A. W., & Schatz, G. (1986). The presequences of two imported mitochondrial proteins contain information for intracellular and intramitochondrial sorting. Cell 44, 801–812.

Van Loon, A. P. G. M., & Schatz, G. (1987). Transport of proteins to the mitochondrial intermembrane space: The "sorting" domain of the cytochrome c_1 presequence is a stop-transfer sequence specific for the inner membrane. EMBO J. 6, 2441–2448.

Van Loon, A. P. G. M., Brändli, A. W., Pesold,-Hurt, B., Blank, D., & Schatz, G. (1987). Transport of proteins to the mitochondrial intermembrane space: The "matrix-targeting" and the "sorting" domains in the cytochrome c_1 presequence. EMBO J. 8, 2433–2439.

Voos, W., Gambill, B. D., Guiard, B., Pfanner, N., & Craig, E. A. (1993). Presequence and mature part of preproteins strongly influence the dependence of mitochondrial protein import on Hsp70 in the matrix. J. Cell Biol. 123, 119–126.

Wachter, C., Schatz, G., & Glick, B. S. (1992). Role of ATP in the intramitochondrial sorting of cytochrome c_1 and the adenine nucleotide translocator. EMBO J. 11, 4787–4794.

Xia, Z-X., & Matthews, F. S. (1990). Molecular structure of flavocytochrome b_2 at 2.4A resolution. J. Mol. Biol. 212, 837–863.

ENERGETICS OF MITOCHONDRIAL PROTEIN IMPORT AND INTRAMITOCHONDRIAL PROTEIN SORTING

Sabine Rospert

I.	A Thumbnail Sketch of Mitochondrial Import	222
II.	The Membrane Potential	223
III.	The ATP Requirement	226
IV.	Cytosolic Factors Involved in Import into Mitochondria	228
	A. The Cytosolic Hsp70 Chaperone System	228
	B. Cytosolic Factors Specifically Interacting with Mitochondrial Proteins Confer Import Competence	229
V.	Is Mitochondrial Hsp70 an ATP-Driven Import Motor?	232
VI.	ATP Is Not Needed for Import Across the Outer Membrane—A Rule with Interesting Exceptions	234
VII.	Outlook	239
	Acknowledgments	239
	References	240

Advances in Molecular and Cell Biology,
Volume 17, pages 221–245.

ISBN: 0-7623-0144-9

I. A THUMBNAIL SKETCH OF MITOCHONDRIAL IMPORT

Mitochondria are indispensable organelles of the eukaryotic cell (Schatz, 1993b). According to the commonly accepted endosymbiont hypothesis, these organelles are derived from ancient prokaryotic organisms (Gray, 1989). Indeed, mitochondria show many typical eubacterial features: (1) They are surrounded by two membranes, as are Gram-negative bacteria; (2) their inner membrane contains a respiratory chain similar to the one from enterobacteria; (3) mitochondria, like bacteria, contain a covalently closed, circular DNA that is not packaged into protein-containing chromosomes; (4) mitochondrial ribosomes, unlike cytosolic ribosomes, are sensitive to several antibacterial antibiotics, and (5) the mitochondrial form of an isozyme pair is usually much more bacteria-like than the cytosolic isoform.

Evolution of mitochondria probably involved extensive transfer of genetic information from the endosymbiont to the host nucleus. As a result, present-day mitochondria are totally dependent on their host cell. The vast majority of mitochondrial proteins are encoded by nuclear genes, synthesized in the cytosol and then imported into the mitochondria. These proteins contain specific signals that target them to the correct location (Hurt and van Loon, 1986; Roise and Schatz, 1988; Hartl et al., 1989). The signals are probably additions to what once were prokaryotic proteins. The targeting signals usually reside at the N-terminus of the mitochondrial "precursor protein" and lack any consensus in primary sequence. Their main common feature is the potential to form a positively charged, amphiphilic structure such as an α-helix (Roise et al., 1986; von Heijne, 1986; Roise, 1993). In most cases, the targeting sequences are proteolytically removed upon import. As mitochondria are subdivided into four compartments (matrix, inner membrane, intermembrane space, and outer membrane), the targeting signal must contain additional information for delivering the attached protein to its correct submitochondrial location (Hurt and van Loon, 1986; Schatz, 1987; Glick et al., 1992a; Segui-Real et al., 1992).

Import of proteins into mitochondria is a multistep process, involving a complex machinery (Hannavy et al., 1993; Horst et al., 1993; Neupert and Pfanner, 1993). Although many components of the import machinery have been identified, it is likely that additional components are involved (see Kronidou et al., this volume). In the narrow sense, the import-machinery consists of components in the outer and inner membrane. These components include outer membrane proteins functioning as receptors for precursor recognition (Verner, 1992), and proteins forming a multisubunit transport channel in each of the two membranes. At sites where the two membranes are in close contact, the two channels may link up to allow precursor proteins to pass across both membranes (Glick et al., 1991). However, protein import into mitochondria also requires soluble proteins in the cytosol and in the mitochondrial matrix (Schatz, 1993a). In general, these soluble proteins are either chaperones that modulate the conformation of imported precursors, or proteases that remove targeting signals.

Most proteins cannot traverse membranes in their native, folded conformation. Consequently, unfolding and refolding are necessary during protein import. Protein folding *in vivo* requires molecular chaperones (Craig et al., 1993) that mediate the folding of proteins without becoming part of the final structure. The two best-studied classes of chaperones involved in import into mitochondria are hsp70 and hsp60. Distinct hsp70-type proteins are present in the cytosol and in the mitochondrial matrix, whereas hsp60 is present exclusively in the mitochondrial matrix. Hsp60 seems to be restricted to prokaryotes and organelles derived from prokaryotic ancestors. At least in some cases, it is involved in refolding of protein precursors after import into the mitochondrial matrix (Cheng et al., 1989; Ostermann et al., 1989; Manning-Krieg et al., 1991).

Protein transport into mitochondria requires energy. The energy is supplied as $\Delta\Psi$ and as ATP (Beasley et al., 1992; Glick et al., 1992c). $\Delta\Psi$ is required for precursors whose import involves permanent or transient integration into the mitochondrial inner membrane. The import of most mitochondrial precursors, and of all precursors transported into the matrix space, requires ATP. ATP is probably required as a substrate for the action of hsp70, of hsp60, and perhaps of other chaperones as well.

The *in vitro* import systems described in the following sections are assays of post-translational import. These assays have been successfully used to solve many important questions concerning the biogenesis of mitochondria. Post-translational import is efficient and does take place *in vivo* (Nelson and Schatz, 1979; Wienhues et al., 1991). However, there is evidence to suggest that import of some precursors *in vivo* can also be co-translational (Ades and Butow, 1980a, 1980b; Suissa and Schatz, 1982; Verner, 1993).

II. THE MEMBRANE POTENTIAL

The electrochemical potential across the mitochondrial inner membrane consists of a pH gradient and an electrical potential ($\Delta\Psi$). Precursor import into isolated mitochondria requires only the electrical potential. Pfanner and Neupert showed that the import of the ADP-ATP translocator into isolated mitochondria could be driven by a valinomycin-induced potassium diffusion potential, and that this import was not inhibited by eliminating the pH gradient with a protonophore (Pfanner and Neupert, 1985; see Table 1).

The current model suggests that the potential, which is negative inside, electrophoretically drives the insertion of the positively charged presequence into the import channel of the inner membrane. Indeed, the more positive charges there are in a precursor's targeting signal, the lower is the threshold potential for import (Martin et al., 1991). Interestingly, the threshold is unaffected by the overall length of the precursor, by the amount of "mature" region of the precursor which is

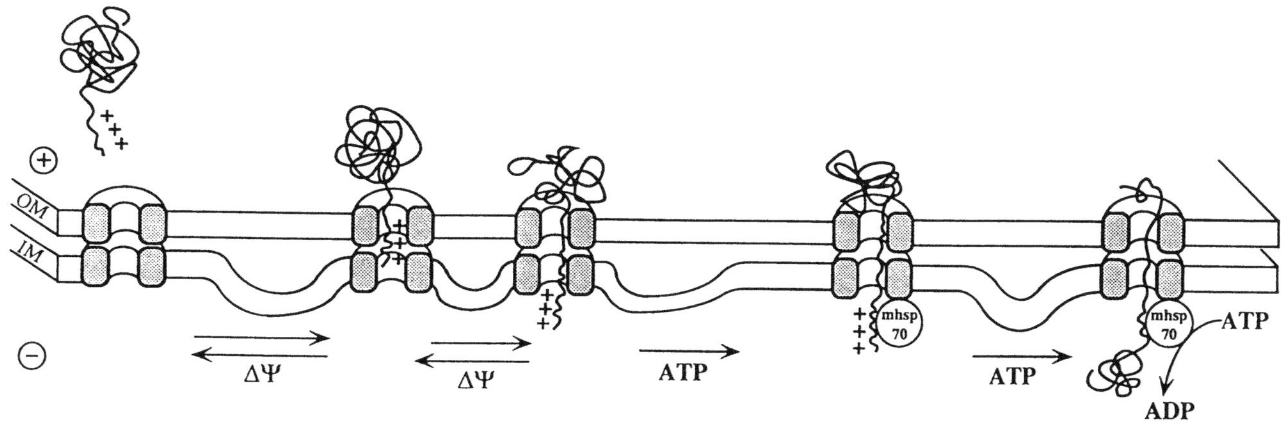

Figure 1. Membrane potential-dependent insertion of the presequence of a mitochondrial precursor into the inner membrane. The presequence of the mitochondrial protein (indicated by the positive charges) interacts with components of the import channel in the inner membrane. This interaction is reversible until interaction with matrix-localized mHsp70 occurs.

Table 1. Commonly Used Inhibitors of Mitochondrial Protein Import

Inhibitor	*Target*	*Mechanism*
Apyrase	ATP level	Cleaves the terminal phosphoryl group from ATP and ADP
Oligomycin	Fo portion of the ATP synthase	Blocks H^+-transport through Fo
Efrapeptin	F_1 portion of the ATP synthase	?
Carboxyatractyloside (CAT)	ADP-ATP translocator	Inhibits by binding to the translocator from the outside of the inner membrane
Carbonyl cyanide p-trifluoromethoxyphenylhydrazone (FCCP)	Δ pH	Increases the permeability of the inner membrane for protons
Valinomycin + K^+	$\Delta\Psi$	Carrier for K^+, collapses $\Delta\Psi$ by carrying K^+ down the electric potential across the inner membrane

attached to the presequence, or by the folding state of the precursor (Martin et al., 1991). The potential thus acts exclusively on the presequence and not on the mature part of the protein (Schleyer and Neupert, 1985; Roise and Schatz, 1988).

A wealth of additional observations supports the view that the membrane potential is only required for an early step of protein translocation across the inner membrane. When precursors of matrix proteins are presented to mitochondria at low temperature (Schleyer and Neupert, 1985), or to ATP-depleted mitochondria (Hwang et al., 1991), they remain stuck across one or both mitochondrial membranes. These import intermediates can then be chased into the matrix either by raising the temperature (Schleyer and Neupert, 1985) or by adding ATP (Hwang et al., 1991). In both cases, the chase does not require a membrane potential. Lowering intramitochondrial ATP levels (see following) traps a very early translocation intermediate that depends on $\Delta\Psi$ to remain stuck. $\Delta\Psi$ is also required to enable these translocation intermediates to be chased into the matrix by ATP (Cyr et al., 1993; Glick et al., 1993). The targeting signal of these intermediates has probably already inserted into the inner membrane, but the insertion remains reversible as long as matrix ATP levels remain low. Thus, the membrane potential is sufficient to drive the insertion of the presequence into the inner membrane, but ATP in the matrix is required to make the insertion irreversible (see Figure 1).

A membrane potential is required for the import of all proteins that insert into, or traverse the inner mitochondrial membrane. In contrast, transport of proteins into the mitochondrial outer membrane is potential-independent (Freitag et al., 1982;

Mihara et al., 1982; Gasser and Schatz, 1983; Pfaller et al., 1990). The same is true for cytochrome *c*, a protein located in the intermembrane space. Apocytochrome *c* appears to be exceptional in that it can diffuse across the outer membrane without the aid of a proteinaceous translocation machinery. Its unidirectional import is probably driven by the covalent attachment of heme in the intermembrane space and is independent of either $\Delta\Psi$ or ATP (Stuart and Neupert, 1990).

III. THE ATP REQUIREMENT

The ATP requirement for protein import into the matrix of isolated mitochondria was demonstrated in three independent studies (Pfanner and Neupert, 1986; Chen and Douglas, 1987; Eilers et al., 1987). These studies established the current view that mitochondrial protein import requires both a membrane potential and ATP. However, they did not settle the question of whether ATP was required outside the mitochondria, inside the mitochondria, or at both locations. Initially, it was generally felt that ATP acted only outside the matrix in order to establish or maintain a loose, translocation-competent conformation of the precursor protein (discussed in Wachter et al., 1994). However, Hwang and Schatz (1989) showed that transport of a protein into the matrix requires ATP inside the inner membrane. If ATP in the matrix is depleted, a matrix-targeted precursor accumulates as a translocation intermediate that is stuck across the inner membrane. This intermediate can be chased into the matrix by addition of ATP. As the chase is blocked by carboxyactractyloside (Table 1), the added ATP must be transported into the matrix in order to act. Whereas the requirement for external ATP can be overcome by destabilizing the precursor (Pfanner et al., 1987, 1988; Verner and Schatz, 1987; Chen and Douglas, 1988), this is not the case for the ATP requirement inside. However, translocation of matrix-targeted proteins across the outer membrane is possible in the absence of matrix ATP (Hwang et al., 1991; Rassow and Pfanner, 1991). In the absence of matrix ATP, matrix-targeted precursors are imported across the outer membrane but do not cross the inner membrane. The translocation intermediate generated in the absence of matrix ATP is membrane-bound and exposed to the intermembrane space. As it can be chased by addition of ATP, it is on its correct import pathway (Figure 2).

It seems likely that earlier reports on ATP-independent import into the matrix reflected incomplete depletion of matrix ATP (Chen and Douglas, 1988) or translocation of the precursor across the outer, but not the inner membrane (Verner and Schatz, 1987; Ostermann et al., 1989; Pfanner et al., 1990). Wachter and colleagues (1994) also showed that the requirement for external ATP could not be met by other nucleotides and that it was restricted to a subset of precursors (Table 2). Import of some precursors was independent of external ATP whereas that of others was partly or strongly dependent.

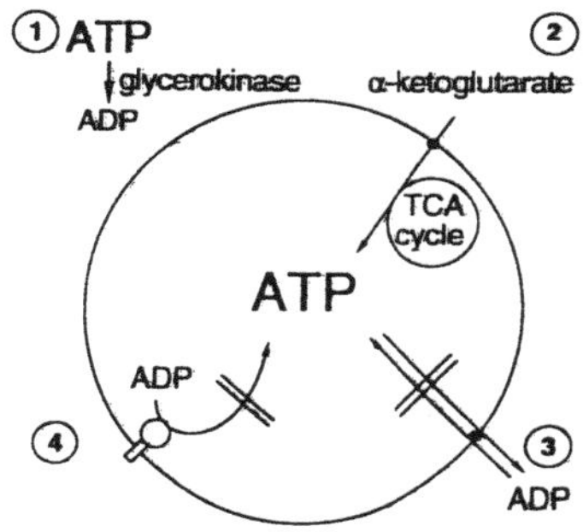

Figure 2. Experimental system for studying the ATP requirements for mitochondrial protein import. In this schematic drawing only the mitochondrial inner membrane is represented because the outer membrane is permeable to small molecules. (1) ATP is either added to the incubation medium, or is converted to ADP by an enzymic trap such as glycerokinase/glycerol. (2) To generate matrix ATP, α-ketoglutarate is added. This metabolite is transported into the matrix by a specific carrier, and then produces ATP by substrate-level phosphorylation in the tricarboxylic acid (TCA) cycle. (3) Carboxyatractyloside (CAT) is added to inhibit the adenine nucleotide translocator and prevent passage of ATP and ADP across the inner membrane. (4) The mitochondrial ATP synthase is blocked by adding a combination of oligomycin and efrapeptin.

Table 2. ATP Requirements for Import of Authentic and Artificial Mitochondrial Precursor Proteins

Precursor	*Destination*	*Matrix ATP*	*External ATP*	*References*
$F_1\beta$	matrix	+	+	Wachter et al. (1994)
MPPα[a]	matrix	+	+	Wachter et al. (1994)
Fe/S protein[a]	matrix	+	+	Wachter et al. (1994)
ADHIII[a]	matrix	+	+	Wachter et al. (1994)
hsp60	matrix	+	–	Wachter et al. (1994)
CoxIV	matrix	+	–	Wachter et al. (1994)
$F_1\beta$-Δ(94-381)	matrix	+	–	Wachter et al. (1994)
CoxIV-DHFR[b]	matrix	+	–	Hwang and Schatz (1989)
cytochrome b_2 [c]	IMS	+	–	Glick et al. (1993)
CoxVa	IM	+	–	Miller and Cumsky (1991) Cyr et al. (1993)
cytochrome c_1	IMS	–	+	Wachter et al. (1992)
adenine nucleotide translocator	IM	–	+	Wachter et al. (1992)
porin[d]	OM	–	+	Hwang and Schatz (1989)
b_2(1–167)-DHFR	IMS	–	–	Glick et al. (1992a, 1993)
c_1(1–64)-DHFR	IMS	–	–	Glick et al. (1992a)
cytochrome *c* heme lyase	IMS	–	–	Lill et al. (1992)

Notes: [a]The requirement for external ATP can be overcome by urea-denaturation (Wachter et al., 1994).
[b]Import of other matrix-targeted DHFR fusion proteins also requires only matrix ATP (Wachter et al., 1992; C. Wachter, unpublished observations).
[c]Like authentic cytochrome b_2, fusion protein derivatives of cytochrome b_2 containing the intact heme-binding domain require only matrix ATP (Glick et al., 1993; Voos et al., 1993).
[d]The requirement for external ATP can be overcome by acid-base treatment of the precursor (Pfanner et al., 1988).
OM: outer membrane; IMS: intermembrane space; IM: inner membrane.

IV. CYTOSOLIC FACTORS INVOLVED IN IMPORT INTO MITOCHONDRIA

The requirement for external ATP has been traditionally explained by cytosolic proteins directly interacting with mitochondrial precursors. However, it is not yet clear how and why this interaction occurs (Lithgow et al., 1993a).

A. The Cytosolic Hsp70 Chaperone System

In 1982 the eukaryotic family of 70 kDa heat shock proteins (Hsp70s; Ingolia and Craig, 1982; Ingolia et al., 1982) was discovered. Functional studies on this class of proteins suggested that they might have an unfolding function: after heat shock, inducible members of this protein family underwent hydrophobic binding to nuclear proteins, which could then be rapidly and specifically released by adding ATP (Lewis and Pelham, 1985). As it became apparent that proteins are transported across membranes in an incompletely folded state (Eilers and Schatz, 1986), it was suggested that cytosolic hsp70 proteins might facilitate import by functioning as "unfoldases" (Rothman and Kornberg, 1986).

Binding of newly synthesized proteins to cytosolic Hsp70s was first shown in HeLa cells (Beckmann et al., 1990). The binding was transient and it was dependent on ATP. Direct binding of mitochondrial precursors to cytosolic Hsp70 has not been demonstrated so far. However, a yeast strain in which the level of cytosolic Hsp70s could be regulated indicated the involvement of Hsp70 in mitochondrial import (Deshaies et al., 1988). Depleting three of the cytosolic Hsp70s led to an accumulation of the precursor of F_1-ATPase β-subunit in the cells. The requirement of cytosolic Hsp70 (and/or other heat shock proteins) for import was also indicated by the finding that a temperature-sensitive allele of the yeast heat-shock factor (HSF) gene, which is responsible for the heat-shock response in yeast, caused accumulation of the precursors of $F_1\beta$ and citrate synthase at elevated temperatures (Smith and Yaffe, 1991). Blobel and coworkers showed that at least two factors in the yeast cytosol were indispensable for post-translational import of delta-pyrroline-5-carboxylate dehydrogenase into mitochondria. One factor was a 70 kDa heat shock-related protein, the other a protein sensitive to *N*-ethylmaleimide. Both proteins acted synergistically (Murakami et al., 1988).

The NEM-sensitive protein in the yeast cytosol could be identical to a cytosolic DnaJ-like protein, which is encoded by the YDJ1/MAS5 gene. This protein participates in mitochondrial protein import in yeast (Atencio and Yaffe, 1992). DnaJ-type proteins generally function together with Hsp70 by regulating the latter's ATPase activity. This was shown with a cytosolic Hsp70 homologue (SSA1p) and the purified cytosolic DnaJ (Ydj1p). Ydj1p and ATP significantly decreased the affinity of Hsp70 for carboxymethylated α-lactalbumin as well as for prebound substrate (Cyr et al., 1992). The participation of DnaJ *in vivo* was shown with the aid of a temperature-sensitive YDJ1 allele. Import of several mitochondrial proteins was defective at the nonpermissive temperature (Caplan et al., 1992).

The studies mentioned so far imply that hsp70 is the chaperone interacting directly with nascent polypeptides and that a DnaJ homologue modulates the ATPase activity and/or the substrate affinity of hsp70. However, recent experimental evidence shows that DnaJ itself, when added to a wheat germ translation system, can bind to nascent polypeptides. This interaction was shown by co-translationally cross-linking DnaJ to a polypeptide chain as short as 55 residues (Hendrick et al., 1993). In contrast to previous findings (Beckmann et al., 1990; Chirico, 1992; Nelson et al., 1992), no stable interaction of nascent polypeptides with Hsp70 was observed in this study. An interaction of cytosolic Hsp70 or DnaJ homologue with precursors of authentic mitochondrial proteins has so far not been reported.

The original proposal that precursors are unfolded actively by chaperone-like proteins has not been disproven, but has become less attractive. None of the numerous *in vitro* studies detected an "unfoldase" activity of Hsp70 proteins. The only exception appears to be the RNA polymerase from *E. coli*, which can be disaggregated by either groEL/ES or by the Hsp70/DnaJ/GrpE system. However, not even in this case can an indirect role for the chaperones be ruled out (Ziemienowicz et al., 1993). Very recently, a cytosolic chaperone (MSF) has been identified that can apparently disaggregate a mitochondrial precursor (Hachiya et al., 1993; see also following). However, the best-documented general function of cytosolic chaperones appears to be prevention (rather than reversal) of aggregation. Cytosolic chaperones would thus mediate mitochondrial protein import by preventing folding or aggregation, thereby preserving precursors in an import-competent state (Figure 3).

Interaction of precursor proteins with cytosolic chaperones possessing ATPase function probably accounts for the requirement for external ATP: in the absence of external ATP, precursors are not released from chaperones. Conversely, a precursor that does not require ATP in the cytosol probably does not require the mediation of an ATP-dependent chaperone (Figure 3). For example, DHFR folds into its native, three-dimensional structure after synthesis in the reticulocyte lysate (Eilers and Schatz, 1986) and does not require external ATP for import. Thus, unfolding of the DHFR moiety can occur independently of cytosolic ATP, most likely on the surface of the outer membrane (Endo and Schatz, 1988).

B. Cytosolic Factors Specifically Interacting with Mitochondrial Proteins Confer Import Competence

Starting in 1983, several laboratories reported that import of proteins into isolated mitochondria was stimulated by cytosolic proteins (Argan et al., 1983; Miura et al., 1983; Ohta and Schatz, 1984). This requirement is particularly striking if import of a purified precursor is being studied. Ohta and Schatz (1984) purified the precursor

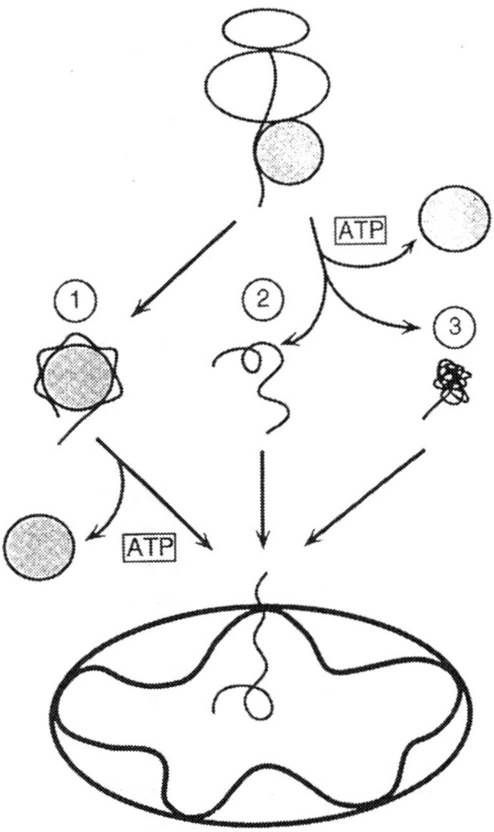

Figure 3. Import of different precursor proteins differs in the requirement for external ATP. Cytosolic chaperones (filled circles) bind to nascent precursor chains on free ribosomes. Precusors can then follow one of three pathways. (1) Cytosolic chaperones remain bound to the precursors until import occurs. Release from the chaperones requires ATP. (2) The precursor is released from the chaperones in the cytosol and remains loosely folded until it engages the translocation machinery. Import is independent of external ATP. (3) The precursor is released from the chaperone, and then undergoes partial or complete folding. It is again unfolded during import, which is independent of external ATP.

of the F_1-ATPase β subunit from yeast cells that had been inhibited with an uncoupler of the mitochondrial membrane potential, and showed that import was absolutely dependent on a cytosolic fraction from either yeast or reticulocytes. The factor was partially purified and proved to be nondialyzable and trypsin-sensitive, but it was not checked whether it required ATP for activity.

A soluble factor of approximately 200 kDa consisting of 28 kDa subunits was detected in a reticulocyte lysate by Ono and Tuboi (1988, 1990a, 1990b). This cytosolic factor was shown to facilitate binding of authentic mitochondrial precursors to the surface of mitochondria. Binding and import of ornithine aminotransferase and sulfite oxidase were inhibited by IgGs raised against the purified factor. As the cytosolic factor did not bind to an ATP-agarose column, the factor probably functions independently of ATP.

Yet another soluble factor of approximately 400 kDa, composed of identical 50 kDa subunits, was purified. The high-molecular weight complex binds to presequences of a subset of mitochondrial precursors and was identified in reticulocytes by Mori and coworkers (Murakami and Mori, 1990; Murakami et al., 1992). Sequencing of the human cDNA corresponding to this factor showed no homology to known proteins (M. Mori, personal communication). The factor (termed presequence binding factor or PBF) binds to the precursor of ornithine carbamoyltransferase (OTC) and to the corresponding presequence peptide, but not to mature OTC. Reticulocyte lysate depleted of PBF failed to import pOTC and some other precursors and purified PBF restored import. Hsp70 enhanced the stimulating effect but did not stimulate import of these precursors in the absence of PBF. There is no indication that PBF utilizes ATP, although the stimulation of its activity by Hsp70 might suggest that it is part of a more complex targeting pathway that requires ATP. PBF might possibly be identical to the NEM-sensitive factor identified in reticulocyte lysate by Sheffield and colleagues (1990). The factor is similar to PBF in that it maintains pOTC in an import-competent state, and that its activity seems also to be presequence-dependent and independent of ATP. The active component could be partially purified as a complex of approximately 200 kDa which also contained Hsp70. However, hsp70 by itself did not confer import competence to the pOTC (Sheffield et al., 1990).

The latest addition to the group of soluble cytosolic factors involved in mitochondrial protein import is the "mitochondrial import stimulation factor" (MSF) purified from rat liver cytosol (Hachiya et al., 1993). MSF is a heterodimer of a 30 and a 32 kDa subunit. The amino acid sequences of the two subunits shows no homology with the known molecular chaperones belonging to the hsp families, but were highly homologous to the 14-3-3 protein family, whose participation in intracellular protein traffic had not previously been described (T. Omura, personal communication). MSF differs from all chaperones described so far in that it can "de-polymerize" aggregates of an *in vitro*-synthesized precursor such as adrenodoxin. The unfolding activity of this factor might possibly be related to its relatively high ATPase activity, which is about 200-fold higher than that of the peptide-activated ATPase activity of Hsp70 (Flynn et al., 1990). So far, MSF is the only example of a chaperone possessing ATP-dependent "unfoldase" activity. This activity is specific and seems to be restricted to mitochondrial precursors; it fits the criteria initially suggested for proteins facilitating transport across membranes (Rothman and Kornberg, 1986).

Cytosolic proteins involved in mitochondrial protein import thus fall into three groups. Members of the first group require ATP for function and include cytosolic

Hsp70-type proteins, proteins acting in concert with Hsp70 such as DnaJ (Caplan et al., 1993), and perhaps a GrpE homologue and PBF. Members of the second group also require ATP, but are different from Hsp70. This group is represented by MSF. Binding of precursor proteins to factors of one of these two groups probably accounts for the observed ATP requirement outside the mitochondrial inner membrane. A third possible group of chaperones functions independently of ATP. Putative members of this group are PBF—if it functions independently of hsp70—and the cytosolic factor described by Ono and Tuboi. Chaperones of this third group might mediate import of those precursors that do not require external ATP. The first group of chaperones is involved in a multitude of cellular processes, including protein targeting to the ER, and probably plays a general role in recognizing unfolded, mature parts of many precursor proteins. A cytosolic Hsp70 has been detected bound to the surface of the outer mitochondrial membrane (Lithgow et al., 1993b). In contrast, chaperones of the other two groups specifically recognize mitochondrial targeting sequences. Thus, they may only function in mitochondrial protein import by conferring import-competence to the precursor proteins and/or targeting them to mitochondrial receptors.

V. IS MITOCHONDRIAL HSP70 AN ATP-DRIVEN IMPORT MOTOR?

In 1989 it was shown that the protein encoded by the *SSC1* gene is the only essential member of the yeast Hsp70-multigene family, and that this protein is present in the mitochondrial matrix (Craig et al., 1987, 1989). A year later, two groups showed that this mitochondrial Hsp70 (mHsp70) was involved in the import of proteins into mitochondria (Kang et al., 1990; Scherer et al., 1990). Import of proteins into isolated mitochondria containing a temperature-sensitive mhsp70 protein was strongly inhibited after preincubation of the organelles at the nonpermissive temperature. Precursor proteins could be coimmunoprecipitated with antibodies against mHsp70. However, binding of precursors to wild type mhsp70 could not be detected (Kang et al., 1990; Ostermann et al., 1990). A direct interaction of imported precursor with wild type mHsp70 was detected by Scherer and colleagues (1990). mHsp70 could be crosslinked to a translocation-arrested precursor protein that spanned both membranes. Binding of mHsp70 was observed to both precursor and mature form of the imported protein, indicating that binding was not restricted to the presequence. In a subsequent study (Manning-Krieg et al., 1991), it was shown that all matrix-targeted precursors tested became transiently bound to mHsp70. If precursor was added to mitochondria that maintained a membrane potential but were depleted of ATP, it became stuck across the inner membrane (Hwang et al., 1991), but did not bind to mHsp70. Readdition of ATP caused transient binding to mHsp70 and complete import into the matrix. That ATP is necessary for binding of mHsp70 to precursor proteins was also indicated by the observation that under "low-ATP conditions"—in presence of apyrase but absence of ATPase inhibitors—a stable binding of mHsp70 to translocation-arrested precursor proteins could be shown (S. Rospert, unpublished results). These findings indicate that low levels of ATP are needed to establish binding of a precursor to mHsp70 and that the release of the precursor requires higher levels of ATP.

Manning-Krieg and colleagues (1991) reported that precursors are released from mHsp70 as incompletely folded chains. In contrast, Kang and colleagues suggested that mHsp70 mediated the folding of imported proteins. Their suggestion was based on the observation that denaturation of a matrix-targeted precursor appeared to bypass the requirement of import for mHsp70. However, the temperature-sensitive mHsp70 used for their studies was still partially active at the restrictive temperature. Indeed, import into the matrix is not possible in the absence of ATP or of functional mHsp70 even if the precursor is denatured (Eilers et al., 1988; Gambill et al., 1993; Glick et al., 1993; Wachter et al., 1994). Thus, it is not possible to separate the "import function" from the "folding function" by studying import into mitochondria whose mutated mHsp70 can still mediate import. A mutant mHsp70 defective in folding but not in import is not yet available. It would be interesting to see if such a mutation (if it can be obtained) is lethal for cell growth, especially as the mHsp70 homologue of *E. coli* is not essential for growth of this bacterium at low temperature.

So far, the best indication that mHsp70 is the "import-motor" has come from two recent studies by Pfanner and coworkers (Gambill et al., 1993; Voos et al., 1993). The authors made use of the already described temperature-sensitive mHsp70 mutant (ssc1-2) but also used a second temperature-sensitive mutant of mHsp70, ssc1-3. Each of these mHsp70 alleles contains a single point mutation, but the location of the mutations is different. The mutation in ssc1-2 is located in the peptide-binding domain of mHsp70. Import of precursors into mitochondria isolated from this mutant is retained, but precursors are imported into the matrix and processed to the mature form. Thus, this mutant mHsp70 behaves like a slowed-down wild type. In contrast, the mutation in ssc1-3 maps in the ATPase domain of the mHsp70. The phenotype of this mutation is more severe and mimics the situation observed upon ATP-depletion: proteins can still cross the mitochondrial outer membrane, but fail to enter the matrix. In contrast to wild type and ssc1-2 mHsp70, the ssc1-3p mHsp70 was not bound to imported proteins. This study shows that import of proteins across the inner membrane requires not only ATP in the matrix but also mHsp70 with a functional ATPase domain.

It has been proposed that mHsp70 drives protein import into mitochondria by binding to translocating precursors. In one version of this model, precursors oscillate in the translocation channel due to Brownian motion; binding of mhsp70 to a precursor would prevent the reverse oscillation, thereby driving the import into the matrix. Thus, ATP-dependent binding would convert the random movement of an unfolded polypeptide chain into a directed movement into the matrix (Simon et al., 1992; Neupert and Pfanner, 1993). However, this model does not readily explain the observation that mHsp70 is involved in unfolding the tightly folded domain of cyt b_2 on the outside of the mitochondrion (Gambill et al., 1993; Glick et al., 1993). It is difficult to envision how passive binding of mHsp70 to the cyt b_2 presequence could pull this folded structure across the outer membrane. Perhaps mHsp70 has a more active function in translocation as it has been proposed for import into the ER (Brodsky and Schekman, 1993).

Some proteins imported into the matrix depend on a second chaperonin system that consists of two components, Hsp60 (McMullin and Hallberg, 1988; Cheng et al., 1989; Reading et al., 1989) and cpn10 (Lubben et al., 1990; Hartman et al., 1992; Rospert et al., 1993a, 1993b). These two components are homologous to groEL and groES from prokaryotes and, like their bacterial counterparts, mediate protein folding (Hendrick and Hartl, 1993; Lorimer et al., 1993). Both chaperonins are essential for viability of yeast (Cheng et al., 1989; Rospert et al., 1993b), implying that their function in folding of proteins in the matrix is essential. Indeed, a yeast mutant expressing a temperature-sensitive Hsp60 protein accumulated mitochondrial precursors in the cytosol at the nonpermissive temperature (Cheng et al., 1989). Also, some proteins imported into the matrix bind transiently (Manning-Krieg et al., 1991) or stably (Rospert et al., 1994) to Hsp60. As Hsp60 possesses ATPase activity, proteins that are imported into the matrix and require Hsp60 and cpn10 for folding depend on matrix ATP not only for their translocation but also for their correct folding.

VI. ATP IS NOT NEEDED FOR IMPORT ACROSS THE OUTER MEMBRANE—A RULE WITH INTERESTING EXCEPTIONS

So far, we have discussed the energy requirements of proteins transported into the mitochondrial matrix. Most of them are soluble proteins, but some of them are components of the inner membrane that assemble into that membrane from the matrix side. The best-studied example is the Fe/S protein that is first completely imported into the matrix in a reaction requiring a membrane potential and matrix ATP, and is then assembled within the other components of complex III in the inner membrane. The matrix-localized chaperone Hsp60 is required for this sorting process (Hartl et al., 1986; Cheng et al., 1989; Kang et al., 1990). In mammalian mitochondria, the precursor of the Fe/S protein is cleaved once by the general matrix peptidase (Brandt et al., 1993). In yeast mitochondria, it is first cleaved by the general matrix peptidase, and then by the mitochondrial intermediate peptidase. Both cleavages occur in the matrix (Fu et al., 1990). Thus, the sorting of the Fe/S protein resembles the sorting of matrix-targeted proteins in every respect. Although the protein is imported into the inner membrane, its import pathway requires matrix ATP.

The soluble intermembrane space protein cytochrome *c* heme lyase (CCHL) uses the same outer membrane import channel as matrix-localized proteins, but does not require the components of the inner membrane channel. As CCHL is only transported across the outer membrane, it requires neither matrix ATP nor a membrane potential. The force driving import of this protein is unknown, but could be the folding of the protein in the intermembrane space (Lill et al., 1992). Import of the ADP-ATP translocator, like that of CCHL, is independent of matrix ATP, suggesting that it assembles into the inner membrane from the outer face of that membrane without a "detour" through the matrix (Mahlke et al., 1990; Wachter et al., 1992). As discussed earlier, cytochrome *c* is yet another intermembrane space protein whose import is independent of components in the inner membrane or the matrix. As expected, import of cytochrome *c* is independent of matrix ATP.

A third import mechanism is used by a group of intermembrane space proteins characterized by their complex targeting signal. The best-studied examples are cytochrome c_1 (cyt c_1), a subunit of complex III that is anchored by its C-terminus in the inner membrane, and cytochrome b_2 (cyt b_2), a soluble enzyme of the intermembrane space. This group probably also includes cytochrome *c* peroxidase, creatine kinase, and nucleoside diphosphokinase. The targeting signals of these proteins are bipartite. The amino-terminal portion resembles a matrix-targeting signal; the carboxy-terminal portion contains a hydrophobic region and acts as sorting signal for the intermembrane space. Mutations that decrease the hydrophobicity or the helix-forming potential of this hydrophobic stretch cause missorting of the precursor to the matrix (Jensen et al., 1992; Beasley et al., 1993; Schwarz et al., 1993). Import of these proteins depends on an energized inner membrane and is accompanied by the proteolytic removal of the presequence in two successive steps. The first proteolytic cleavage takes place in the matrix (Yang et al., 1988); the second takes place on the outer face of the inner membrane and generates the mature form of the protein in the intermembrane space (Schneider et al., 1991, 1994; Nunnari et al., 1993; Figure 4).

The study of this complex pathway has been technically challenging and, perhaps not surprisingly, has led to two different models. In both models, the proteins use the common translocation machinery of the outer and the inner membranes. According to the "stop-transfer" model, the amino-terminal part of the presequence is imported into the matrix, but the intermembrane space targeting portion of the signal arrests further translocation through the inner membrane. Dissociation of the inner and outer membrane components then allows the mature part of the protein to cross the outer membrane and to reach the intermembrane space (Figure 4). The alternative pathway postulates that the entire protein is first transported into the matrix where it is processed to the intermediate form. This matrix-localized intermediate is then re-exported to the intermembrane space. This second model has recently been modified by suggesting that import and re-export occur simultaneously, generating an intermediate spanning three membranes. According to this second model, the second half of the targeting signal functions as a signal for an as yet unidentified export machinery in the inner membrane. This hypothetical re-export machinery would have a function similar to the export machinery in the plasma membrane of bacteria. Because of this evolutionary aspect, the hypothetical path-

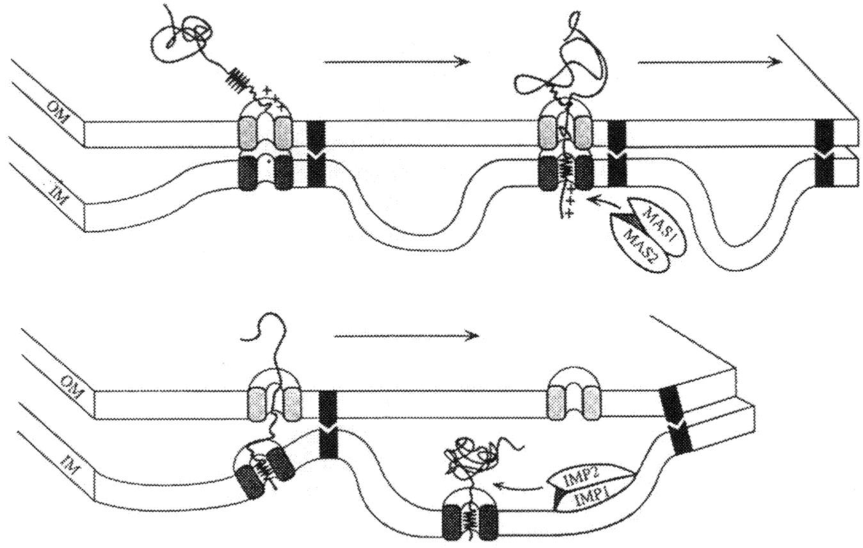

Figure 4. Model for the sorting of intermembrane space proteins containing a bipartite signal sequence. The precursor protein targeted to the intermembrane space inserts into the inner membrane. Translocation is arrested by the stop-transfer signal residing in the carboxy-terminal portion of the targeting signal. Thus, only the N-terminal portion of the protein is exposed to the matrix compartment where the matrix processing peptidase (MAS1/MAS2) cleaves off the matrix-targeting portion of the signal. Components of the inner membrane translocation machinery (dark gray symbols) diffuse out of the contact site (black symbols) pulling the intermediate into the intermembrane space where the second processing event by the innermembrane protease I (IMP1/IMP2) generates the mature protein. Light gray symbols represent components of the outer membrane translocation machinery. OM: outer membrane; IM: inner membrane.

way was termed "conservative sorting." A more detailed discussion of intramitochondrial sorting can be found in several recent reviews (Hartl et al., 1989; Beasley et al., 1992; Glick et al., 1992a, 1992c; Segui-Real et al., 1992, and also this volume).

The existence of a stop-transfer mechanism was first shown using a fusion protein comprising the presequence of cyt c_1 fused to DHFR. This fusion protein was targeted to the intermembrane space; the unprocessed fusion protein and the intermediate were bound to the inner membrane and were always exposed to the intermembrane space. At no time during the import reaction were intermediates detected in the matrix (van Loon et al., 1986 1987). However, subsequent publications did report that import intermediates of cyt c_1 and cyt b_2 could be detected in the matrix (Hartl et al., 1987; Nicholson et al., 1989) and that Hsp60 function was needed for the maturation of cyt b_2 *in vivo* (Cheng et al., 1989), as well as for the import of cyt b_2 fusion proteins into isolated mitochondria (Cheng et al., 1989; Koll et al., 1992). These findings were cited as strong support for the conservative sorting model. Indeed, the presence of genuine import intermediates in the matrix and the transient binding of these intermediates to matrix-localized Hsp60 would be incompatible with a stop-transfer mechanism. However, reinvestigation of the question by repeating and extending the earlier experiments, and by using new approaches to test predictions of the two models showed that there is no experimental support for the hypothesis that cyt c_1 and cyt b_2 are sorted to the intermembrane space via the matrix. Intermediates of authentic cyt b_2 or of cyt c_1 were never detected in the matrix and the same was true of several cyt b_2-DHFR fusion proteins. Two independent groups found that Hsp60 is not required for sorting of intermembrane space proteins *in vivo* and *in vitro*. Most of the published results (Hartl et al., 1987; Cheng et al., 1989; Koll et al., 1992) could be reproduced, but the additional information gained by the reinvestigations showed that the earlier results had to be reinterpreted (Glick et al., 1992b; Hallberg et al., 1993; Rospert et al., 1994).

In sum, all of the currently available evidence indicates that the mature parts of cytochrome c_1 and cytochrome b_2 do not pass through the matrix. Of course, this negative conclusion cannot exclude that only trace amounts of intermediate are present in the matrix at steady-state level. However, there is also positive evidence for the stop-transfer model: matrix ATP is not required for importing these proteins into the intermembrane space. Indeed, the energetics of this import pathway provide one of the most convincing arguments against a re-export pathway.

As described above, complete import into the matrix always requires ATP in the matrix; depletion causes matrix-targeted precursors to get stuck across the inner membrane. Therefore, a protein that passes through the matrix—as suggested by the conservative sorting mechanism—should not be able to reach the intermembrane space when ATP in the matrix is depleted.

In fact, cyt c_1 is imported and processed normally in the absence of matrix ATP. The efficiency of import into ATP depleted mitochondria equals that into fully energized mitochondria. The protein requires ATP only outside the inner membrane, suggesting that it interacts with ATP-utilizing chaperones in the cytosol (Wachter et al., 1992, 1994). Thus, import of cyt c_1 into the intermembrane space occurs normally under conditions in which mHsp70-driven import into the matrix is impossible.

In contrast, import of cyt b_2 does require matrix ATP. At first sight, this finding is inconsistent with the stop-transfer model described above. However, a closer study revealed that the matrix ATP was not used to pull the precursor completely into the matrix, but to unfold a tight heme binding domain near the precursor's amino terminus. This tightly folded domain is already present in the *in vitro*-synthesized precursor and blocks translocation across the outer membrane (Table 2 and Figure 5). When the heme-binding domain is disrupted by mutation or by urea denaturation, import and correct processing take place even with ATP-depleted

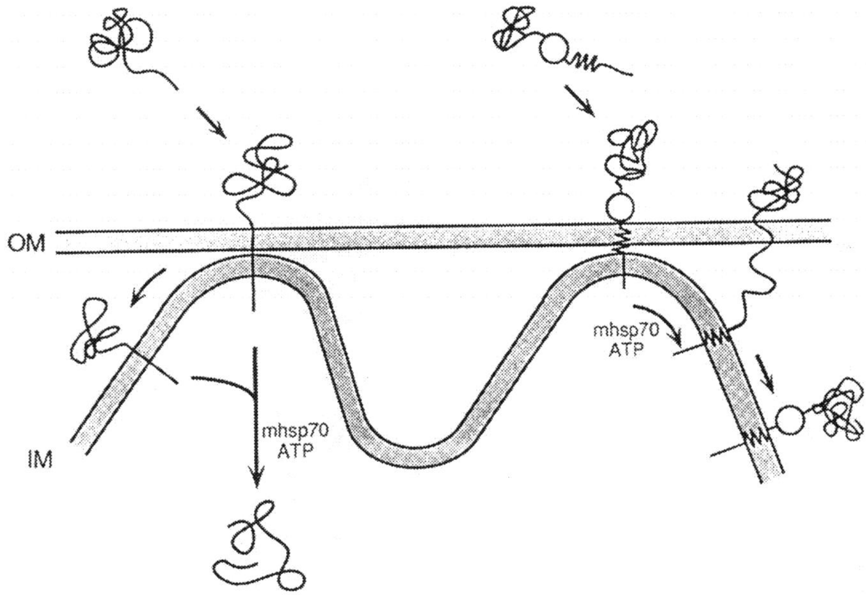

Figure 5. Proposed functions of matrix ATP in mitochondrial protein import. Left: With matrix-targeted precursors, mHsp70 uses matrix ATP to drive translocation across the inner membrane. In the absence of matrix ATP, translocation across the inner membrane is arrested; the arrested precursors can often cross the outer membrane to yield translocation intermediates facing the intermembrane space. Addition of ATP then allows completion of import into the matrix.

Right: In the case of cytochrome *b*$_2$ mHsp70 uses matrix ATP to exert a pulling force on the presequence; this force causes the heme-binding domain in the precursor to unfold and to cross the outer membrane. The stop-transfer signal in the cytochrome b$_2$ presequence (zig-zag line) prevents further translocation across the inner membrane. The mechanism of ATP-dependent mHsp70 function is presumably the same for the complete import of matrix-targeted precursors and for the partial translocation of the cytochrome *b*$_2$ presequence across the inner membrane. OM: outer membrane; IM: inner membrane.

mitochondria. Cyt b_2 then behaves like cyt c_1, indicating that the two proteins are sorted by the same stop-transfer mechanism and do not pass through the matrix. In contrast to cyt c_1, cyt b_2 does not require external ATP (Glick et al., 1993). With cytochrome b_2, unfolding is thus driven by the action of a component in the matrix. This was also shown by studies of Pfanner and coworkers (Voos et al., 1993). They showed that cyt b_2-DHFR fusion proteins that lack the heme-binding domain were imported in the absence of functional mhsp70. Thus, mitochondria containing an import-defective mHsp70 still sort cyt b_2 to the intermembrane space. In contrast, fusion proteins containing a tightly folded full-length heme-binding domain required mHsp70 for correct sorting.

What drives the translocation of cyt c_1 and cyt b_2 across the outer membrane? This question cannot yet be answered. One possible driving force could be the diffusion of inner membrane import channel in the inner membrane (Figure 4). The stuck precursor would then be pulled across the outer membrane channel by diffusion, resulting in an increase in entropy. Much further work will be required to support or refute this hypothesis.

VII. OUTLOOK

It is now clear that ATP is essential for protein import into the matrix and that some precursors also need ATP outside the inner membrane. The requirement for external ATP can usually be overcome by denaturating the precursor, whereas import across the inner membrane always requires ATP in the matrix regardless of the conformation of the precursor protein. It follows that proteins that do not require matrix ATP are not completely imported across the inner membrane. Progress has also been made in identifying the ATP-utilizing proteins involved in import. The most important one is mHsp70. By analogy with bacterial systems, mHsp70 would be expected to perform its functions together with partner proteins resembling bacterial DnaJ (Rowley et al., 1994) and GrpE (Bolliger et al., 1994). Such proteins have indeed been identified, but the mechanism of their interaction with mHsp70 during import remains to be defined. It is also an open question if other matrix-localized ATPases are involved in mitochondrial protein import.

The precise role of ATP outside the mitochondria is still poorly understood. Defining the roles of the different cytosolic factors in mitochondrial protein import will require the use of a homologous import system, and the use of purified components, and purified precursor proteins. The available insights into the requirement for external ATP should make it possible (1) to reconstitute a homologous, Hsp70-dependent pathway for a protein which requires external ATP, (2) to address the question of whether proteins imported independently of external ATP are bound to ATP-independent chaperones, and (3) to ask if the release in this case is caused by the pulling force of mHsp70 in the matrix.

ACKNOWLEDGMENTS

I wish to thank Jeff Schatz for helpful suggestions and criticism. Thanks also to Kevin Hannavy and Trevor Lithgow for critically reading the manuscript. Most of the figures were kindly provided by Clemens Wachter and Ben Glick. I was supported by a Long-term Fellowship from the European Molecular Biology Organization. Literature referenced in this article includes publications up to the beginning of 1994.

REFERENCES

Ades, I. Z., & Butow, R. A. (1980a). The products of mitochondria-bound cytoplasmic polysomes in yeast. J. Biol. Chem. 255, 9918–9924.

Ades, I. Z., & Butow, R. A. (1980b). The transport of proteins into yeast mitochondria. J. Biol. Chem. 255, 9925–9935.

Argan, C. A., Lusty, C. J., & Shore, G. C. (1983). Membrane & cytosolic components affecting transport of the precursor for ornithine carbamyltransferase into mitochondria. J. Biol. Chem. 258, 6667–6670.

Atencio, D. P., & Yaffe, M. P. (1992). MAS5, a yeast homolog of DnaJ involved in mitochondrial protein import. Mol. Cell. Biol. 12, 283–291.

Beasley, E. M., Wachter, C., & Schatz, G. (1992). Putting energy into mitochondrial protein import. Curr. Op. Cell Biol. 4, 646–651.

Beasley, E. M., Müller, S., & Schatz, G. (1993). The signal that sorts yeast cytochrome b_2 to the mitochondrial intermembrane space contains three distinct functional regions. EMBO J. 12, 2303–2311.

Beckmann, R. P., Mizzen, L. A., & Welch, W. J. (1990). Interaction of hsp70 with newly synthesized proteins: Implications for protein folding and assembly. Science 248, 850–854.

Bolliger, L., Deloche, O., Glick, B. S., Georgopoulos, C., Jenö, P., Kronidou, N., Horst, M., Morishima, N., & Schatz, G. (1994). A mitochondrial homolog of bacterial GrpE interacts with mitochondrial hsp70 and is essential for viability. EMBO J. 13, 1998–2006.

Brandt, U., Yu, L., Yu, C-A., & Trumpower, B. L. (1993). The mitochondrial targeting presequence of the Rieske iron-sulfur protein is processed in a single step after insertion into the cytochrome bc_1 complex in mammals and retained as a subunit in the complex. J. Biol. Chem. 268, 8387–8390.

Brodsky J. L., & Schekman, R. (1993). A Sec63p-BiP complex from yeast is required for protein translocation in a reconstituted proteoliposome. J. Cell Biol. 123, 1355–1363.

Caplan, A. J., Cyr, D. M., & Douglas, M. G. (1992). Ydj1p facilitates polypeptide translocation across different intracellular membranes by a conserved mechanism. Cell 71, 1143–1155.

Caplan, A. J., Cyr, D. M., & Douglas, M. G. (1993). Eukaryotic homologues of *Escherichia coli* DnaJ: A diverse protein family that functions with hsp70 stress proteins. Mol. Biol. Cell 4, 555–563.

Chen, W-J., & Douglas, M. G. (1987). Phosphodiester bond cleavage outside mitochondria is required for the completion of protein import into the mitochondrial matrix. Cell 49, 651–658.

Chen, W-J., & Douglas, M. G. (1988). An F1-ATPase β-subunit precursor lacking an internal tetramer-forming domain is imported into mitochondria in the absence of ATP. J. Biol. Chem. 263, 4997–5000.

Cheng, M. Y., Hartl, F-U., Martin, J., Pollock, R. A., Kalousek, F., Neupert, W., Hallberg, E. M., Hallberg, R. L., & Horwich, A. L. (1989). Mitochondrial heat-shock protein hsp60 is essential for assembly of proteins imported into yeast mitochondria. Nature 337, 620–625.

Chirico, W. J. (1992). Dissociation of complexes between 70 kDa stress proteins and presecretory proteins is facilitated by a cytosolic factor. Biochem. Biophys. Res. Commun. 189, 1150–1156.

Craig, E. A., Kramer, J., & Kosic-Smithers J. (1987). *SSC1*, a member of the 70-kDa heat shock protein multigene family of *Saccharomyces cerevisiae*, is essential for growth. Proc. Natl. Acad. Sci. 84, 4156–4160.

Craig, E. A., Kramer, J., Shilling, J., Werner-Washburne, M., Holmes, S., Kosic-Smithers, J., & Nicolet, C. M. (1989). *SSC1*, an essential member of the yeast hsp70 multigene family, encodes a mitochondrial protein. Mol. Cell. Biol. 9, 3000–3008.

Craig, E. A., Gambill, D. B., & Nelson, J. R. (1993). Heat shock proteins: Molecular chaperones of protein biogenesis. Microbiol. Rev. 57, 402–414.

Cyr, D. M., Lu, X., & Douglas, M. G. (1992). Regulation of hsp70 function by a eukaryotic DnaJ homolog. J. Biol. Chem. 267, 20927–20931.

Cyr, D. M., Stuart, R. A., & Neupert, W. (1993). A matrix ATP requirement for presequence translocation across the inner membrane of mitochondria. J. Biol. Chem. 268, 23751–23754.

Deshaies, R. J., Koch, B. D., Werner-Washburne, M., Craig, E. A., & Schekman, R. (1988). A subfamily of stress proteins facilitates translocation of secretory and mitochondrial precursor polypeptides. Nature 332, 800–805.

Eilers, M., & Schatz, G. (1986). Binding of a specific ligand inhibits import of a purified precursor protein into mitochondria. Nature 322, 228–232.

Eilers, M., Oppliger, W., & Schatz, G. (1987). Both ATP and an energized inner membrane are required to import purified precursor protein into mitochondria. EMBO J. 6, 1073–1077.

Eilers, M., Hwang, S., & Schatz, G. (1988). Unfolding and refolding of a purified precursor protein during import into isolated mitochondria. EMBO J. 7, 1139–1145.

Endo, T., & Schatz, G. (1988). Latent membrane perturbation activity of a mitochondrial precursor protein is exposed by unfolding. EMBO J. 7, 1153–1158.

Flynn, G. C., Pohl, J., Flocco, M. T., & Rothman, J. E. (1990). Peptide-binding specificity of the molecular chaperone BiP. Nature 353, 726–730.

Freitag, H., Janes, M., & Neupert, W. (1982). Synthesis of mitochondrial porin and insertion in the outer membrane of *Neurospora crassa*. Eur. J. Biochem. 126, 197–202.

Fu, W., Japa, S., & Beattie, D. S. (1990). Import of the iron-sulfur protein for the cytochrome bc_1 complex into yeast mitochondria. J. Biol. Chem. 265, 16541–16547.

Gambill, B. D., Voos, W., Kang, P. J., Miao, B., Langer, T., Craig, E. A., & Pfanner, N. (1993). A dual role for mitochondrial heat shock protein 70 in membrane translocation of preproteins. J. Cell Biol. 123, 109–117.

Gasser, S. M., & Schatz, G. (1983). Import of proteins into mitochondria. J. Biol. Chem. 258, 3427–3430.

Glick, B., Wachter, C., & Schatz, G. (1991). Protein import into mitochondria: Two systems acting in tandem? TICB 1, 99–103.

Glick, B. S., Beasley, E. M., & Schatz, G. (1992a). Protein sorting into mitochondria. TIBS 17, 453–459.

Glick, B. S., Brandt, A., Cunningham, K., Müller, S., Hallberg, R. L., & Schatz, G. (1992b). Cytochromes c_1 and b_2 are sorted to the intermembrane space of yeast mitochondria by a stop-transfer mechanism. Cell 69, 809–822.

Glick, B. S., Wachter, C., & Schatz, G. (1992c). The energetics of protein import into mitochondria. Biochim. Biophys. Acta 1101, 249–251.

Glick, B. S., Wachter, C., Reid, G. A., & Schatz, G. (1993). Import of cytochrome b_2 to the mitochondrial intermembrane space: The tightly folded heme-binding domain makes import dependent upon matrix ATP. Prot. Sci. 2, 1901–1917.

Gray, M. W. (1989). The evolutionary origins of organelles. TIG 5, 294–299.

Hachiya, N., Alam, R., Sakasegawa, Y., Sakaguchi, M., Mihara, K., & Omura, T. (1993). A mitochondrial import factor purified from rat liver cytosol is an ATP-dependent conformational modulator for precursor proteins. EMBO J. 12, 1579–1586.

Hallberg, E. M., Shu, Y., & Hallberg, R. L. (1993). Loss of mitochondrial hsp60 function: Nonequivalent effects on matrix-targeted and intermembrane-targeted proteins. Mol. Cell. Biol. 13, 3050–3057.

Hannavy, K., Rospert, S., & Schatz, G. (1993). Protein import into mitochondria: A paradigm for the translocation of polypeptides across membranes. Curr. Op. Cell Biol. 5, 694–700.

Hartl, F.-U., Schmidt, B., Wachter, E., Weiss, H., & Neupert, W. (1986). Transport into mitochondria and intramitochondrial sorting of the Fe/S protein of ubiquinol-cytochrome c reductase. Cell 47, 939–951.

Hartl, F.-U., Ostermann, J., Guiard, B., & Neupert, W. (1987). Successive translocation into and out of the mitochondrial matrix: Targeting of proteins to the intermembrane space by a bipartite signal peptide. Cell 51, 1027–1037.

Hartl, F.-U., Pfanner, N., Nicholson, D. W., & Neupert, W. (1989). Mitochondrial protein import. Biochim. Biophys. Acta 988, 1–45.

Hartman, D. J., Hoogenraad, N. J., Condron, R., & Hoj, P. B. (1992). Identification of a mammalian 10-kDa heat shock protein, a mitochondrial chaperonin 10 homologue essential for assisted folding of trimeric ornithine transcarbamoylase *in vitro*. Proc. Natl. Acad. Sci. USA 89, 3394–3398.

Hendrick, J. P., & Hartl, F-U. (1993). Molecular chaperone functions of heat-shock proteins. Annu. Rev. Biochem. 62, 349–384.

Hendrick, J. P., Langer, T., Davis, T. A., Hartl, F-U., & Wiedmann, M. (1993). Control of folding and membrane translocation by binding of the chaperone DnaJ to nascent polypeptides. Proc. Natl. Acad. Sci. USA 90, 10216–10220.

Horst, M., Kronidou, N. G., & Schatz, G. (1993). Through the mitochondrial inner membrane. Curr. Biol. 3, 175–177.

Hurt, E. C., & van Loon, A. P. G. M. (1986). How proteins find mitochondria and intramitochondrial compartments. TIBS 11, 204–207.

Hwang, S. T., & Schatz, G. (1989). Translocation of proteins across the mitochondrial inner membrane but not into the outer membrane, requires nucleoside triphosphates in the matrix. Proc. Natl. Acad. Sci. USA 86, 8432–8436.

Hwang, S. T., Wachter, C., & Schatz, G. (1991). Protein import into the yeast mitochondrial matrix. J. Biol. Chem. 266, 21083–21089.

Ingolia, T. D., & Craig, E. A. (1982). A *Drosophila* gene related to the major heat shock induced gene is transcribed at normal temperatures and not induced by heat shock. Proc. Natl. Acad. Sci. USA 79, 525–529.

Ingolia, T. D., Slater, M. R., & Craig, E. A. (1982). *Saccharomyces cerevisiae* contains a complex multigene family related to the major heat shock-inducibile gene of *Drosophila*. Mol. Cell. Biol. 2, 1388–1398.

Jensen, R. E., Schmidt, S., & Mark, R. J. (1992). Mutations in a 19-amino-acid hydrophobic region of the yeast cytochrome c_1 presequence prevent sorting to the mitochondrial intermembrane space. Mol. Cell. Biol. 12, 4677–4686.

Kang, P-J., Ostermann, J., Shilling, J., Neupert, W., Craig, E. A., & Pfanner, N. (1990). Requirement for hsp70 in the mitochondrial matrix for translocation and folding of precursor proteins. Nature 348, 137–142.

Koll, H., Guiard, B., Rassow, J., Ostermann, J., Horwich, A. L., Neupert, W., & Hartl, F-U. (1992). Antifolding activity of hsp60 couples protein import into the mitochondrial matrix with export to the intermembrane space. Cell 68, 1163–1175.

Lewis, M. J., & Pelham, H. R. B. (1985). Involvement of ATP in the nuclear and nucleolar functions of the 70 kDa heat shock protein. EMBO J. 4, 3137–3143.

Lill, R., Stuart, R. A., Drygas, M. E., Nargang, F. E., & Neupert, W. (1992). Import of cytochrome c heme lyase into mitochondria: A novel pathway into the intermembrane space. EMBO J. 11, 449–456.

Lithgow, T., Hoj, P. B., & Hoogenraad, N. J. (1993a). Do cytosolic factors prevent promiscuity at the membrane surface? FEBS Lett. 329, 1–4.

Lithgow, T., Ryan, M., Anderson, R. L., Hoj, P. B., & Hoogenraad, N. J. (1993b). A constitutive form of heat-shock protein 70 is located in the outer membranes of mitochondria from rat liver. FEBS Lett. 332, 277–281.

Lorimer, G. H., Todd, M. J., & Viitanen, P. V. (1993). Chaperonins and protein folding: Unity and disunity of mechanisms. Phil. Trans. R. Soc. Lond. B 339, 297–304.

Lubben, T. H., Gatenby, A. A., Gail, K. D., Lorimer, G. H., & Viitanen, P. V. (1990). Identification of a groES-like chaperonin in mitochondria that facilitates protein folding. Proc. Natl. Acad. Sci. USA 87, 7683–7687.

Mahlke, K., Pfanner, N., Martin, J., Horwich, A. L., Hartl, F-U., & Neupert, W. (1990). Sorting pathways of mitochondrial inner membrane proteins. Eur. J. Biochem. 192, 551–555.

Manning-Krieg, U. C., Scherer, P. E., & Schatz, G. (1991). Sequential action of mitochondrial chaperones in protein import into the matrix. EMBO J. 10, 3273–3280.

Martin, J., Mahlke, K., & Pfanner, N. (1991). Role of an energized inner membrane in mitochondrial protein import. J. Biol. Chem. 266, 18051–18057.

McMullin, T. W., & Hallberg, R. L. (1988). A highly evolutionarily conserved mitochondrial protein is structurally related to the protein encoded by the *Escherichia coli* groEL gene. Mol. Cell. Biol. 8, 371–380.

Mihara, K., Blobel, G., & Sato, R. (1982). *In vitro* synthesis and integration into mitochondria of porin, a major protein of the outer mitochondrial membrane of *Saccharomyces cerevisiae*. Proc. Natl. Acad. Sci. USA 79, 7102–7106.

Miura, S., Mori, M., & Tatibana, M. (1983). Transport of ornithine carbymoyltransferase precursor into mitochondria. J. Biol. Chem. 258, 6671–6674.

Murakami, H., Pain, D., & Blobel, G. (1988). 70-kD heat shock-related protein is one of at least two distinct cytosolic factors stimulating protein import into mitochondria. J. Cell Biol. 107, 2051–2057.

Murakami, K., & Mori, M. (1990). Purified presequence binding factor (PBF) forms an import-competent complex with a purified mitochondrial precursor protein. EMBO J. 9, 3201–3208.

Murakami, K., Tanase, S., Morino, Y., & Mori, M. (1992). Presequence binding factor-dependent and -independent import of proteins into mitochondria. J. Biol. Chem. 267, 13199–13122.

Nelson, N., & Schatz, G. (1979). Energy-dependent processing of cytoplasmically made precursors to mitochondrial proteins. Proc. Natl. Acad. Sci. USA 76, 4365–4369.

Nelson, R. J., Ziegelhoffer, T., Nicolet, C., Werner-Washburne, M., & Craig, E. A. (1992). The translation machinery and 70 kd heat shock protein cooperate in protein synthesis. Cell 71, 97–105.

Neupert, W., & Pfanner, N. (1993). Roles of molecular chaperones in protein targeting to mitochondria. Phil. Trans. R. Soc. Lond. B 339, 355–362.

Nicholson, D. W., Stuart, R. A., & Neupert, W. (1989). Biogenesis of cytochrome c_1. J. Biol. Chem. 264, 10156–10168.

Nunnari, J., Fox, T. D., & Walter, P. (1993). A mitochondrial protease with two catalytic subunits of nonoverlapping specificities. Science 262, 1997–2003.

Ohta, S., & Schatz, G. (1984). A purified precursor polypeptide requires a cytosolic protein fraction for import into mitochondria. EMBO J. 3, 651–657.

Ono, H., & Tuboi, S. (1988). The cytosolic factor required for import of precursors of mitochondrial proteins into mitochondria. J. Biol. Chem. 263, 3188–3193.

Ono, H., & Tuboi, S. (1990a). Presence of the cytosolic factor stimulating the import of precursor of mitochondrial proteins in rabbit reticulocytes and rat liver cells. Arch. Biochem. Biophys. 277, 368–373.

Ono, H., & Tuboi, S. (1990b). Purification and identification of a cytosolic factor required for import of precursors of mitochondrial proteins into mitochondria. Arch. Biochem. Biophys. 280, 299–304.

Ostermann, J., Horwich, A. L., Neupert, W., & Hartl, F-U. (1989). Protein folding in mitochondria requires complex formation with hsp60 and ATP hydrolysis. Nature 341, 125–129.

Ostermann, J., Voos, W., Kang, P. J., Craig, E. A., Neupert, W., & Pfanner, N. (1990). Precursor proteins in transit through mitochondrial contact sites interact with hsp70 in the matrix. FEBS Lett. 277, 281–284.

Pfaller, R., Kleene, R., & Neupert, W. (1990). Biogenesis of mitochondrial porin: The import pathway. Experientia 46, 153–161.

Pfanner, N., & Neupert, W. (1985). Transport of proteins into mitochondria: a potassium diffusion potential is able to drive the import of ADP/ATP carrier. EMBO J. 4, 2819–2825.

Pfanner, N., & Neupert, W. (1986). Transport of F_1-ATPase subunit β into mitochondria depends on both a membrane potential and nucleoside triphosphates. FEBS Lett. 209, 152–156.

Pfanner, N., Tropschug M., & Neupert, W. (1987). Mitochondrial protein import: Nucleoside triphosphates are involved in conferring import-competence to precursors. Cell 49, 815–823.

Pfanner, N., Pfaller, R., Kleene, R., Ito, M., Tropschug, M., & Neupert, W. (1988). Role of ATP in mitochondrial protein import. J. Biol. Chem. 263, 4049–4051.

Pfanner, N., Rassow, J., Guiard, B., Söllner, T., Hartl, F.-U., & Neupert, W. (1990). Energy requirements for unfolding and membrane translocation of precursor proteins during import into mitochondria. J. Biol. Chem. 265, 16324–16329.

Rassow, J., & Pfanner, N. (1991). Mitochondrial preproteins en route from the outer membrane to the inner membrane are exposed to the intermembrane space. FEBS Lett. 293, 85–88.

Reading, D. S., Hallberg, R. L., & Myers, A. M. (1989). Characterization of the yeast *HSP60* gene coding for a mitochondrial assembly factor. Nature 337, 655–659.

Roise, D., Horvath, S. J., Tomich, J. M., Richards, J. H., & Schatz, G. (1986). A chemically synthesized mitochondrial presequence of an imported protein can form an amphiphilic helix and perturb natural and artificial phospholipid bilayers. EMBO J. 5, 1327–1334.

Roise, D., & Schatz, G. (1988). Mitochondrial presequences. J. Biol. Chem. 263, 4509–4511.

Roise, D. (1993). The amphipathic helix in mitochondrial targeting sequences. In: *The Amphipathic Helix* (Epand, R. M., Ed.). CRC Press, Inc., Boca Raton, Fl., pp. 258–279.

Rospert, S., Glick, B. S., Jenö, P., Schatz, G., Todd, M. J., Lorimer, G. H., & Viitanen, P. V. (1993a). Identification and functional analysis of chaperonin 10, the groES homolog from yeast mitochondria. Proc. Natl. Acad. Sci. USA 90, 10967–10971.

Rospert, S., Junne, T., Glick, B. S., & Schatz, G. (1993b). Cloning and disruption of the gene encoding yeast mitochondrial chaperonin 10, the homolog of *E. coli* groES. FEBS Lett. 335, 358–360.

Rospert, S., Müller, S., Schatz, G., & Glick, B. S. (1994). Fusion proteins containing the cytochrome b_2 presequence are sorted to the mitochondrial intermembrane space independently of hsp60. J. Biol. Chem. 269, 17279–17288.

Rothman, J. E., & Kornberg, R. D. (1986). An unfolding story of protein translocation. Nature 322, 209–210.

Rowley, N., Prip-Buus, C., Westermann, B., Brown, C., Schwarz, E. Barrell, B., & Neupert, W. (1994). Mdj1p, a novel chaperone of the DnaJ family, is involved in mitochondrial biogenesis and protein folding. Cell 77, 249–259.

Schatz, G. (1987). 17. Sir Hans Krebs lecture: Signals guiding proteins to specific intramitochondrial locations. Eur. J. Biochem. 165, 1–6.

Schatz, G. (1993a). The protein import machinery of mitochondria. Prot. Sci. 2, 141–146.

Schatz, G. (1993b). From "Granules" to organelles: How yeast mitochondria became respectable. In: *The Early Days of Yeast Genetics* (Hall, M. N., & Linder, P., Eds.). Cold Spring Harbor Laboratory Press, New York, pp. 241–246.

Scherer, P. E., Krieg, U. C., Hwang, S. T., Vestweber, D., & Schatz, G. (1990). A precursor protein partly translocated into yeast mitochondria is bound to a 70 kd mitochondrial stress protein. EMBO J. 9, 4315–4322.

Schleyer, M., & Neupert, W. (1985). Transport of proteins into mitochondria: Translocational intermediates spanning contact sites between outer and inner membranes. Cell 43, 339–350.

Schneider, A., Behrens, M., Scherer, P., Pratje, E., Michaelis, G., & Schatz, G. (1991). Inner membrane protease I, an enzyme mediating intramitochondrial protein sorting in yeast. EMBO J. 10, 247–254.

Schneider, A., Oppliger, W., & Jenö P. (1994). Purified inner membrane protease I of yeast mitochondria is a heterodimer. J. Biol. Chem. 269, 8635–8638.

Schwarz, E., Seytter, T., Guiard, B., & Neupert, W. (1993). Targeting of cytochrome b_2 into the mitochondrial intermembrane space: Specific recognition of the sorting signal. EMBO J. 12, 2295–2302.

Segui-Real, B., Stuart, R., & Neupert, W. (1992). Transport of proteins into the various subcompartments of mitochondria. FEBS Lett. 313, 2–7.

Sheffield, W. P., Shore, G. C., & Randall, S. K. (1990). Mitochondrial precursor protein. J. Biol. Chem. 265, 11069–11076.

Simon, S. M., Peskin, S. P., & Oster, G. F. (1992). What drives the translocation of proteins? Proc. Natl. Acad. Sci. USA 89, 3770–3774.
Smith, B. J., & Yaffe, M. P. (1991). A mutation in the yeast heat-shock factor gene causes temperature-sensitive defects in both mitochondrial protein import and the cell cycle. Mol. Cell. Biol. 11, 2647–2655.
Stuart, R. A., & Neupert, W. (1990). Apocytochrome c: An exceptional mitochondrial precursor protein using an exceptional import pathway. Biochimie 72, 115–121.
Suissa, M., & Schatz, G. (1982). Import of proteins into mitochondria. J. Biol. Chem. 257, 13048–13055.
van Loon, A. P. G. M., Brändli, A. W., & Schatz, G. (1986). The presequence of two imported mitochondrial proteins contain information for intracellular and intramitochondrial sorting. Cell 44, 801–812.
van Loon, A. P. G. M., & Schatz, G. (1987). Transport of proteins to the mitochondrial intermembrane space: the 'sorting' domain of the cytochrome c_1 presequence is a stop-transfer sequence specific for the mitochondrial inner membrane. EMBO J. 6, 2441–2448.
Verner, K., & Schatz, G. (1987). Import of an incompletely folded precursor protein into isolated mitochondria requires an energized inner membrane, but no added ATP. EMBO J. 6, 2449–2456.
Verner, K. (1992). Early events in yeast mitochondrial protein targeting. Mol. Microbiol. 6, 1723–1728.
Verner, K. (1993). Co-translational protein import into mitochondria: An alternative view. TIBS 18, 366–371.
von Heijne, G. (1986). Mitochondrial targeting sequences may form amphiphilic helices. EMBO J. 5, 1335–1342.
Voos, W., Gambill, B. D., Guiard, B., Pfanner, N., & Craig, E. A. (1993). Presequence and mature part of preproteins strongly influence the dependence of mitochondrial protein import on heat shock protein 70 in the matrix. J. Cell Biol. 123, 119–126.
Wachter, C., Schatz, G., & Glick, B. (1992). Role of ATP in the intramitochondrial sorting of cytochrome c_1 and the adenine nucleotide translocator. EMBO J. 11, 4787–4794.
Wachter, C., Schatz, G., & Glick, B. S. (1994). Protein import into mitochondria: The requirement for external ATP is precursor-specific whereas intramitochondrial ATP is universally needed for translocation into the matrix. Mol. Biol. Cell. 5, 465–474.
Wienhues, U., Becker, K., Schleyer, M., Guiard, B., Tropschug, M., Horwich, A. L., Pfanner, N., & Neupert, W. (1991). Protein folding causes an arrest of preprotein translocation into mitochondria in vivo. J. Cell Biol. 115, 1601–1609.
Yang, M., Jensen, R. E., Yaffe, M. P., Oppliger, W., & Schatz, G. (1988). Import of proteins into yeast mitochondria: The purified matrix processing protease contains two subunits which are encoded by the nuclear *MAS1* and *MAS2* genes. EMBO J. 7, 3857–3862.
Ziemienowicz, A., Skowyra, D., Zeilstra-Ryalls, J., Fayet, O., Georgopoulos, C., & Zylicz, M. (1993). Both the *Escherichia coli* chaperone systems, groEL/ES and dnaK/DnaJ/GrpE, can reactivate heat-treated RNA polymerase. J. Biol. Chem 268, 25425–25431.

EXPORT OF PROTEINS FROM MITOCHONDRIA

Robert O. Poyton, Kevin A. Sevarino,
Edward E. McKee, David J. M. Duhl,
Vicki Cameron, and Bradley Goehring

I. Introduction 248
II. Pathways for Protein Export in Overview 249
III. Insertion of Mitochondrially-encoded Proteins into the Inner Membrane . . . 250
A. Mitochondrial Gene Products 250
B. Preprotein Precursors 250
C. Leader Peptides 254
D. The Leader Peptide on Presubunit II is Not Present in All Species 255
E. Evidence for Co-translational Insertion 259
F. Energy Requirements for Insertion 259
IV. Export of Mitochondrially-encoded Proteins to the Plasma Membrane 262
V. Export of Nuclear-encoded Proteins to the Intermembrane Space 263
VI. Export of Nuclear-encoded Proteins to the Cytosol 265
A. Cytochrome *c* 265
B. Cytochrome *c* peroxidase 266
C. Aspartate Aminotransferase and Malate Dehydrogenase 266

Advances in Molecular and Cell Biology,
Volume 17, pages 247–280.

ISBN: 0-7623-0144-9

VII. The Machinery for Export 266
A. Mitochondrially-encoded Proteins 267
B. Nuclear-coded Proteins 271
VIII. Relationship Between Import and Export 272
IX. Concluding Remarks 273
Acknowledgments 274
References 274

I. INTRODUCTION

Mitochondria are semi-autonomous organelles that communicate with the rest of the cell via a variety of transport phenomena. This communication is essential not only for mitochondrial function but also for cell metabolism and viability (Baker and Schatz, 1991). Historically, the first transport pathways studied were the import and export pathways followed by protons, electrons, ions, and metabolites across the inner mitochondrial membrane (Racker, 1976; Boyer et al., 1977). These pathways are essential for energy generation and conservation (Boyer et al., 1977). The protein catalysts that participate in these pathways as well as the energetics that underlie them are now well understood. The next transport pathways to be examined were those for protein import from the cytosol to the mitochondrion. These studies were initiated 20 years ago (Kellems et al., 1974; Hallermayer and Neupert, 1976). They have not only elucidated how nuclear-coded proteins are localized to mitochondria, but also have clarified the communication pathway by which the nuclear genome "talks to" the mitochondrion. In addition, these studies have contributed significantly to our general understanding of protein transport across membranes (Verner and Schatz, 1988). As discussed in other articles in this volume, protein import is a multi-event process followed by preprotein precursors that are initially translated on free cytosolic ribosomes. It involves preprotein unfolding in the cytosol, binding to surface receptors, translocation into the organelle, preprotein processing, protein refolding, and submitochondrial sorting. These studies have also led to the discovery and initial characterization of channels (or pores) through the inner and outer mitochondrial membranes (Inoue et al., 1991; Szabo and Zoratti, 1991; Glick et al., 1991).

Recently, a third type of transport has come into focus. It has become clear that mitochondria are capable of exporting (or releasing) proteins (Poyton et al. 1992), RNA (Nugent and Palmer, 1991; Covello and Gray, 1993; Kobayashi et al. 1993), and DNA (Thorsness and Fox, 1990, 1993). Studies on the pathways and mechanisms by which these export processes occur are in their early phases. This review will consider protein export, and its relationship(s) to the other transport pathways of mitochondria.

II. PATHWAYS FOR PROTEIN EXPORT IN OVERVIEW

Proteins encoded in the mitochondrion as well as proteins encoded in the nucleus are exported from the mitochondrial matrix into or across the inner membrane, a process we have defined previously (Poyton et al. 1992) as "protein export." Exported proteins are localized to at least four cellular compartments: the cell surface, the cytosol, the intermembrane space, and the inner membrane. The pathways followed by these proteins are outlined in Figure 1. Proteins encoded by mitochondrial genes are translated on endogenous mitochondrial ribosomes, which are bound to the matrix face of the inner membrane (Kuriyama and Luck, 1973; Spithill et al., 1978), and are inserted into it co-translationally (Figure 1, pathways 4 and 5). Insofar as most mitochondrially encoded proteins are transmembranous, with domains exposed on the outer surface of the membrane, these proteins were the first recognized to be exported from the matrix (Figure 1, Pathway 4). This type of export may be analogous to the secretory pathway for proteins of the bacterial cytoplasmic membrane. Although most mitochondrially encoded proteins reside in the inner membrane (Attardi and Schatz, 1988; Pon and Schatz, 1991), short domains of mitochondrial gene products have been found in the plasma membrane

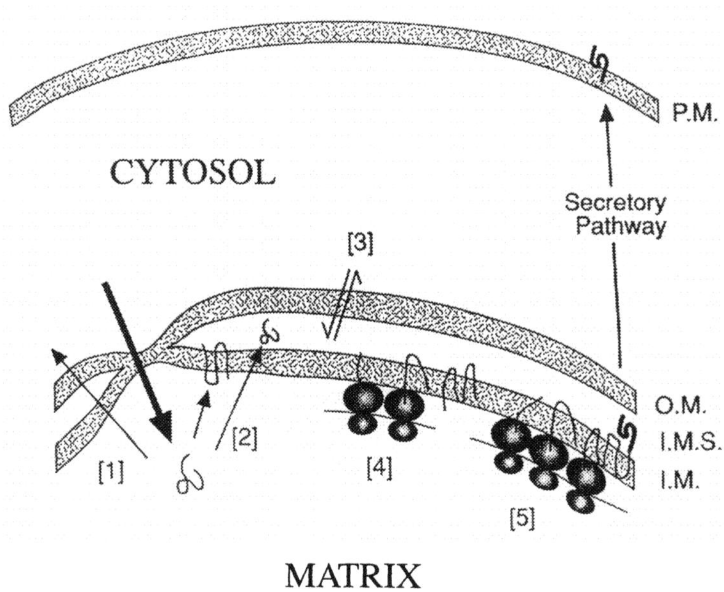

Figure 1. Mitochondrial export pathways. Proteins that follow pathways 1, 2, and 3 are encoded by the nuclear genome; proteins that follow pathways 4 and 5 are encoded by the mitochondrial genome. See text for details. P.M., plasma membrane; O.M., outer membrane; I.M.S., intermembrane space; I.M., inner membrane. The heavy arrow indicates import.

of some eucaryotes (Loveland et al., 1990; Fischer-Lindahl et al., 1991) (Figure 1, pathway 5). Thus, mitochondrial gene products may be localized to two different cellular compartments: inner mitochondrial membrane and plasma membrane.

Nuclear gene products are exported from the mitochondrial matrix to the inner membrane space by a process termed "conservative sorting" (Hartl and Neupert, 1990; Figure 1, pathway 2). Proteins that follow this pathway are translated in the cytosol and localized to the matrix by standard import pathways. Nuclear gene products may also be released from the mitochondrial matrix or intermembrane space into the cytosol (Kaput et al. 1989; Igbavboa et al., 1989; Hakvoort et al. 1990; Passarella et al., 1990; Figure 1, pathways 1 and 3). Thus, nuclear gene products may be exported to either the intermembrane space or to the cytosol.

III. INSERTION OF MITOCHONDRIALLY-ENCODED PROTEINS INTO THE INNER MEMBRANE

A. Mitochondrial Gene Products

The mitochondrial genome in all eukaryotes encodes a small subset of proteins. These are subunits of multimeric protein complexes that contain nuclear-encoded subunit polypeptides as well. Three of these genetic chimeric complexes, common to all eukaryotes, are coenzyme Q cytochrome *c* reductase, cytochrome *c* oxidase, and the $F_1 F_0$ ATPase (Poyton et al., 1992; Poyton and McEwen, 1996). Additional, but not universal, chimeric protein complexes include NADH dehydrogenases of vertebrates, plants, and some fungal mitochondria, and the RNA splicing enzymes of fungal mitochondria (Attardi and Schatz, 1988; Poyton and McEwen, 1996). The polypeptides encoded by the mitochondrial genomes of most organisms include: cytochrome b, ATPase subunits 6 and 8, and cytochrome *c* oxidase subunits I, II, and III. In some plants, the gene for cytochrome *c* oxidase subunit II may be expressed in the nucleus (Nugent and Palmer, 1991; Covello and Gray, 1993). Some mitochondrial genomes encode additional polypeptides. These include: NADH dehydrogenase subunits 1, 2, 3, 4, 4L, 5, and 6 in plants, vertebrates, and some fungi; ATPase subunit 9 in *Saccharomyces cerevisiae*, and ATPase subunit α in maize. These polypeptides encoded by mitochondrial DNA are extremely hydrophobic. Many of them are predicted to have multiple transmembrane helices (Figure 2) and domains that are exposed on the outer surface of the inner membrane. Hence, they qualify as exported proteins.

B. Preprotein Precursors

Currently, most information on preprotein precursors comes from studies with yeast mitochondrial gene products. By using isolated yeast mitochondria optimized for protein synthesis (McKee and Poyton, 1984), it has been possible to sequence most of the primary translation products encoded by the yeast mitochondrial

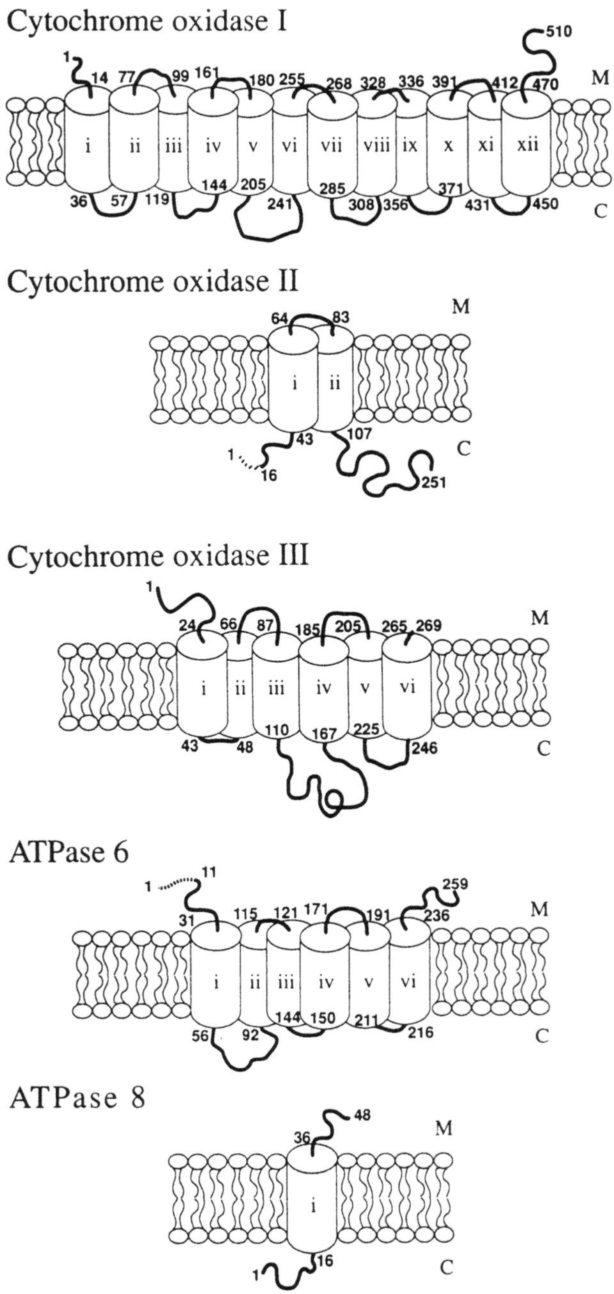

(continued)

Figure 2. Topogenic models for the disposition of proteins encoded by the yeast mitochondrial genome across the inner mitochondrial membrane. The dashed line shown for cytochrome oxidase II and ATPase 6 designates their leader peptides. See Poyton et al. (1992) for details.

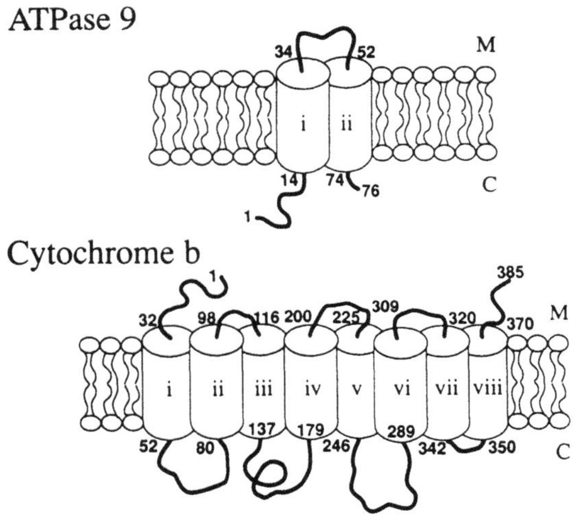

Figure 2. (Continued)

genome (Sevarino, 1983). These studies revealed that cytochrome *c* oxidase pre-subunit II is a preprotein that has a 15 amino acid N-terminal leader peptide. The other primary translation products synthesized by isolated yeast mitochondria (i.e., cytochrome *c* oxidase subunits I and III, cytochrome *b*, and ATPase subunits 8 and 9) are not preprotein precursors, as determined by aligning their N-terminal sequences with N-terminal sequences of the mature subunits themselves and with the sequences predicted by their genes. Moreover, the mature subunits have blocked N-termini, as deduced from low yields during Edman degradation sequencing reactions. The block appears to be due to the retention of formylmethionine at their N-termini because it is removed when these proteins are deformylated (with acidic methanol) prior to sequencing (Sevarino and Poyton, unpublished data). These findings are consistent with the conclusion that these mature subunits are equivalent to primary translation products and retain formylmethionine, their initiating amino acid, at their N-terminus. This is supported by the finding that the majority of yeast mitochondrial translation products can be radiolabeled *in vivo* with formate (Mahler et al., 1972; Feldman and Mahler, 1974). More recently, it has been proposed from protein sequence analysis that ATPase presubunit 6 is also derived from a preprotein, which has a 10 amino acid N-terminal leader peptide (Michon et al., 1988).

Normally, pre-subunit II is processed co-translationally and is usually not seen. However, it has been observed in four different experimental situations.

1. Presubunit II accumulates in isolated mitochondria that have been incubated with the triphenylmethane drug aurintricarboxylic acid (Sevarino and Poy-

ton, 1980). Because aurintricarboxylic acid inhibits translational initiation it has not been possible to determine whether presubunit II accumulates because of a direct effect on the processing protease itself, or an indirect effect resulting from an alteration in the mode of membrane insertion in a run-off system that is disconnected from translational initiation. This question has been addressed with a variety of triphenylmethane drug analogs of aurintricarboxylic acid (Sevarino, 1983). One of these, pyrocatechol violet, promotes the accumulation of presubunit II (Figure 3) but does not inhibit translation (Sevarino, 1983). Hence, it is unlikely that the accumulation of presubunit II is the result of an inhibition of polypeptide chain initiation. Moreover, other initiation inhibitors, such as pactamycin, do not lead to the accumulation of presubunit II in isolated yeast mitochondria. Thus, it seems likely that both aurintricarboxylic acid and pyrocatechol violet inhibit presubunit processing per se, probably by inhibiting inner membrane protease I (Imp1p; Pratje et al., 1983; Schneider et al., 1991), the processing enzyme for presubunit II. Although presubunit II is normally processed co-translationally, presubunit II that has been accumulated in the presence of aurintricarboxylic acid can be processed post-translationally upon removal of the drug. This post-translational processing is inhibited by the cold (4°C) or by the absence of an energy generating system.

2. Presubunit II accumulates in isolated mitochondria when it is translated under conditions that prevent the formation of a membrane potential (Clarkson and Poyton, 1989). When accumulated under these conditions, presubunit II is inserted into the inner membrane and behaves like an integral protein, indicating that the membrane potential per se is not required for membrane insertion but is required for processing. This finding, together with the observation that the post-translational processing of presubunit II accumulated in aurintricarboxylic acid poisoned mitochondria requires an energy generating system, raises the interesting possibility that the catalytic activity of Imp1p may depend on a membrane potential.
3. Presubunit II accumulates *in vivo* in strains that carry mutations in either Imp1p or Imp2p, a recently discovered protease that interacts with Imp1p (Nunnari et al., 1993). Although Imp1p is the protease that processes presubunit II, Imp2p appears to be essential also, probably because it complexes with Imp1p and apparently stabilizes it.
4. Finally, presubunit II accumulates *in vivo* in a strain that carries a mutation in the coding region of mature subunit II (Cameron et al., 1996).

In contrast to presubunit II, ATPase presubunit 6 does not accumulate under any of the above conditions, suggesting that it is not processed by Imp1p. This is supported by the fact that the sequence preceding the cleavage site in presubunit 6, YIT, is different than the consensus sequence, IMN, in the two proteins (presubunit II and an intermediate in the biogenesis of cytochrome b_2) known to be processed

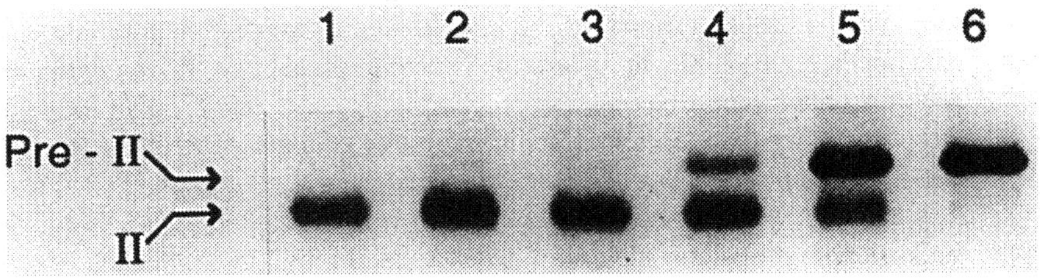

Figure 3. Synthesis of presubunit II of cytochrome *c* oxidase by isolated yeast mitochondria incubated in the presence of different concentrations of pyrocatechol violet. Isolated mitochondria were allowed to synthesize protein (McKee et al., 1984) in the presence of [^{35}S]-methionine and increasing concentrations of pyrocatechol violet (PCV), then solubulized with SDS and analyzed by SDS-PAGE. PCV had no effect on mitochondrial protein synthesis at concentrations up to 1 mM. Lanes: 1 = No PCV, 2 = 50 uM, 3 = 100uM, 4 = 250 uM, 5 = 500 uM, 6 = 1mM.

by Imp1p (Pratje and Guiard, 1986). It also appears unlikely that presubunit 6 is processed by Imp2p because this enzyme seems to process substrates in which amino acids with small side chains are present at the positions -1 and -3 relative to the cleavage site (Nunnari et al., 1993). Furthermore, unlike the leader peptide of presubunit II, which is exposed on the cytosolic face of the inner membrane, the leader peptide of presubunit 6 is predicted to be exposed on the matrix face of the inner membrane (Figure 2). Thus, it is possible that this precursor is processed by a matrix-localized protease; perhaps *MIP1* Protease (Isaya et al. 1994) or the matrix processing protease that processes imported proteins (Glick and Schatz, 1991).

C. Leader Peptides

The leader peptides on presubunit II and presubunit 6 share partial primary sequence identity (Figure 4) and both are predicted to form α-helices (Poyton, 1983). They differ, however, in overall charge. It is not clear if the other mitochondrial gene products from yeast have uncleaved N-terminal domains that function as "leader peptides." However, it is clear that there is little sequence homology between the leader peptides on presubunit II and presubunit 6 and the N-termini of all other mitochondrial gene products (Poyton, 1983). Furthermore, some of these N-terminal regions are predicted to form α-helices while others are predicted to form β sheets (Poyton, 1983). Although these considerations argue against the presence of an uncleaved leader peptide on other yeast mitochondrial gene products, they do not rule out the possibility that these polypeptides have internal domains that perform leader peptide functions, whatever they might be.

Compared to the majority of leader peptides on proteins that are inserted into or through membranes co-translationally, the leader peptide on yeast cytochrome *c* oxidase subunit II appears to be somewhat unusual. Like most other leader peptides

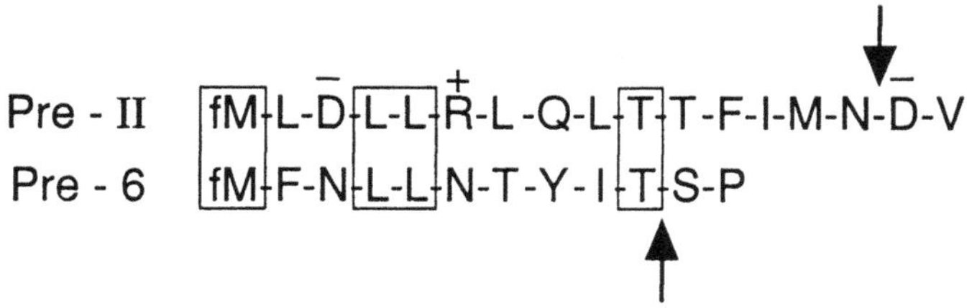

Figure 4. Comparison of the leader peptides of presubunit II of cytochrome c oxidase and presubunit 6 of ATPase from *Saccharomyces cerevisiae.* The junction of each leader peptide with its mature polypeptide is designated by an arrow.

it contains an uncharged core, from asn-(-1) to leu-(-9), which is preceded by a positively charged residue, arg -(-10). However, unlike most other leader peptides, the hydrophobic core is short and only weakly hydrophobic. In view of the emphasis placed on overall hydrophobicity of the uncharged cores of other leader peptides in formulating hypotheses for their insertion into membranes (von Heijne and Blomberg, 1979; Engelman and Steitz, 1981), it is important to note that the uncharged core in the leader peptide of presubunit II is predicted to yield a favorable energy of transfer from an aqueous phase into a membrane of only 47.3 kd/mole—about one-half the minimum energy gain calculated for the majority of other leader peptide cores. This low free energy gain for the leader peptide on presubunit II is inconsistent with its partition into the inner membrane by a "direct transfer" mechanism (von Heijne and Blomberg, 1979). Moreover, the uncharged core of this leader peptide is too short to fit the "helical hairpin" hypothesis (Engelman and Steitz, 1981). The low free energy gain, together with the short length of the uncharged core of the presubunit II leader peptide, suggests that it may not insert into the hydrophobic milieu of the inner membrane at all, and that perhaps it transverses the inner membrane via a pore or channel.

Although the structural characteristics of the presubunit II leader peptide differ from those of the majority of other known leader peptides, they are not completely unique. For example, the uncharged core of the leader peptide on S_2 ovine pre-casein is nine amino acids long (Mercier et al., 1978) and has a hydrophobicity index of 1.89, and that of *Halobacterium halobium* bacteriorhodopsin is 6 amino acids long (Seehra and Khorana, 1984) and has a hydrophobicity index of 2.17. The latter leader peptide is especially interesting because its N-terminus exhibits partial primary sequence homology with the presubunit II leader peptide.

D. The Leader Peptide on Presubunit II Is Not Present in All Species

The DNA sequences of the structural gene, *COX2*, for cytochrome *c* oxidase from many different species of eukaryotes are now known (Figure 5). In addition, the N-terminal sequences of the mature subunit II polypeptide have been determined

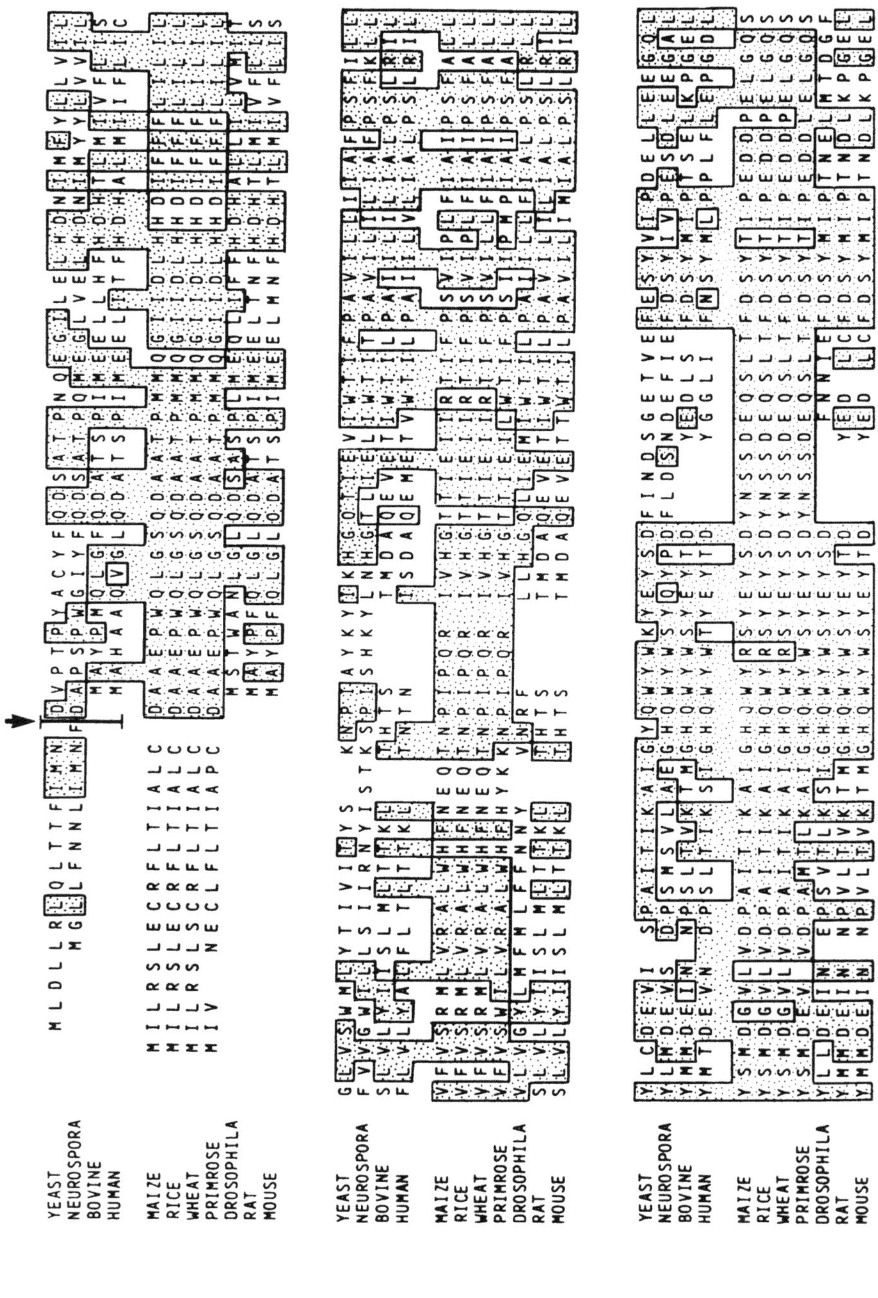

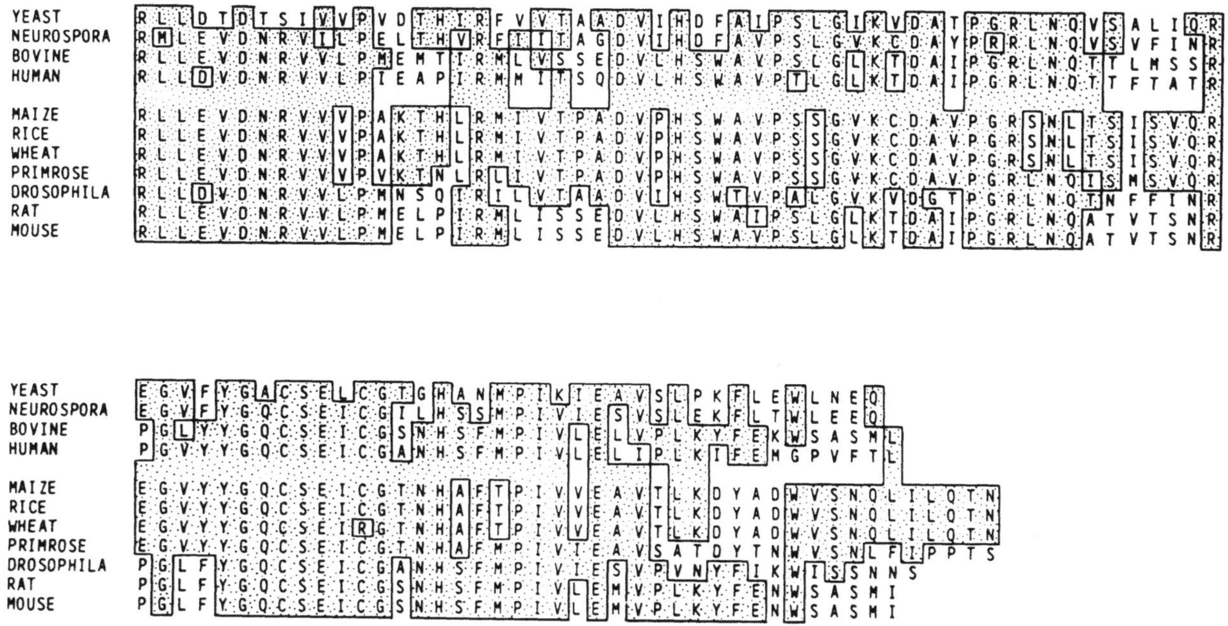

Figure 5. Primary sequence homologies of cytochrome c oxidase subunit II polypeptides from different species. The amino acid sequences are those predicted by the DNA sequences of the COX2 structural genes in the designated species. The DNA sequences were taken from the following references: Yeast (Coruzzi and Tzagoloff, 1979); *Neurospora* (Macino and Morelli, 1983); bovine (Anderson et al., 1982); human (Anderson et al., 1981); maize (Fox and Leaver, 1981); rice (Kao et al., 1984); wheat (Bonen et al., 1984); primrose (Hiesel and Brennicke, 1983); *Drosophila* (de Bruijn, 1983); rat (Brown and Simpson, 1982); and mouse (Bibb et al., 1981). Those four sequences for which the NH_2-terminal sequence of the subunit II protein has been determined are shown at the top. The amino acid residue to the immediate right of the vertical line is the NH_2-terminus of the mature protein. All sequences have been aligned and boxed for optimal homology.

in four of these species (*Saccharomyces cerevisiae, Neurospora crassa*, cow, and human). Both mammalian subunit II polypeptides have formulated N-terminal methionine residues (Steffans and Buse, 1979; Chomyn et al., 1981) and both begin at the first possible initiation codon on their structural gene. Therefore, neither mammalian subunit II is derived from a preprotein. The mature subunit II polypeptide from *N. crassa* cytochrome *c* oxidase has neither an N-terminal methionine nor a blocked N-terminus (Machleidt and Werner, 1979). It starts at codon 13 of its structural gene. Even though no precursor polypeptide has been observed yet, these findings strongly suggest that *N. crassa* subunit II is also derived from a preprotein. The predicted leader peptide for *N. crassa* has several features in common with the *S. cerevisiae* subunit II leader peptide including: a hydrophobic core, the sequence IMN near the cleavage site, and an α helical predicted secondary structure. As yet, it is not known if other species besides *N. crassa* and *S. cerevisiae* have preprotein precursors for subunit II. However, based on the alignments for the coding regions of mature subunit II it seems likely that preprotein precursors exist in plants. The N-terminal sequences of all four plant species shown in Figure 5 extend upstream from the determined N-terminal amino acid of mature subunit II from *S. cerevisiae, N. crassa*, cow, and human.

Why is subunit II from yeast, *N. crassa*, and possibly plants, derived from a preprotein precursor while subunit II from human, cow (and probably rat, mouse, and *Drosophilia*; see Figure 5) not derived from a preprotein? This question is of significance because the subunit II polypeptides from different species, irrespective of whether they are derived from preproteins, have a high degree of structural homology. Moreover, yeast subunit II, a polypeptide derived from a preprotein, and bovine subunit II, a polypeptide that is not derived from a preprotein, have similar topographical orientations across the inner membrane (Figure 2). In view of this it is hard to argue that the leader peptide on yeast subunit II is required for targeting or inserting the mature subunit into the inner membrane. Therefore, it is likely that this structure has some other role. One possibility is that it functions to assemble one or more of the other subunits into a holoenzyme. This role could be species dependent since holocytochrome *c* oxidases in different species have slightly different subunit compositions (Patterson et al., 1987). Another possibility is that the leader peptide functions to keep the mature subunit II polypeptide chain in an insertion-competent conformation. In mammals, this function may have been acquired by a nuclear-coded subunit. A third possibility is that the leader peptide functions to regulate the synthesis of subunit II in plants and fungi, perhaps at the level of translation, by interacting with the mitochondrial equivalent of a signal recognition particle. This possibility is intriguing because it would imply that this leader peptide has only one of the two generic functions (i.e., protein targeting and translational arrest) currently attributed to leader peptides on proteins that are inserted into membranes co-translationally. Moreover, it is interesting because mitochondrial gene expression is regulated mostly post-transcriptionally in fungal mitochondria (Costanzo and Fox, 1990).

E. Evidence for Co-translational Insertion

Several lines of evidence indicate that mitochondrially-encoded proteins are inserted into the inner mitochondrial membrane co-translationally. The most direct evidence comes from studies with cytochrome *c* oxidase presubunit II in yeast. The leader peptide of this preprotein is processed by Imp1p, which is located on the external surface of the inner membrane (Schneider et al., 1991). Hence, processing is a convenient measure of insertion because the precursor becomes accessible to the protease only after its insertion into the inner membrane. Because presubunit II is processed co-translationally, it follows that it is also inserted into the membrane co-translationally. The evidence that the other mitochondrial gene products in *S. cerevisiae* are also inserted into the inner membrane co-translationally is more indirect, summarized as follows:

1. Mitochondrial proteins translated *in vivo* and *in vitro* appear in the inner membrane fraction concomitantly with synthesis and there are no detectable pools of these proteins in the matrix fraction (Sevarino and Poyton, 1980; McKee and Poyton, 1984; Clarkson and Poyton, 1989).
2. When nascent polypeptide chains are released from mitochondrial ribosomes with puromycin they are membrane bound and are not released with Na_2CO_3 at pH 11.5 (G.H.D. Clarkson, unpublished), conditions that release peripheral membrane proteins (Fujiki et al., 1982).
3. Growing nascent polypeptide chains that are translated by isolated mitochondria accumulate in the inner membrane fraction. In the presence of pactamycin, an inhibitor of mitochondrial translational initiation (Sevarino and Poyton, 1980; McKee et al., 1984), these nascent polypeptide chains are elongated within the inner membrane to full length polypeptides (Figure 6). Virtually no nascent chains are released from these membranes by sonication or treatment with Na_2CO_3 at pH 11.5. Hence, the nascent polypeptides are integral proteins, like mature mitochondrial gene products. This argues against mitochondrial gene products being inserted into the membrane via an aqueous pore as proposed for integral proteins of the endoplasmic reticulum (Simon et al., 1992).

These observations, together with those for presubunit II, suggest a co-translational pathway for insertion of most, if not all, mitochondrially-translated proteins in yeast. It is not yet known whether all of these proteins follow a single pathway for membrane insertion. This may not be the case because translation of some of these proteins require mRNA-specific translational activators that may function in membrane insertion (see following).

F. Energy Requirements for Insertion

The export pathway for mitochondrially-encoded proteins, like the import pathway for nuclear-coded proteins, requires an electrochemical potential across the

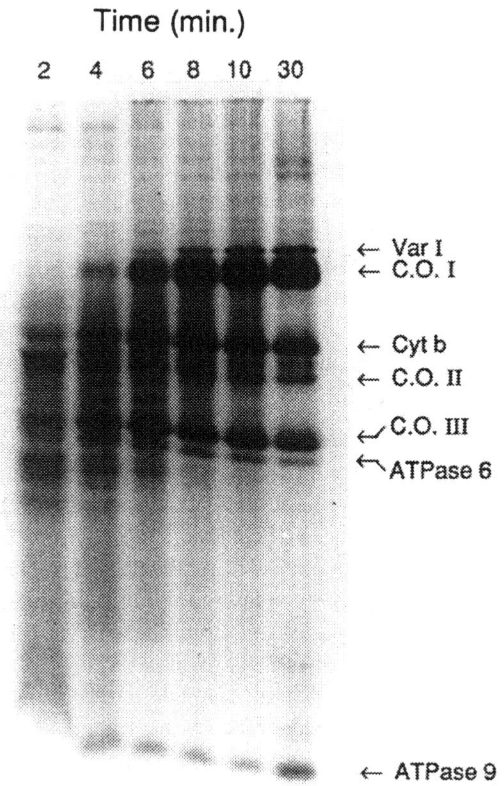

Figure 6. Growing nascent polypeptide chains translated by yeast mitochondria in a run-off system are membrane bound. Isolated yeast mitochondria were incubated in the presence of [^{35}S] methionine in an optimized protein synthesizing medium (McKee and Poyton, 1984) but in the presence of 2uM pactamycin, an inhibitor of translational initiation (McKee et al., 1984). Samples were withdrawn at the times indicated and then sonicated. The membrane fraction was collected by centrifugation and then analyzed by SDS-PAGE. Under the run-off conditions used for this experiment, nascent polypeptide chains are dispersed in size and appear as background labelling in the gel at early times (2–8 min). They are elongated to full-length polypeptides by 30 min. In this experiment more than 99% of all TCA-precipitable counts co-fractionated with the membrane fraction at each time interval. Hence, growing nascent polypeptide chains encoded by the mitochondrial genome are membrane associated. Full-length mitochondrial gene products are identified by the following abbreviations: var1 for polypeptide var1 of the small ribosomal subunit; C.O.I, II, and III for subunits I, II, and III of cytochrome *c* oxidase; cyt b for the cytochrome b binding subunit of coenzyme Q cytochrome *c* reductase; and ATPase 6 and 9 for subunits 6 and 9 of oligomycin-sensitive ATPase.

inner membrane and the cleavage of nucleoside triphosphates both in the cytosol and matrix (Glick and Schatz, 1991). However, the use of these energy sources are different for import and export.

1. Membrane Potential

Studies with isolated yeast mitochondria have revealed that cytochrome *c* oxidase presubunit II is accumulated when mitochondrial translation occurs under conditions that prevent the formation of a membrane potential across the inner membrane (Clarkson and Poyton, 1989). The accumulated presubunit II is located in the inner membrane and behaves like an integral membrane protein (Clarkson and Poyton, 1989). The other yeast mitochondrial gene products synthesized in the absence of a membrane potential behave like integral proteins as well. Thus, a membrane potential is not required for insertion of yeast mitochondrial gene products into the inner membrane but is required for the processing of presubunit II. This reveals a fundamental difference between import and export pathways. For import, a membrane potential functions to initiate translocation of preproteins into the matrix, perhaps via an electrophoretic effect on the positively charged domains of the leader peptide or mature regions of these polypeptides (Hartl et al., 1989). For export, a membrane potential appears to function only in processing a preprotein and is not required for translocation of polypeptides into the membrane.

2. Nucleoside Triphosphates

So far, it has been difficult to establish that nucleoside triphosphates per se are required for insertion of mitochondrial gene products into the inner membrane because insertion occurs co-translationally and as such is linked to translation. Because translation requires nucleoside triphosphates, it is difficult to readily distinguish an effect on insertion from an effect on translation. Nevertheless, studies with isolated yeast mitochondria have revealed that the mechanism of ATP generation effects translation and/or insertion of some mitochondrial gene products into the inner membrane. These studies have shown that both an exogenous 'ATP-regenerating' system as well as endogenous oxidative phosphorylation (driven by added oxidizable substrates) are required for optimal synthesis of mitochondrial gene products (McKee and Poyton, 1984). Either ATP generating system by itself supports some protein synthesis. However, the ATP generated exogenously by the ATP-regenerating system does not support the synthesis of full-length cytochrome *c* oxidase subunits I and II. In the absence of added oxidizable substrates, these are replaced by truncated forms (of 19.5 and 18kDa) less than one-half of the length of the full-length cytochrome *c* oxidase subunits I and II. In contrast, other mitochondrial gene products are produced at normal levels under these conditions.

Why is the synthesis and/or membrane insertion of cytochrome *c* oxidase subunits I an II differentially affected by oxidative phosphorylation? There are at least two possibilities. First, electron transport itself (driven by oxidizable substrate) may be required for the insertion or synthesis of full-length subunit I and II polypeptides. These two subunits bind the four prosthetic groups (hemes a and a_3, Cu_A and Cu_B) of holocytochrome c oxidase. It is possible that they are added co-translationally and electron transport itself is required to alter their redox state for addition to the growing nascent chain. Second, the synthesis or insertion of these two polypeptides may require higher levels of intramitochondrial ATP than the other mitochondrial gene products. This might occur, for example, if these two mitochondrial gene products alone require heat shock proteins for folding into the inner membrane.

IV. EXPORT OF MITOCHONDRIALLY-ENCODED PROTEINS TO THE PLASMA MEMBRANE

Studies with the MTF component of Mta, a mouse minor histocompatibility antigen (Fischer-Lindahl et al., 1991), have provided evidence for an export pathway for mitochondrially-encoded peptides to the plasma membrane (Figure 1, pathway 5). MTF is a maternally inherited peptide that is identical to the N-terminus of NADH dehydrogenase subunit 1 (ND1) (Loveland et al., 1990), a protein encoded by the mitochondrial genome in vertebrates. It is presented on the cell surface by HMT, a major histocompatibility complex class I protein. Cells pre-treated with a peptide containing the first 17 amino acids of ND1 are lysed by Mta-specific cytotoxic T lymphocytes. An N-terminal formylmethionine, the initiating amino acid for mitochondrial translation, is essential for MTF function in the assay. Also, if Mta is removed from the cell surface enzymatically its resynthesis is inhibited by chloramphenicol, an inhibitor of mitochondrial translation (Han et al., 1989). Together, these findings establish that a peptide derived from a mitochondrial gene product and translated on mitochondrial ribosomes is present on the surface of mouse cells.

Several questions concerning how the MTF peptide is produced and routed to the cell surface remain open. First, how does this fragment of a mitochondrial protein arise and how large is it? In theory, it may be generated by premature translational termination during synthesis or by proteolytic processing of mature ND1. If it is generated by proteolytic processing, is the protease a mitochondrial or cytosolic enzyme? Some likely candidates for proteases that could generate a fragment of ND1 are the mammalian homoloques of yeast YTA10p, an ATP-dependent inner membrane protease (Pajic et al., 1994), or yeast MSP1p, an ATP-dependent outer membrane protease (Nakai et al., 1993). Second, how is MTF exported from the mitochondrion? As discussed ahead, this may occur by diffusion through pores or channels either in the inner membrane, the outer membrane, or contact sites between both membranes. Third, how is MTF routed to the cell surface once it is released from the mitochondrion? Available evidence suggests that MTP is routed through the endoplasmic reticulum to the cell surface via the secretory pathway. This evidence comes from the finding that the presentation of the MTF peptide on the cell surface is diminished in RMA-S mutant cells (Hermel et al.,

1991). These cells are defective in ATP-dependent ABC transport proteins that have been implicated in the transport of antigens from the cytosol to the lumen of the endoplasmic reticulum (Attaya et al., 1992).

The export of mitochondrially-coded proteins to the plasma membrane of other organisms is not well documented yet. However, several lines of indirect evidence suggests that it occurs. First, rat cells also have a surface antigen that is encoded by the mitochondrial genome (Fischer-Lindahl et al., 1991). This antigen is unrelated to MTP and is probably not derived from ND1. Second, some auto-immune sera in humans contain antibodies against mitochondria (Mackay and Gershwin, 1989). Currently, it is not known if these are produced in response to mitochondrial proteins that are secreted or by cell lysis. Finally, cell surface functions (e.g., sugar uptake, cell agglutination, and floculation) in yeast are affected by the mitochondrial genome (Wilkie and Evans, 1982). It is not yet known if these changes are direct, resulting from protein export to the cell surface, or indirect, resulting from an effect of mitochondrial genes on the expression of nuclear genes for cell surface proteins.

V. EXPORT OF NUCLEAR-ENCODED PROTEINS TO THE INTERMEMBRANE SPACE

The export of nuclear-encoded proteins from the matrix to the intermembrane space (Figure 1, pathway 2) has been referred to as "conservative sorting" (Hartl and Neupert, 1990). This pathway has been implicated in the sorting of a limited set of proteins to either the intermembrane space or the outer surface of the inner membrane. These include the inner membrane proteins cytochrome c_1, the Rieske Fe-S subunit of coenzyme Q cytochrome *c* reductase, subunit IV of cytochrome *c* oxidase, and subunit 9 of F_1F_0 ATPase from *N. crassa* (Hartl et al., 1986, 1987; Fu et al., 1990; Mahlke et al., 1990; Hurt et al., 1984) and the intermembrane space protein cytochrome b_2 (Hartl et al., 1987).

The precursors for these proteins fall into two classes, based upon the nature of their presequences. One class, represented by the Fe-S protein, cytochrome *c* oxidase subunit IV, and ATPase subunit 9, is characterized by a hydrophobic and basic presequence that functions as a matrix-targeting sequence. This simple presequence is removed in the matrix by a two step process. The first step, catalyzed by matrix protease, results in the generation of a soluble intermediate. In the second step this intermediate is processed by MIP1 protease (Isaya et al., 1994) to the mature form, which is then exported in a process apparently directed by internal domains within the body of the mature protein. The second class, represented by cytochrome c_1 and cytochrome b_2, has a complex bipartite presequence that contains an N-terminal hydrophobic basic matrix-targeting signal followed by a domain of hydrophobic amino acids. This presequence is also processed in two steps: the N-terminal portion is removed in the matrix by the matrix protease, while the remainder is processed by the inner membrane protease whose catalytic sites face the intermembrane space. Imp1p catalyses the second step in cytochrome b_2 processing, while Imp2p catalyzes the second step in cytochrome c_1 processing (Nunnari et al., 1993).

Currently, there is general agreement that the class of proteins with a simple presequence follows a conservative sorting pathway. However, the route taken by members of the class of proteins with bipartite presequences is controversial. Two different pathways have been proposed for the sorting of cytochrome c_1 and cytochrome b_2 to the intermediate space. One of these is the conservative sorting pathway. The other, a "stop-transfer" pathway (Glick et al., 1992), proposes that precursors with bipartite presequences are not completely translocated across the inner membrane and that the hydrophobic region of the presequence acts as a "stop-transfer" sequence. In both models, the matrix protease processes the N-terminal region of the presequence, and the inner membrane protease complex (Imp1p and Imp2p) processes the intermediate form to the mature protein. The primary difference between these models is the intramitochondrial location where the preprotein is processed to the intermediate form and whether the intermediate form is completely translocated to the matrix. In addition, it has been suggested that proteins that follow the conservative sorting route interact with Hsp60 in the mitochondrial matrix (Hartl and Neupert, 1990; Mahlke et al., 1990; Koll et al., 1992) and that ATP is required for the sorting of their intermediate forms to the intermembrane space. The ATP is used to release the preprotein from Hsp60.

Neupert and his colleagues have identified intermediates of cytochrome c_1 and cytochrome b_2 in the matrix and a dependency on Hsp60 for the processing of these intermediates to their mature forms (Hartl et al., 1987; Cheng et al., 1989; Koll et al., 1992). In addition, mutations in the hydrophobic region of the cytochrome c_1 presequence prevent sorting to the intermembranes space but not import into the matrix (Jensen et al., 1992). These findings support the "conservative sorting" model. However, attempts to repeat some of these studies have produced conflicting results. Glick and colleagues (1992) have reported that the processing of precursors to cytochrome b_2 and cytochrome c_1 does not require Hsp60 and have failed to find intermediates in the matrix. It is not clear why matrix-localized intermediates were found in one set of studies but not in the other. It has been suggested that the different methods used for mitochondrial fractionation in these two sets of studies may account for the observed differences by altering the distribution of intermediates between mitochondrial subfractions (Glick et al., 1992). It is also possible that both pathways are operable and that the different import conditions used in the two sets of studies favored one pathway over the other. Finally, as discussed ahead, the buffer systems commonly used in both sets of studies do not support mitochondrial translation (Duhl, 1991; Duhl et al., 1996). This also may have effected the export pathway for nuclear-coded proteins.

VI. EXPORT OF NUCLEAR-ENCODED PROTEINS TO THE CYTOSOL

Evidence for the reversible transport and/or exchange of proteins into and out of the mitochondrial matrix or intermembrane space compartments has been available for several years (Waksman et al., 1977; Passarella et al., 1982). Although this type of transport involves import as well as export (Figure 1, pathways 1 and 3), it is generally ignored in reviews on mitochondrial import. Here, we will consider evidence for the existence of this type of transport for four proteins: cytochrome *c* and cytochrome *c* peroxidase, which reside in the intermembrane space, and malate dehydrogenase and aspartate aminotransferase, which reside in the matrix.

A. Cytochrome *c*

Cytochrome *c* is an intermembrane space protein that binds to the outer surface of the inner membrane by ionic interactions with either acidic phospholipids, coenzyme Q cytochrome *c* reductase, or cytochrome *c* oxidase. Hence, it can exist as either a soluble protein or a peripheral membrane protein. Its import to the intermembrane space does not involve receptors on the mitochondrial surface and does not require ATP (Hartl and Neupert, 1990). Moreover, because localization of cytochrome *c* to the intermembrane space does not require a membrane potential across the inner membrane, it appears unlikely that it is imported first into the matrix and then sorted to the intermembrane space by Pathway 2 (Figure 1).

Several studies have demonstrated that the covalent attachment of heme to apocytochrome *c* is important for its accumulation in the intermembrane space. The reaction is catalyzed by heme lyase, which is localized at the outer surface of the inner membrane (Dumont et al., 1991). Early attempts to elucidate the import pathway for cytochrome *c* lead to a model in which the apoprotein remains at the mitochondrial surface until it is converted to holocytochrome *c* by the covalent attachment of reduced heme c (Nicolson et al., 1987). This model presupposes that heme lyase can add heme while the apoprotein is either bound at the outer mitochondrial surface or is transversing the outer membrane. More recent studies have lead to a two step model for apocytochrome *c* import (Hakvoort et al., 1990; Dumont et al., 1991). In the first step, apocytochrome *c* equilibrates across the outer membrane in a process that is freely reversible (Hakvoort et al., 1990; Figure 1, pathway 3). In the second step, apocytochrome *c* binds to heme lyase at the inner membrane and is converted to holocytochrome *c* (Dumont et al., 1991). It is the second step that confers irreversibility to the transport process and which leads to net accumulation of cytochrome *c* in the intermembrane space. The reversible translocation of apocytochrome *c* across the outer membrane is relevant to export and raises several questions concerning the mechanism of translocation. Does it occur through the lipid bilayer of the outer membrane itself (Rietveld and deKruijff, 1984) or through protein aqueous pores (see following) in the outer membrane?

B. Cytochrome *c* Peroxidase

The import pathway followed by cytochrome c peroxidase to the intermembrane space is poorly understood. Although it has a bipartite presequence like cytochrome b_2 and cytochrome c_1, it does not appear to be processed by the matrix protease (Reid et al., 1982) and does not require a membrane potential for localization to the intermembrane space (Kaput et al., 1989). Hence, unlike precytochrome b_2 and precytochrome c_1, precytochrome *c* peroxidase does not appear to be exposed to the matrix at any time during import. It is possible, therefore, that it may follow an import pathway like cytochrome *c*. This possibility is supported by the finding that mature cytochrome *c* peroxidase can be released from the intermembrane space after it is processed from precytochrome c peroxidase *in vitro* (Kaput et al., 1989).

C. Aspartate Aminotransferase and Malate Dehydrogenase

Both aspartate aminotransferase and malate dehydrogenase are released from isolated intact rat liver mitochondria. This release requires an external effector. External aspartate aminotransferase is an effector for malate dehydrogenase release, while external malate dehydrogenase is the effector for aspartate aminotransferase release. Both enzymes are released from the mitochondrial matrix (Figure 1, pathway 1). This release can be considered a form of export because it occurs in intact mitochondria and is selective. No other proteins are released in the presence of each effector. Moreover, this type of export from the matrix appears to be linked to import of the effector because: (1) the kinetics of export are similar to the kinetics of import of the effector; (2) it is inhibited by bathophenanthroline and β-mercaptoethanol, which are also inhibitors of import of the effectors; (3) import of the effector and export occur concomitantly; and (4) export, like import, requires a pH gradient across the inner membrane. Thus, it seems likely that there is an exchange transport pathway for the release of proteins from the mitochondrial matrix. Currently, it is not clear if the proteins are exported directly to the outside (perhaps via a zone of adhesion between inner and outer membranes) or if they are released first into the intermembrane space and follow Pathway 3 (Figure 1). In this regard, it is noteworthy that Waksman and colleagues (1977) found that aspartate aminotransferase could localize from the matrix to the intermembrane space under certain metabolic conditions.

VII. THE MACHINERY FOR EXPORT

Currently, little is known about the molecular machinery that is required for protein export across either the inner or outer mitochondrial membranes. However, several aspects of this are presently under study. For mitochondrial gene products, experimental emphasis has been placed on factors required for translation, membrane insertion, processing, or stability. For nuclear gene products, the possibility that pores or channels involved in import may also be involved in export is being addressed.

A. Mitochondrially-encoded Proteins

It is not known if mitochondrial gene products are inserted directly into the hydrophobic milieu of the inner membrane or if they first gain entry to the inner membrane via a proteinaceous pore, as has been proposed for integral proteins of the endoplasmic reticulum (Simon et al., 1992) and for the localization of proteins to the outer surface of the inner membrane by a "stop-transfer" pathway (Glick et al., 1991). Also unclear is whether protein factors are required for other aspects of export: for example, folding mitochondrial gene products into export competent structures, coupling translation with membrane insertion, or conferring stability to inserted proteins. Protein factors that may be involved in these processes have been sought in two ways: by genetic studies with yeast nuclear PET mutants and by biochemical studies with isolated mitochondria and mitochondrial subfractions (Poyton et al., 1996).

Genetic Studies

In view of the limited coding capacity of mitochondrial DNA, it seems likely that any protein factors required for this process will be encoded by the nuclear genome. Genetic studies with yeast nuclear PET mutants have revealed several proteins that are involved in the translation, insertion, or stability of mitochondrially-translated proteins. Some of these act globally on the expression of all mitochondrial genes. They include mitochondrial ribosomal proteins, tRNA synthetases, and initiation or elongation factors (Pon and Schatz, 1991). Others act on the expression of specific mitochondrial genes. These include: translational activators (Kloeckener-Griussem et al., 1987; Kloeckener-Griussem et al., 1988; Valencik et al., 1989; Valencik and McEwen, 1991; Costanzo and Fox, 1990), the inner membrane protease I complex (Michaelis et al., 1982; Pratje et al., 1993; Nunnari et al., 1993), and the nuclear-coded subunits of respiratory complexes that contain mitochondrial gene products.

Translational activators. Gene specific translation activators affect the translation of at least three mitochondrial mRNAs (COXII, COXIII, and COB). *CBS1* and *CBS2* specifically affect COB mRNA translation (Roedel et al., 1985, 1986; Roedel, 1986); *PET111* and *PET112* specifically affect COX2 mRNA translation (Poutre and Fox, 1987; Costanzo et al., 1986; Strick and Fox, 1987); and *PET54*, *PET122*, and *PET494* specifically affect translation of COX3 mRNA (Muller et al., 1984; McEwen et al., 1986; Kloeckener-Griussem et al., 1987; Costanzo and Fox, 1986). Each of these activators modulate translation through sites in the 5′-untranslated regions of their target mRNAs (Mulero and Fox, 1993; Constanzo and Fox, 1988; Roedel and Fox, 1987). The best understood of these are the products of the *PET54*, *PET122*, and *PET494* genes. Mutations in one of them, *PET122*, can be suppressed by mutations in ribosomal protein genes (Haffter et al., 1990, 1991), indicating an interaction between the PET122 protein and the mitochondrial ribosome. All three of these activator proteins are associated with the inner membrane (McMullin and Fox, 1993; Brown et al., 1994). Similarly, the COB-mRNA specific activator proteins CBS1p and CBS2p, as well as the COX2-mRNA specific activator protein, PET111p, are also membrane localized (Michaelis et al., 1991; Strick, 1988). Both CBS1p and PET111p appear to be integral proteins while CBS2p is a peripheral protein. In view of their membrane association and, in the case of PET122p, interaction with the mitochondrial ribosome, it has been suggested that these proteins may function in membrane localization of specific mitochondrial mRNAs (Brown et al., 1994) or in the insertion of cytochrome *c* oxidase subunits II, III, and cytochrome *b* into the inner membrane (McMullin and Fox, 1993). The precise mechanism by which these activators facilitate translation or insertion is unknown; its elucidation will require the development of a mitochondrial mRNA-dependent translation system (see below).

Inner membrane protease. Inner membrane protease I consists of two catalytic subunits, Imp1p and Imp2p, of nonoverlapping specificities (Nunnari et al., 1993). Imp1p processes cytochrome *c* oxidase pre-subunit II as well as the nuclear-encoded protein cytochrome b_2. Imp2p processes the nuclear-encoded protein cytochrome c_1. Mutations in the Imp1p protease affect the expression of cytochrome *c* oxidase subunit II but not other yeast mitochondrial gene products because this protease only processes pre-subunit II. The first mutant characterized with a defect in Imp1p was a temperature sensitive petite, pet^{ts} 2858 (Michaelis et al., 1982; Pratje et al., 1983). This strain lacks cytochrome aa_3 and is incapable of growth on nonfermentable carbon sources at 36°C, the nonpermissive temperature. However, it has normal levels of cytochrome aa_3 and grows on nonfermentable carbon sources at 23°C, the permissive temperature. Interestingly, pre-subunit II is not processed at either temperature. Thus, processing is not required for membrane insertion or assembly into a functional cytochrome *c* oxidase, at least at the permissive temperature. It is not clear how the presence of presubunit II affects assembly of the holoenzyme at the nonpermissive temperature. Perhaps it affects the stability of the complex. Mutations in the Imp2p protease also prevent the processing of pre-subunit II (Nunnari et al., 1993). Even though this protease does not process pre-subunit II by itself, it appears to be required for the stable expression of Imp1p. Mutants in Imp2p also differ from those in Imp1p insofar as they affect the expression of cytochrome oxidase subunit I (see Figure 1 in Nunnari et al., 1993). The level of subunit I synthesized *in vivo* during a pulse is greatly reduced relative to wild type. Imp2p may, therefore, affect the synthesis or insertion of subunit I, as well as the processing of pre-subunit II.

Nuclear-coded subunits of chimeric respiratory complexes. The third class of specific effectors are the nuclear-coded subunits of the respiratory complexes themselves. Nuclear mutations in the structural genes for some subunits of

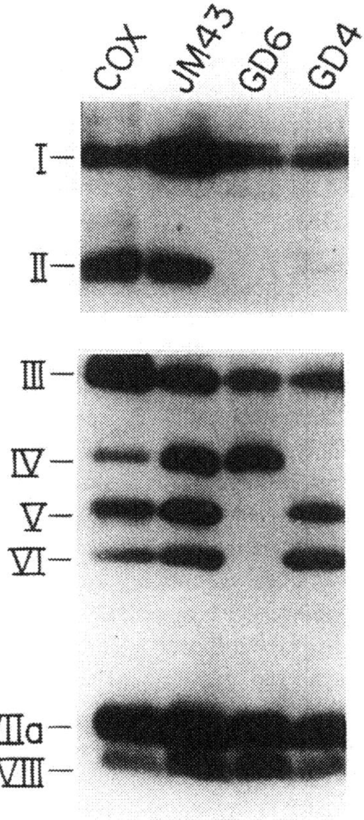

Figure 7. Null mutants that lack subunits IV or VI of cytochrome c oxidase lack or have reduced levels of the mitochondrially coded subunits I and II of the holoenzyme. A Western immunoblot of mitochondria from strains JM43 (wild type), GD6 (a COX6-*null* mutant), and GD4 (a COX4-*null* mutant). A cytochrome *c* oxidase sample (COX) was run to localize the positions of the subunits (indicated by Roman numerals).

cytochrome *c* oxidase and coenzyme Q cytochrome *c* reductase show substantial decreases in the level of the mitochondrially-encoded subunits of these complexes. *Null* mutants that carry a gene disruption in yeast *COX6*, the nuclear gene for cytochrome *c* oxidase subunit VI, or in *COX4*, the nuclear gene for subunit IV, have reduced levels of subunits I and III and lack subunit II entirely (Figure 7). *Null* mutants with gene disruptions in *COX5a* and *COX5b*, the structural genes for the Va and Vb isoforms of cytochrome *c* oxidase, have less than half of the normal levels of subunits I and II and lack subunit III entirely (L.E. Farrell and C.R. Trueblood, unpublished). And a *null* mutant that carries a deletion in *COX7*, the gene for subunit VII of cytochrome *c* oxidase, has reduced levels of subunit I

(Calder and McEwen, 1991). The results of *in vivo* pulse-chase studies with this latter strain suggest that there is either a decreased rate of translation of the mRNA for *COX1* (possibly due to a defect in insertion of subunit I into the inner membrane) or increased rates of subunit I degradation. Similar studies with *null* mutants in the nuclear-coded subunits of coenzyme Q cytochrome *c* reductase have revealed that mutations in *COR1*, *COR2*, *COR4*, and *COR5* lead to decreased levels of cytochrome *b* (Crivellone et al., 1988).

Together, these studies imply that the nuclear-coded subunits of these complexes may be required for synthesis, membrane insertion, or stability of their mitochondrially-encoded subunit counterparts.

Other factors. It is not known if protein factors (e.g., SRP, SEC, trigger factor, heat shock proteins, ABC transporters, ATP-dependent proteases, and pores) required for insertion of proteins into other membranes (e.g., endoplasmic reticulum, bacterial cytoplasmic membrane) are also required for protein export from mitochondria. There is suggestive evidence that hsp60 functions in the biogenesis of the subunit of F_1 F_0-ATPase (Prasad et al., 1990), a mitochondrial gene product in maize. However, this subunit is part of the F_1 component of the enzyme and as such it assembles into a peripheral protein complex on the matrix face of the inner membrane. Hence, it is not exported. Recently, it has been reported that SRP deficient yeast rapidly acquire a rho$^-$ phenotype (Hann and Walter, 1991). This effect may be due to an effect on import or export of mitochondrial gene products. The latter seems unlikely because attempts to detect SRP in mitochondria have been unsuccessful (Walter, personnel communication).

Finally, a growing number of ATP-dependent proteases are being identified in mitochondrial subfractions. At least two of these are membrane proteins. MSP1p is an integral protein of the outer membrane (Nakai et al., 1993), and YTA10p is an integral protein of the inner membrane (Pajic et al., 1994). The latter is involved in the degradation of unassembled subunits of ATP synthase and respiratory chain complexes. It belongs to a novel class of ATPases that have diverse cellular functions. Insofar as some of these functions included facilitating the insertion of proteins into membranes, it is possible that YTA10p functions in the export pathway for mitochondrially-coded proteins.

Biochemical Studies

Two biochemical approaches have been used to study the export machinery. One makes use of isolated yeast mitochondria and the other uses inverted inner membrane preparations. Previously, it has been shown that protein synthesis in isolated yeast mitochondria is stimulated by the addition of cytosolic factors that are proteinaceous (Poyton and Kavanagh, 1976; McKee and Poyton, 1984), and that endogenous pools of protein required for translation can be depleted by preincubating cells in cycloheximide prior to isolating mitochondria. These findings suggest that nuclear-coded proteins are required for mitochondrial protein synthesis or stability and that mitochondria accumulate a pool of these proteins in exponentially growing cells. More recently, Black-Schaefer and colleagues (1991) have found that proteins made by isolated yeast mitochondria define two distinct pools with respect to turnover: one that is rapidly degraded ($t_{1/2}$ of minutes) and one that is much more resistant to proteolysis ($t_{1/2}$ of hours). Currently, isolated yeast mitochondria optimized for protein synthesis and import (Duhl et al., 1996; Poyton et al., 1996) are being used to address the relationships between import, export, and assembly pathways. By using isolated mitochondria capable of simultaneously importing nuclear-coded proteins while they are synthesizing mitochondrial gene products, it has been possible to show that neither the translation nor insertion of mitochondrial gene products into the inner membrane are coupled to import and that mitochondrially encoded subunits of respiratory protein complexes are unstable in the absence of nuclear-coded subunits. This suggests that the rapidly-degraded pool observed by Black-Schaefer and colleagues (1991) may be mitochondrial gene products that have not assembled with their nuclear-coded subunit counterparts. The degradation of these proteins is ATP dependent and is probably catalyzed, in part, by YTA10p.

Inverted inner membrane preparations are being used to provide further refinement to studies on the mechanism of insertion. So far, these preparations have been shown to be capable of run-off translation and to be translocation competent (Poyton, unpublished). These are currently being used to try to develop a mitochondrial mRNA-dependent translation/translocation system that can be used to assay for "insertion factors."

B. Nuclear-coded Proteins

As discussed previously, nuclear gene products may be exported to the intermembrane space or the cytosol. Hence, they must cross either the inner membrane or both the inner and outer membranes. Studies on protein import has revealed that both the inner and outer membranes possess protein translocation channels (for review see Glick et al., 1991; Pfanner et al., 1992) that are distinct and different in composition from one another. Currently, it is thought that these two translocation channels localize to the zones of adhesion between inner and outer membranes and collaborate to import proteins to the mitochondrial matrix. However, several lines of evidence suggest that these two translocation channels may function independently of one another. It is not known what triggers the opening or closing of these channels. Also, it is not known if they can work in reverse to effect protein export.

The inner membrane channel deduced from import studies may be identical to the giant channel that has been identified in electrophysiology experiments (Petronilli et al., 1989; Moran et al., 1990; Szabo and Zoratti, 1991; Nicolli et al., 1996). The latter has an extremely high conductance (up to 1.3 nS), behaves as if it is composed of units that can assemble with variable stoichiometries, and may also be located at zones of adhesion. Recent studies suggest that this channel is identical to a Ca^{++} dependent permeability transition channel (Szabo and Zoratti, 1991; Nicolli et al., 1996). Both are inhibited by cyclosporin A. The Ca^{++} dependent permeability transition channel has been implicated in the release of mitochondrial matrix proteins (Igbavboa et al., 1989). Therefore, it may be operative in the export pathways shown in Figure 1. The finding that this channel is inhibited by cyclosporin A provides a way of testing this hypothesis experimentally.

VIII. RELATIONSHIP BETWEEN IMPORT AND EXPORT

The assembly of several protein complexes of the inner mitochondrial membrane requires both the import and export of protein subunits across and/or into the inner membrane. Currently, it is not clear if import and export pathways are interconnected or independent of one another during the assembly of these multimeric proteins. This aspect of mitochondrial inner membrane assembly is interesting in light of the fact that import and export require a membrane potential. Are the requirements for a membrane potential independent of one another or does one pathway require the other? This question is best illustrated with yeast cytochrome *c* oxidase. This enzyme is composed of stoichiometric amounts of three subunits (I, II, and III) encoded by the mitochondrial genome and six subunits (IV, Va or Vb, VI, VII, VIIa, and VIII) encoded by the nuclear genome (Power et al., 1984; Poyton et al., 1995). An important question for this enzyme, and other respiratory complexes that are genetic chimeras, is when do the import and export pathways (Figure 8) converge for assembly of the holoenzyme? We have begun to address this question with an *in vitro* system with isolated yeast mitochondria capable of simultaneous mitochondrial import and translation (Duhl et al., 1996; Poyton et al., 1996). Unlike other *in vitro* mitochondrial import systems (Pfaller et al., 1988; Eilers et al., 1989) this system has been optimized for both import and mitochondrial translation. So far, we have used pre-subunit Va to follow import and subunits I, II, and III to follow export in this system. These studies have shown that the import of at least one nuclear-coded subunit, Va, does not affect the rate of translation of any of the mitochondrially-coded subunits nor does it affect the processing of pre-II. Moreover, the rate of import of subunit Va is not affected by synthesis of subunits I, II, or III. Thus, it appears that the import and export pathways for these subunits are not coupled and that assembly occurs after the subunits have been synthesized and localized to the inner membrane. It is not yet known if the import pathways for the other nuclear-coded subunits of cytochrome *c* oxidase or those for the nuclear-coded subunits of other complexes are coupled to export.

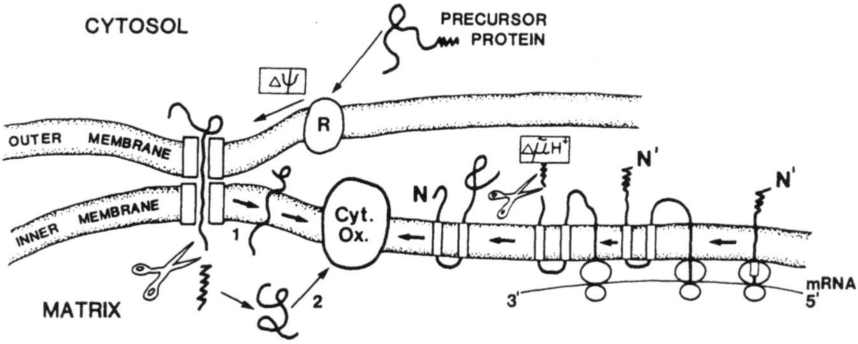

Figure 8. Model illustrating the convergence of nuclear and mitochondrially-coded subunits in the assembly of yeast cytochrome *c* oxidase. Shown here are key steps during post-translational import of a presequence-containing subunit and during export of subunit II as it is synthesized on membrane bound mitochondrial ribosomes. Zig-zag lines designate cleavable presequences. For import, a presequence-containing precursor protein binds to a receptor (R) and then moves to a contact zone between inner and outer membranes where it initiates translocation into the matrix in a $\Delta\Psi$ dependent step. Once inside the matrix the leader peptide is removed by the matrix processing protease (large scissors) and the mature protein moves to cytochrome *c* oxidase (cyt. ox.) either by lateral diffusion (path 1) or re-export from the matrix (path 2). Path 1 is a "stop-transfer" pathway and is followed by subunit Va whereas path 2 is "conservative sorting" (or "conservative assembly") from the matrix and is followed by subunit IV. For insertion of subunit II into the inner membrane we have assumed a loop model to explain the insertion of both transmembrane helices (open boxes) as well as the topological placement of the N and C termini on the cytoplasmic face of the inner membrane. The presequence of pre-II is removed by inner membrane protease I (small scissors) and requires a membrane potential ($\Delta\mu H^+$). It is not clear if pre-II assembles before or after assembly with other subunits into cytochrome *c* oxidase.

IX. CONCLUDING REMARKS

The export of proteins into or across the inner mitochondrial membrane is now well documented. However, our understanding of this phenomenon is still very rudimentary. Many questions remain. Is the same molecular machinery used for export and import? How are mitochondrially-synthesized proteins inserted into the inner membrane? Do exported proteins follow one pathway or multiple pathways? Do other macromolecules (e.g., DNA and RNA) leave mitochondria via protein export pathways? What factors regulate export? Answers to these questions should provide a mechanistic underpinning for our understanding of the export process itself. In addition, they may provide insight concerning the communication pathway by which the mitochondrial genome "talks to" the nuclear genome.

ACKNOWLEDGMENTS

Some of the work discussed here, done in the author's laboratory, was supported by grants GM39324 and GM30228 from the National Institutes of Health and a gift from Leslie F. Bailey.

REFERENCES

Anderson, S., Bankier, A. T., Barrell, B. G., de Bruijn, M. H. L., Coulson, A. R., Drouin, A., Eperon, E. C., Nierlich, D. P., Roe, B. A., Sanger, F., Schreier, P. H., & Young, E. G. (1981). Sequence and organization of the human mitochondrial genome. Nature 290, 459–465.

Anderson, S., deBruijn, M. H. L., Coulson, A. R., Eperon, E. C., Sanger, F., & Young, E. G. (1982). Conserved features of the mammalian mitochondrial genome. J. Mol. Biol. 156, 683–717.

Attardi, G., & Schatz, G. (1988). Biogenesis of mitochondria. Annu. Rev. Cell. Biol. 4, 289–333.

Attaya, M., Jameson, S., Martinez, C. K., Hermel, E., Aldrich, C., Forman, J., Fischer Lindahl, K., Bevan, M. J., & Monaco, J. J. (1992). Ham-2 corrects the class 1 antigen-processing defect in RNA-S cells. Nature 355, 647–649.

Baker, K. P., & Schatz G. (1991). Mitochondrial proteins essential for viability mediate protein import into yeast mitochondria. Nature 349, 205–208.

Bibb, M. J., vanEtten, R. R., Wright, C. T., Walberg, M. W., & Clayton, D. A. (1981). Sequence and gene organization of mouse mitochondrial DNA. Cell 26, 167–180.

Black-Shaefer C. L., McCourt, J. D., Poyton, R. O., & McKee, E. E. (1991). Mitochondrial gene expression in *Saccharomyces cerevisiae*. Proteolysis of nascent chains in yeast mitochondria optimized for protein synthesis. J. Biochem. 274, 199–205.

Bonen, L., Boer, P. H., & Gray, M. W. (1984). The wheat cytochrome *c* oxidase subunit II gene has an intron insert and three radical amino acid changes relative to maize. EMBO J. 3, 2531–2536.

Boyer, P. D., Chance, B., Ernster, L., Mitchell, P., Racker, E., & Slater, E. C. (1977). Oxidative phosphorylation and photophosphorylation. Annu. Rev. Biochem. 46, 955–1026.

Brown, G. G., & Simpson, M. V. (1982). Novel features of animal mtDNA evolution as shown by sequences of two rat cytochrome c oxidase subunit II genes. Proc. Natl. Acad. Sci. USA 79, 3246–3250.

Brown, N. G., Costanzo, M. C., & Fox, T. D. (1994). Interactions among three proteins that specifically activate translation of the mitochondrial COX3 mRNA in *Saccharomyces cerevisiae.* Mol. Cell Biol. 14, 1045–1053.

Calder, K. M., & McEwen, J. E. (1991). Deletion of the *COX7* gene in *Saccharomyces cerevisiae* reveals a role for cytochrome *c* oxidase subunit VII in assembly of remaining subunits. Mol. Microbiol. 5, 1769–1777.

Cameron, V. L., McKee, E. E. Goehring, B., & Poyton, R. O. (1996). Protein export from mitochondria. A mutation at a conserved site within the coding sequence of yeast cytochrome *c* oxidase subunit II prevents processing of its presequence. J. Biol. Chem. Submitted.

Cheng, M. Y., Hartl, F-U., Martin, J., Pollock, R. A., Kalousek, F., Neupert, W., Hallberg, E. M., Hallberg, R. L., & Horwich, A. L. (1989). Mitochondrial heat-shock protein hsp60 is essential for assembly of proteins imported into yeast mitochondria. Nature 337, 620–625.

Chomyn, A., Hunkapiller, M. W., & Attardi, G. (1981). Alignment of the amino-terminal amino acid of human cytochrome *c* oxidase subunits I and II with the sequence of their putative mRNAs. Nucleic Acids Res. 9, 867–877.

Clarkson, G. H. D., & Poyton, R. O. (1989). A role for membrane potential in the biogenesis of cytochrome *c* oxidase subunit II, a mitochondrial gene product. J. Biol. Chem. 264, 10114–10118.

Coruzzi, G., & Tzagoloff, A. (1979). Assembly of the mitochondrial membrane system. DNA sequence of subunit 2 of yeast cytochrome c oxidase. J. Biol. Chem. 254, 9324–9330.

Costanzo, M. C., & Fox, T. D. (1986). Product of *Saccharomyces cerevisiae* nuclear gene *PET494* activates translation of a specific mitochondrial mRNA. Mol. Cell. Biol. 6, 3694–3703.

Costanzo, M. C., & Fox, T. D. (1988). Specific translational activation by nuclear gene products occurs in the 5′-untranslated leader of yeast mitochondrial mRNA. Proc. Natl. Acad. Sci. 55, 2677–2681.

Costanzo, M. C., & Fox, T. D. (1990). Control of mitochondrial gene expression in *Saccharomyces cerevisiae*. Annu. Rev. Genetics 24, 91–113.

Costanzo, M. C., Seaver, E. C., & Fox, T. D. (1986). At least two nuclear gene products are specifically required for translation of a single mitochondrial mRNA. EMBO J. 5, 3637–3641.

Covello, P. S., & Gray, M. W. (1992). Silent mitochondrial and active nuclear genes for subunit 2 of cytochrome *c* oxidase (COX2) in soybean: Evidence for RNA-mediated gene transfer. *EMBO J.* 11, 3815–3820.

Crivellone, M. D., Wu, M., & Tzagoloff, A. (1988). Assembly of the mitochondrial membrane system. Analysis of structural mutants of the yeast coenzyme QH_2-cytochrome c reductase complex. J. Biol. Chem. 263, 14323–14333.

de Bruijn, M. H. L. (1983). *Drosphila melanogaster* mitochondrial DNA, a novel organization and genetic code. Nature 304, 234–241.

Duhl, D. J. M. (1991). Protein localization and assembly of cytochrome *c* oxidase subunits in *Saccharomyces cerevisiae*. Ph.D. dissertation, University of Colorado, CO.

Duhl, D. M. J., Clarkson, G. H., Cassidy-Stone, A., & Poyton, R. O. (1996). Mitochondrial translation and mitochondrial import are not coupled events in the assembly of cytochrome *c* oxidase. J. Biol. Chem. submitted.

Dumont, M. E., Cardillo, T. S., Hayes, M. K., & Sherman, F. (1991). Role of cytochrome c heme lyase in mitochondrial import and accumulation of cytochrome c in *Saccharomyces cerevisiae*. Mol. Cell. Biol. 11, 5487–5996.

Eilers, M., Endo, J., & Schatz, G. (1989). Adriamycin, a drug interacting with phospholipids, blocks the import of precursor proteins by isolated yeast mitochondria. J. Biol. Chem. 264, 2945–2950.

Engelman, D. M., & Steitz, T. A. (1981). The spontaneous insertion of proteins into and across membranes: The helical hairpin hypothesis. Cell 77, 2023–2027.

Feldman, F., & Mahler, H. R. (1974). Mitochondrial biogenesis: Retention of terminal formyl methionine in membrane proteins and regulation of their synthesis. J. Biol. Chem. 249, 3702–3709.

Fischer-Lindahl, K., Hermel, E., Loveland, B. E., & Wang, C. R. (1991). Maternally transmitted antigen of mice: A model transplantation antigen. Annu. Rev. Immunol. 9, 357–372.

Fox, T. D., & Leaver, C. J. (1981). The *Zea mays* mitochondrial gene coding cytochrome c oxidase subunit II has an intervening sequence and does not contain TGA codons. Cell 26, 315–323.

Fu, W., Japa, S., & Beattie, D. S. (1990). Import of iron-sulfur protein of the cytochrome bc_1 complex into yeast mitochondria. J. Biol. Chem. 265, 16541–16547.

Fujiki, Y., Fowler, S., Shio, H., Hubbard, A. L., & Lazarow, P. B. (1982). Polypeptide and phospholipid composition of the membrane of rat liver peroxisomes: Comparison with endoplasmic reticulum and mitochondrial membranes. J. Cell. Biol. 93, 103–110.

Glick, B., & Schatz, G. (1991). Import of proteins into mitochondria. Annu. Rev. Genet. 25, 21–44.

Glick, B., Wachter, C., & Schatz, G. (1991). Protein import into mitochondria: Two systems acting in tandem. Trends in Cell Biol. 1, 99–103.

Glick, B., Brandt, A., Cunningham, K., Muller, S., Hallberg, R. C., & Schatz, G. (1992). Cytochrome c_1 and b_2 are sorted to the intermembrane space of yeast mitochondria by a stop-transfer mechanism. Cell 69, 809–822.

Haffter, P., McMullin, T. W., & Fox, T. D. (1990). A genetic link between an mRNA-specific translational activator and the translation system in yeast mitochondria. Genetics 125, 495–503.

Haffter, P., McMullin, T. W., & Fox, T. D. (1991). Functional interactions among two yeast mitochondrial ribosomal proteins and an mRNA-specific translational activator. Genetics 127, 319–326.

Hakvoort, T. B. M., Sprinkle, J. R., & Margoliash, E. (1990). Reversible import of apocytochrome c into mitochondria. Proc. Natl. Acad. Sci. USA 87, 4996–5000.

Hallermayer, G., & Neupert, W. (1976). Studies on the synthesis of mitochondrial proteins in the cytoplasm and on their transport into the mitochondrion. In: *Genetics and Biogenesis of Chloroplasts and Mitochondria* (Bucher, T., Neupert, W., Sebald, W., & Werner, S., Eds.). North-Holland, Amsterdam, pp. 807–813.

Han, A. C., Rodgers, J. C., & Rich, R. R. (1989). An unexpectedly labile mitochondrially encoded protein is required for Mta expression. Immunogenetics 29, 258–264.

Hann, B. C., & Walter, P. (1991). The signal recognition particle in *S. cerevisiae*. *Cell* 67, 131–144.

Hartl, F. U., Schmidt, B., Wachter, E., Weiss, H., & Neupert, W. (1986). Transport into mitochondria and intramitochondrial sorting of the Fe/S protein of ubiquinol-cytochrome c reductase. Cell 47, 939–951.

Hartl, F. U., Ostermann, J., Guiard, B., & Neupert, W. (1987). Successive translocation into and out of the mitochondrial matrix: Targeting of proteins to the intermembrane space by a bipartite signal peptide. Cell 51, 1027–1037.

Hartl, F. U., Pfanner, N., Nicolson, D. W., & Neupert, W. (1989). Mitochondrial protein import. Biochim. Biophys. Acta 988, 1–45.

Hartl, F. U., & Neupert, W. (1990). Protein sorting to mitochondria-evolutionary conservations of folding and assembly. Science 247, 930–938.

Hermel, E., Grigorenko, E., & Fischer Lindahl, K. (1991). Expression of medial class 1 histocompatibility antigens on RNA-S mutant cells. Int. Immunol. 3, 407–412.

Hiesel, R., & Brennicke, A. (1983). Cytochrome c oxidase subunit II gene in mitochondria of *Oenothera* has no intron. EMBO J. 2, 2173–2178.

Hurt, E. C., Pesold-Hurt, B., & Schatz, G. (1984). The amino-terminal region of an imported mitochondrial precursor polypeptide can direct cytoplasmic dihydrofolate reductase into the mitochondrial matrix. EMBO J. 3, 3149–3156.

Igbavboa, U., Zwizinski, C. W., & Pfeiffer, D. R. (1989). Release of mitochondrial matrix proteins through a Ca^{2+}-requiring cyclosporin-sensitive pathway. Biochem. Biophys. Res. Comm. 161, 619–625.

Inoue, I., Nagase, H., Kishi, K., & Higuti, T. (1991). ATP-sensitive K^+ channel in the mitochondrial inner membrane. Nature 352, 244–247.

Isaya, G., Miklos, D., & Rollins, R. A. (1994). *MIP1*, a new yeast gene homologous to the rat intermediate peptidase gene, is required for oxidative metabolism in *Saccharomyces cerevisiae*. Mol. Cell Biol. 14, 5603–5616.

Jensen, R. E., Schmidt, S., & Mark, R. J. (1992). Mutations in a 19-amino acid hydrophobic region of the yeast cytochrome c_1 presequence prevent sorting to the mitochondrial intermembrane space. Mol. Cell. Biol. 12, 4677–4686.

Kao, T-h., Moon, E., & Wu, R. (1984). Cytochrome oxidase subunit II gene from rice has an insertion sequence within the intron. Nucleic Acids Res. 12, 7305–7315.

Kaput, J., Brandriss, M. C., & Prussak-Wieckowska, T. (1989). *In vitro* import of cytochrome c peroxidase into the intermembrane space: Release of the processed form by intact mitochondria. J. Cell. Biol. 109, 101–112.

Kellems, R. E., Allison, U. F., & Butow, R. A. (1974). Cytoplasmic type 80s ribosomes associated with yeast mitochondria II. Evidence for the association of cytoplasmic ribosomes with the outer mitochondrial membrane *in situ*. J. Biol. Chem. 249, 3297–3303.

Kloeckener-Griussem, B., McEwen, J. E., & Poyton, R. O. (1987). Nuclear functions required for cytochrome *c* oxidase biogenesis in *Saccharomyces cerevisiae*: Multiple trans-acting nuclear genes exert specific effects on expression of each of the cytochrome *c* oxidase subunits encoded by mitochondrial DNA. Curr. Genetics. 12, 311–322.

Kloeckener-Griussem, B., McEwen, J. E., & Poyton, R. O. (1988). Identification of a third nuclear protein-coding gene required specifically for post-transcriptional expression of the mitochondrial COX3 gene in *S. cerevisiae*. J. Bacteriol. 170, 1399–1402.

Kobayashi, S., Amikura, R., & Okada, M. (1993). Presence of mitochondrial large ribosomal RNA outside mitochondria in genoplasm of *Drosophila Melanogaster*. Science 260, 1521–1524.

Koll, H., Guiard, B., Rassow, J., Ostermann, J., Horwich, A. L., Neupert, W., & Hartl, F-U. (1992). Antifolding activity of hsp60 couples protein import into the mitochondrial matrix with export to the intermembrane space. Cell. 68, 1163–1175.

Kuriyama, Y., & Luck, D. J. L. (1973). Membrane-associated ribosomes in mitochondria of *Neurospora crassa*. J. Cell. Biol. 59, 776–784.

Loveland, B., Wang, C. R., Yonekawa, H., Hermel, E., & Fischer-Lindahl, K. (1990). Maternally transmitted histocompatibility antigen of mice: A hydrophobic peptide of a mitochondrially encoded protein. Cell 60, 971–980.

Machleidt, W., & Werner, S. (1979). Is the mitochondrially made subunit 2 of cytochrome *c* oxidase synthesized as a precursor in *Neurospora crassa*? FEBS Letters 107, 327–330.

Macino, G., & Morelli, G. (1983). Cytochrome oxidase subunit II gene in *Neurospora crassa* mitochondria. J. Biol. Chem. 258, 13230–13235.

Mackay, I. R., & Gershwin, M. E. (1989). Molecular basis of mitochondrial autoreactivity in primary biliary cirrhosis. Immunol. Today 10, 315–318.

Mahler, H. R., Dawidowicz, K., & Feldman, F. (1972). Formate as a specific label for mitochondrial translation products. J. Biol. Chem. 247, 7439–7442.

Mahlke, K., Pfanner, N., Martin, J., Horwich, A. L., Hartl, F-U., & Neupert, W. (1990). Sorting pathways of mitochondrial inner membrane proteins. Eur. J. Biochem. 192, 551–555.

McEwen, J. E., Ko, C., Kloeckener-Griussem, B., & Poyton, R.O. (1986). Nuclear functions required for cytochrome c oxidase biogenesis in *Saccharomyces cerevisiae*. Characterization of mutants in 34 complementation groups. J. Biol. Chem. 261, 11872–11879.

McKee, E. E., & Poyton, R. O. (1984). Mitochondrial gene expression in *Saccharomyces cerevisiae* I. Optimal conditions for protein synthesis in isolated mitochondria. *J. Biol. Chem.* 259, 9320–9331.

McKee, E. E., McEwen, J. E., & Poyton, R. O. (1984). Mitochondrial gene expression in *Saccharomyces cerevisiae* II. Fidelity of translation in isolated mitochondria from wild type and respiratory-deficient mutant cells. J. Biol. Chem. 259, 9332–9338.

McMullin, T. W., & Fox, T. D. (1993). COX3 mRNA-specific translational activator proteins are associated with the inner mitochondrial membrane in *Saccharomyces cerevisiae*. J. Biol. Chem. 268, 11737–11741.

Mercier, J. C., Haze, G., Gaye, P., & Hue, D. (1978). Amino-terminal sequence of the precursor of ovine beta-lactoglobulin. Biochem. Biophys. Res. Comm. 82, 1236–1245.

Michaelis, G., Mannhaupt, G., Pratje, E., Fischer, E., Naggert, J., & Schweizer, E. (1982). Mitochondrial translation products in nuclear respiration-deficient pet mutants of *Saccharomyces cerevisiae*. In: *Mitochondrial Genes* (Slonimski, P., Borst, P., & Attardi, G., Eds.) Cold Spring Harbor Laboratory Press, pp. 311–321.

Michaelis, U., Korte, A., & Roedel, G. (1991). Association of cytochrome b translational activator proteins with the mitochondrial membrane: implications for cytochrome b expression in yeast. Mol. Gen. Genet. 230, 177–185.

Michon, T., Galante, M., & Velours, J. (1988). NH_2-terminal sequence of isolated yeast ATP synthase subunit 6 reveals post-translational cleavage. Eur. J. Biochem. 172, 621–625.

Moran, O., Sandri, G., Ponfili, E., Stoehmer, W., & Sorgato, M. C. (1990). Electrophysiological characterization of contact sites in brain mitochondria. J. Biol. Chem. 265, 908–913.

Mulero, J. J., & Fox, T. D. (1993). *PET111* acts in the 5′-leader of *Saccharomyces cerevisiae* mitochondrial COX2 mRNA to promote its translocation. Genetics 133, 509–516.

Muller, P. P., Reif, M. K., Zonghou, S., Sengstag, C., Mason, T. L., & Fox, T. D. (1984). A nuclear mutation that post-transcriptionally blocks accumulation of a yeast mitochondrial gene product can be suppressed by a mitochondrial gene arrangement. J. Mol. Biol. 175, 431–452.

Nakai, M., Endo, T., Hase, T., & Matsubara, H. (1993). Intramitochondrial protein sorting. J. Biol. Chem. 268, 24262–24269.

Nicolson, D. W., Kohler, H., & Neupert, W. (1987). Import of cytochrome c into mitochondria: Cytochrome c lyase. Eur. J. Biochem. 164, 147–157.

Nicolli, A., Basso, E., Petronilli, V., Wenger, R. M., & Bernardi, P. (1996). Interactions of cyclophilin with the mitochondrial inner membrane and regulation of the permeability transition pore, a cyclosporin A-sensitive channel. J. Biol. Chem. 271, 2185–2192.

Nugent, J. M., & Palmer, J. D. (1991). RNA-mediated transfer of the gene COXII from the mitochondrion to the nucleus during flowering plant evolution. Cell 66, 473–481.

Nunnari, J., Fox, T. D., & Walter, P. (1993). A mitochondrial protease with two catalytic subunits of nonoverlapping specificities. Science 262, 1997–2004.

Pajic, A., Taver, R., Feldmann, H., Neupert, W., & Langer, T. (1994). Yta10p is required for the ATP-dependent degradation of polypeptides in the inner membrane of mitochondria. FEBS Letts. 353, 201–206.

Passarella, S., Marra, E., Atlante, A., Barile, M., Doonan, S., & Quagliariello, E. (1990). Uptake of aspartate aminotransferase into mitochondria *in vitro* causes efflux of malate dehydrogenase and vice versa. Biochim. Biophys. Acta 1022, 273–282.

Patterson, T. E., Trueblood, C. E., Wright, R. M., & Poyton, R. O. (1987). Polypeptide subunits encoded by nuclear genes are essential components of cytochrome *c* oxidase. In: *Cytochrome Systems: Molecular Biology and Bioenergetics* (Papa, S., Ed.). Plenum Press, New York, pp. 253–260.

Petronilli, V., Szabo, I., & Zoratti, M. (1989). The inner mitochondrial membrane contains ion-conducting channels similar to those found in bacteria. FEBS Lett. 259, 137–143.

Pfaller, R., Steger, H. F., Rassow, J., Pfanner, N., & Neupert, W. (1988). Import pathways of precursor proteins into mitochondria. J. Cell. Biol. 107, 2483–2490.

Pfanner, N., Rassow, J., van der Kle, I. O., & Neupert, W. (1992). A dynamic model of the mitochondrial protein import machinery. Cell 68, 999–1002.

Pon, L., & Schatz, G. (1991). Biogenesis of yeast mitochondria In: *The Molecular Biology of The Yeast Saccharomyces. Genome Dynamics, Protein Synthesis, and Energetics* (Broach, J. R., Pringle, J. R., & Jones, E. W., Eds.). Cold Spring Harbor Press, New York, pp. 333–406.

Poutre, C. G., & Fox, T. D. (1987). *PET111*, a *Saccharomyces cerevisiae* nuclear gene required for translation of the mitochondrial mRNA encoding cytochrome *c* oxidase subunit II. Genetics 115, 637–647.

Power, S. D., Lochrie, M. A., Sevarino, K. A., Patterson, T. E., & Poyton, R. O. (1984). The nuclear-coded subunits of yeast cytochrome c oxidase: I. Fractionation of the holoenzyme into chemically pure polypeptides and the identification of two new subunits using solvent extraction and reversed phase high performance liquid chromatography. J. Biol. Chem. 259, 6564–6570.

Poyton, R. O. (1983). Memory and membranes: The expression of genetic and spatial memory during the assembly of organelle macrocompartments. Modern. Cell. Biol. 2, 15–72.

Poyton, R. O., Duhl, D. M. J., & Clarkson, G. H. D. (1992). Protein export from the mitochondrial matrix. Trends in Cell Biol. 2, 369–375.

Poyton, R. O., Duhl, D. M. J., Sevarino, K. A., Bellus, G., & Goehring, B. (1996). *In organello* RNA and protein synthesis systems from yeast cells. Meth. Enzymol. 264, 36–42.

Poyton, R. O., & Kavanagh, J. (1976). Regulation of mitochondrial protein synthesis by cytoplasmic proteins. Proc. Natl. Acad. Sci. USA 73, 3947–3951.

Poyton, R. O., Goehring, B., Droste, M., Sevarino, K. A., Allen, L. A., & Zhao, X-S. (1995). Cytochrome c oxidase (complex IV) from *Saccharomyces cerevisiae.* Methods Enzymol. 260, 97–117.

Poyton, R. O., & McEwen, J. E. (1996). Crosstalk between nuclear and mitochondrial genomes. Annu. Rev. Biochem. 65, 563–607.

Prasad, T. K., Hack, E., & Hallberg, R. C. (1990). Function of the maize chaperonin hsp60: Specific association between hsp60 and newly-synthesized F_1-ATPase alpha subunits Mol. Cell. Biol. 10, 3979–3986.

Pratje, E., Mannhaupt, G., Michaelis, G., & Beyreuther, K. (1983). A nuclear mutation prevents processing of a mitochondrially encoded membrane protein in *Saccharomyces cerevisiae*. EMBO J. 2, 1049–1054.

Pratje, E., & Guiard, B. (1986). One nuclear gene controls the removal of transient pre-sequences from two yeast proteins: One encoded by the nuclear and the other by the mitochondrial genome. EMBO J. 5, 1313–1317.

Racker. (1976). *A New Look At Mechanisms in Bioenergetics*. Academic Press, New York.

Reid, G. A., Yonetani, Y., & Schatz, G. (1982). Import of proteins into mitochondria: Import and maturation of the mitochondrial intermembrane space enzymes cytochrome b_2 and cytochrome *c* peroxidase in intact yeast cells. J. Biol. Chem. 257, 13068–13074.

Rietveld, A., & Dekruijff, B. (1984). Is the mitochondrial precursor protein apocytochrome c able to pass a lipid bilayer? J. Biol. Chem. 254, 6704–6707.

Roedel, G., Koerte, A., & Kaudewitz, F. (1985). Mitochondrial suppression of a yeast nuclear mutation which affects the translation of the apocytochrome b transcript. Curr. Genet. 9, 641–648.

Roedel, G., Michaelis, U., Forsbach, V., Kreike, J., & Kaudewitz, F. (1986). Molecular cloning of the yeast nuclear genes CBS1 and CBS2. Curr. Genet. 11, 47–54.

Roedel, G. (1986). Two yeast nuclear genes, CBS1 and CBS2, are required for translation of mitochondrial transcripts bearing the 5′-untranslated COB leader. Curr. Genet. 11, 41–45.

Roedel, G., & Fox, T. D. (1987). The yeast nuclear gene CBS1 is required for translation of mitochondrial mRNAs bearing the COB 5′-untranslated leader. Mol. Gen. Genet. 206, 45–50.

Schneider, A., Behrens, M., Scherer, P., Pratje, E., Michaelis, G., & Schatz, G. (1991). Inner membrane protease I, an enzyme mediating intramitochondrial protein sorting in yeast. EMBO J. 10, 247–254.

Seehra, J. S., & Khorana, H. G. (1984). Bacteriorhodopsin precursor J. Biol. Chem. 259, 4187–4193.

Sevarino, K. A. (1983). Ph. D. dissertation. University of Connecticut Health Sciences Center.

Sevarino, K. A., & Poyton, R. O. (1980). Mitochondrial membrane biogenesis. Identification of a precursor to yeast cytochrome *c* oxidase subunit II, an integral polypeptide. Proc. Natl. Acad. Sci. USA 77, 142–146.

Simon, S. M., Peskin, C. S., & Oster, G. F. (1992). What drives the translocation of proteins? Proc. Natl. Acad. Sci. USA 89, 3770–3774.

Spithill, T. W., Trembath, M. K., Lukins, H. B., & Linnane, A. W. (1978). Mutations of the mitochondrial DNA of *Saccharomyces cerevisiae* which affect the interaction between mitochondrial ribosomes and the inner membrane. Mol. Genet. 164, 155–162.

Steffans, G. S., & Buse, G. (1979). Studies on cytochrome c oxidase IV. Primary sequence and function of subunit II. Hoppe-Seyler's Z. Physiol. Chem. 360, 613–619.

Strick, C. A., & Fox, T. D. (1987). *Saccharomyces cerevisiae* regulatory gene *PET111* encodes a mitochondrial protein that is translated from an mRNA with a long 5′ leader. Mol. Cell. Biol. 7, 2728–2734.

Strick, C.A. (1988). Ph. D. dissertation. Cornell University.

Szabo, I., & Zoratti, M. (1991). The giant channel of the inner mitochondrial membrane is inhibited by cyclosporin A. J. Biol. Chem. 266, 3376–3379.

Thorsness, P. E., & Fox, T. D. (1990). Escape of DNA from mitochondria to the nucleus in *Saccharomyces cerevisiae*. Nature 346, 376–379.

Thorsness, P. E., & Fox, T. P. (1993). Nuclear mutations in *Saccharomyces cerevisiae* that affect the escape of DNA from mitochondria to the nucleus. Genetics 134, 21–28.

Valencik, M. L., Kloeckener-Gruissem, B., Poyton, R. O., & McEwen, J. E. (1989). Disruption of the yeast nuclear *PET54* gene blocks excision of mitochondrial intron aI5B from pre m-RNA for yeast cytochrome *c* oxidase subunit I. EMBO J. 8, 3899–3904.

Valencik, M. L., & McEwen, J. E. (1991). Genetic evidence that different functional domains of the *PET54* gene product facilitate expression of the mitochondrial genes *COX1* and *COX3* in *Saccharomyces cerevisiae*. Mol. Cell. Biol. 11, 2399–2405.

Verner, K., & Schatz, G. (1988). Protein translocation across membranes. Science 241, 1307–1313.

Von Heijne, G., & Blomberg, C. (1979). Trans-membrane translocation of proteins—the direct transfer model. Eur. J. Biochem. 97, 175–181.

Waksman, A., Rendon, A., Cremel, G., Pellicone, C., & Goubaultde-Brugier, J-F. (1977). Intramitochondrial intermembranal reversible translocation of aspartate aminotransferase and malate dehydrogenase through the inner mitochondrial membrane. Biochemistry 16, 4703–4707.

Wilkie, D., & Evans, I. (1982). Mitochondria and the yeast cell surface: Implications for carcinogenesis. Trends. Biochem. Sci. 7, 147–151.

PROTEIN FOLDING IN MITOCHONDRIA

Jörg Höhfeld

I. Introduction . . . 282
II. The Function of Mitochondrial Hsp70 . . . 283
A. Requirement of mt-Hsp70 for Protein Import . . . 283
B. The Hsp70 Reaction Cycle . . . 285
III. HSP60—Mediator of Mitochondrial Protein Folding . . . 286
A. Hsp60 Activity is Regulated by the Cochaperonin Hsp10 . . . 287
B. Role of Hsp60/Hsp10 in Intramitochondrial Protein Sorting . . . 289
C. Role of Hsp60/Hsp10 during Heat Stress . . . 290
D. Sequential Action of Hsp70 and Hsp60 in Protein Folding . . . 290
IV. Mitochondrial Protein Folding—A Model for Chaperone-mediated Folding in the Cytosol . . . 291
V. Perspectives . . . 292
Acknowledgments . . . 293
References . . . 293

Advances in Molecular and Cell Biology,
Volume 17, pages 281–297.

ISBN: 0-7623-0144-9

I. INTRODUCTION

Many proteins are able to reach their native conformation spontaneously upon refolding *in vitro*. This finding led to the proposal that the linear amino acid sequence of a polypeptide chain contains all the information necessary to specify its three-dimensional, biologically active structure (Anfinsen, 1973). Accordingly, protein folding within the cell was assumed to be a spontaneous process as well. In recent years the discovery of proteins required to mediate and catalyze intracellular folding reactions has led to a revision of this long-held view. These proteins, aptly named molecular chaperones, are able to interact with non-native conformations of other proteins, thereby preventing incorrect intra- and intermolecular interactions that lead to aggregation. They can mediate folding to the native state by releasing their substrate in a controlled reaction. Of central importance are the chaperones of the Hsp70 and Hsp60 families, which are functionally distinct but can cooperate in a sequential folding pathway (for review see Hartl et al., 1994). Major insights into the function of molecular chaperones and their significant role during protein folding have been gained through the analysis of mitochondrial protein import.

Mitochondrial precursor proteins are synthesized in the cytosol and post-translationally targeted to mitochondria due to the presence of specific targeting sequences (see the chapter by G. V. Heijne). The precursor proteins are translocated across the mitochondrial membranes in an unfolded, probably rather extended conformation. The maintenance of the import-competent conformation after completion of translation is critically dependent on the activity of cytosolic Hsp70s. Hsp70 also plays an important role on the trans-side of the membrane. The mitochondrial Hsp70 (mt-Hsp70), localized in the matrix space, is essential for the translocation reaction (see also the chapter by M. Meijer et al.). The subsequent folding of many, if not most, newly-imported proteins is mediated by Hsp60. It is apparent that mitochondrial protein import requires defined and regulated interactions between members of the different chaperone families and the precursor protein. In this review, I will focus on the principles of these interactions as they occur during protein folding within the mitochondrial matrix. The subsequent steps and components involved in the assembly of oligomeric protein complexes are discussed in the chapter by M. Prescott and colleagues.

The mitochondrial chaperones are highly homologous to the chaperone proteins of the bacterial cytosol, consistent with the proposed origin of mitochondria from prokaryotic ancestors. This homology has dramatically accelerated the biochemical analysis of chaperone function. The analysis of protein folding in mitochondria suggested certain general principles of chaperone action and continues to inspire the investigation of chaperone-mediated protein folding in other cellular compartments.

II. THE FUNCTION OF MITOCHONDRIAL HSP70

A. Requirement of mt-Hsp70 for Protein Import

The Hsp70s are characterized by their ability to bind and thus stabilize unfolded polypeptides in an ATP-dependent manner (Hendrick and Hartl, 1993). The binding involves the recognition of extended segments of the unfolded polypeptide chain, which are seven to eight residues in length and are enriched in hydrophobic residues (Flynn et al., 1991; Landry et al., 1992; Blond-Elguindi et al., 1993). Thus, Hsp70s shield hydrophic regions of an unfolded polypeptide that are normally buried within the native structure, thereby preventing unproductive interactions within and between unfolded chains (Hartl et al., 1994). This function of Hsp70 is crucial during heat stress but also during protein synthesis and translocation under normal growth conditions when such hydrophobic regions are exposed.

Mt-Hsp70 has been proposed to have a dual role in translocation and subsequent folding of precursor proteins (Figure 1). The direct participation of mt-Hsp70 in protein import into mitochondria has been demonstrated (Kang et al., 1990; Scherer et al., 1990; Ostermann et al., 1990; Gambill et al., 1993). Crosslinking studies revealed that mt-Hsp70 binds to the unfolded precursor proteins as they enter the matrix space (Scherer et al., 1990; Ostermann et al., 1990). Moreover, the inactivation of mt-Hsp70 in temperature-sensitive yeast mutants resulted in the accumulation of precursor proteins spanning both mitochondrial membranes as translocation intermediates (Kang et al., 1990; Gambill et al., 1993; Voos et al., 1993). This phenotype is similar to the situation observed upon import into mitochondria that have been depleted of matrix ATP (Stuart et al., 1994; Wachter et al., 1994) and the ATP-dependency of the import reaction has been attributed to the activity of mt-Hsp70. It appears that mt-Hsp70 undergoes ATP-dependent cycles of binding and release of the unfolded precursor, which pulls the polypeptide chain into the matrix compartment. Recent findings indicate mt-Hsp70 cooperates in this function with MIM44, a membrane protein that is part of the translocation machinery of the inner membrane (Schneider et al., 1994; Rassow et al., 1994; Figure 1). Mt-Hsp70 thus seems to provide the motor for the translocation reaction. Noteworthy, this principle is not only realized in mitochondria but probably also provides the mechanistic basis for the translocation of proteins into the endoplasmic reticulum, based on the activity of the Hsp70 homolog in the ER lumen, BiP (Brodsky and Schekman, 1993). To recruit a chaperone protein as part of a translocation machinery also ensures that the unfolded polypeptide chain is stabilized as it emerges from the inner surface of the membrane and is properly channeled into the folding machinery of the mitochondrial matrix. As will be discussed later, a similar principle applies to the folding of polypeptide chains emerging from ribosomes in the cytosol.

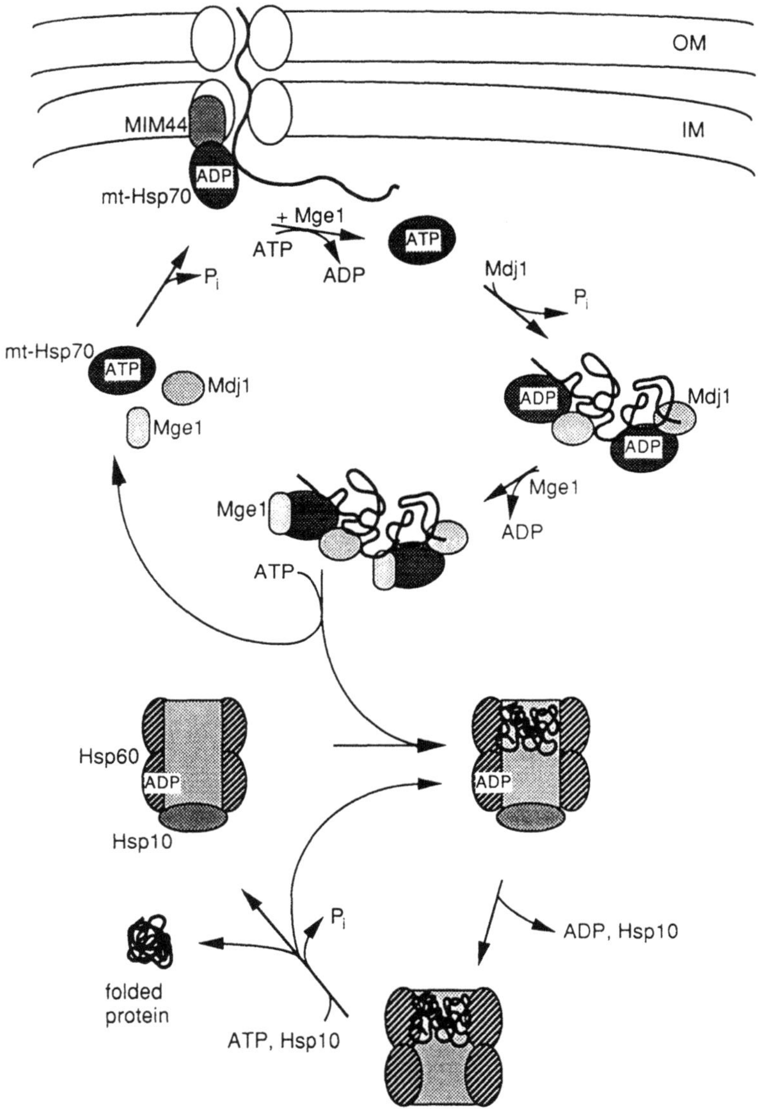

Figure 1. Model of the import and folding of proteins in mitochondria. See text for details. OM, mitochondrial outer membrane; IM, mitochondrial inner membrane; MIM44, inner membrane protein of 44 kDa; mt-Hsp70, mitochondrial 70 kDa heat shock protein; Mdj1, mitochondrial DnaJ homolog; Mge1, mitochondrial GrpE homolog; Hsp60, mitochondrial 60 kDa heat shock protein; Hsp10, mitochondrial 10 kDa heat shock protein; ADP, the tightly bound ADP in the Hsp10-associated ring of Hsp60; the opposite ring of Hsp60 can bind and hydrolyze ATP (not shown).

B. The Hsp70 Reaction Cycle

The mechanism of action of mt-Hsp70 is probably very similar to that of its bacterial homolog DnaK. The activity of DnaK is regulated by DnaJ and GrpE, additional constitutively expressed stress proteins of *E. coli* (for review see Hartl et al., 1994). Both proteins influence the ATPase activity of DnaK. Moreover, DnaJ, a homodimer of 41 kDa subunits, has been shown to bind to unfolded polypeptides itself, characterizing DnaJ as a molecular chaperone on its own (Langer et al., 1992a; Hendrick et al., 1993). The reconstitution of chaperone-mediated folding *in vitro*, using unfolded monomeric rhodanese and firefly luciferase as model substrate proteins, revealed details of the function of DnaK (Langer et al., 1992a; Schröder et al., 1993; Szabo et al., 1994). Only at very high concentrations did DnaK prevent the aggregation of rhodanese that had been diluted from denaturant into chaperone-containing buffer solution. In contrast, DnaJ efficiently bound unfolded rhodanese and strongly potentiated the stabilizing effect of DnaK, resulting in the ATP-dependent formation of a stable ternary complex of unfolded protein, DnaK, and DnaJ (Langer et al., 1992a). The ternary complex dissociated in the presence of GrpE, which is active as a monomer of 21 kDa, and the folding of rhodanese proceeded when the bacterial Hsp60 homolog GroEL was present in the reconstituted system.

Firefly luciferase has been a useful model substrate in these studies because, in contrast to rhodanese, its refolding does not depend on GroEL, at least *in vitro* (Szabo et al., 1994). DnaJ, DnaK, and GrpE mediate the efficient ATP-dependent reactivation of denatured luciferase, whereas the protein aggregates in the absence of chaperones. As observed for rhodanese, unfolded luciferase forms a ternary complex with DnaK and DnaJ in the presence of MgATP. The DnaK in the complex contains bound ADP (Szabo et al., 1994), which dissociates upon addition of GrpE. This dissociation is induced by the interaction of GrpE with the nucleotide binding domain of DnaK (Buchberger et al., 1994). Subsequent rebinding of ATP by DnaK releases the substrate protein. ATP-hydrolysis per se is not required to mediate the release reaction (Palleros et al., 1993) and addition of AMP-PNP, a nonhydrolyzable analogue of ATP, also triggers the release of the substrate protein (Szabo et al., 1994). However, only about 30% of the luciferase molecules were able to refold upon a single release event induced by AMP-PNP, whereas a significant amount aggregated. It appears that repeated cycles of binding and release are required for efficient reactivation of luciferase.

Based on these results, a reaction cycle for the function of DnaK, DnaJ and GrpE was proposed (Szabo et al., 1994). Under physiological conditions, free DnaK is in the ATP-bound state which has a low affinity for polypeptide substrate. In the presence of DnaJ and unfolded polypeptides, ATP hydrolysis by DnaK is induced (Liberek et al., 1991). In the ADP-bound state, DnaK has a high affinity for the substrate protein and a stable ternary complex of substrate, DnaK and DnaJ is formed. While aggregation of the protein is prevented in this complex, an efficient release reaction is necessary for subsequent folding to the native state or transfer to a "downstream" chaperone system. GrpE is required at this step as a nucleotide exchange factor to induce the dissociation of ADP from DnaK. Subsequently, DnaK rebinds ATP and thus returns to the low affinity state. As a consequence, the bound substrate protein is released.

This reaction cycle defines the bacterial Hsp70 protein and its regulators as a chaperone team for the stabilization of unfolded polypeptides. The recent identification of DnaJ- and GrpE-like proteins in the mitochondrial matrix, Mdj1p and Mge1p, respectively, implies that a highly homologous chaperone system is present in this organelle (Rowley et al., 1994; Ikeda et al., 1994; Laloraya et al., 1994; Bolliger et al., 1994) (Figure 1). The GrpE homologue Mge1p was purified on the basis of its stable binding to mt-Hsp70 in the absence of ATP (Bolliger et al., 1994). The analysis of specific yeast mutants revealed that Mge1p is an essential component of the mitochondrial translocation machinery (Laloraya et al., 1994). Mge1p depletion resulted in the accumulation of unprocessed mitochondrial precursor proteins. These results are in agreement with the model of Hsp70 function outlined previously, and indicate that Mge1p acts as an ADP/ATP exchange factor during mt-Hsp70-catalyzed protein translocation (Figure 1). In contrast, deletion of the *MDJ1* gene in yeast was not lethal (Rowley et al., 1994). The *mdj1* null mutant shows a petite phenotype at normal growth temperatures, but fails to grow at 37°C. Further analysis revealed that a lack of Mdj1p left mitochondrial protein import unaffected, whereas the folding of newly-imported proteins in the matrix compartment was drastically impaired (Rowley et al., 1994). These results highlight the importance of the Hsp70 chaperone system for mitochondrial protein folding, although there is yet no evidence for a direct interaction between Mdj1p and mt-Hsp70. However, it appears that Mdj1p does not assist mt-Hsp70 during protein translocation. Recent biochemical and genetic evidence shows an interaction of mt-Hsp70 with MIM44, a component of the inner membrane translocase (Schneider et al., 1994; Rassow et al., 1994). MIM44 may thus contribute to the regulation of mt-Hsp70 activity during protein translocation into mitochondria (Figure 1).

III. HSP60—MEDIATOR OF MITOCHONDRIAL PROTEIN FOLDING

Mitochondrial Hsp60 belongs to a family of molecular chaperones, termed chaperonins (Hemmingsen et al., 1988), which are specialized in mediating intracellular protein folding reactions (reviewed in Hendrick and Hartl, 1993). The chaperonin family includes the Hsp60-homologous proteins GroEL in bacteria (Bochkareva et al., 1988; Hemmingsen et al., 1988), the rubisco subunit-binding protein in chloroplasts (Hemmingsen et al., 1988; Goloubinoff et al., 1989), and the functionally related TRiC complex in the eukaryotic cytosol (Gao et al., 1992; Frydman et al., 1992; Horwich and Willison, 1993). All chaperonins form large oligomeric protein complexes with a ring-like structure. In the case of bacterial

GroEL, the complex consists of fourteen identical subunits of 60 kDa that are arranged in two stacked heptameric rings forming a central cavity (Langer et al., 1992b; Saibil et al., 1993; Braig et al., 1994). A similar oligomeric structure was described for Hsp60 of *Neurospora crassa* (Hutchinson et al., 1989). The distinctive quaternary structure of the chaperonins appears to be directly related to their function in mediating protein folding. It has been shown that substrate proteins are bound within the central cavity of the GroEL complex (Langer et al., 1992b; Saibil et al., 1993; Chen et al., 1994). The chaperonins might thus provide a compartment in which folding can proceed isolated form the cellular surroundings, thereby preventing unproductive interactions.

The essential role of Hsp60 in mitochondrial protein folding was established by the analysis of the yeast mutant strain *mif4* which is defective in the gene encoding Hsp60 (Cheng et al., 1989). The *mif4* strain shows a temperature-sensitive phenotype due the aggregation of mutant Hsp60 at 37°C. Notably, the inactivation of Hsp60 does not affect the translocation of proteins into mitochondria. Hence, *in vitro* import experiments using isolated *mif4* mitochondria provided an excellent system for the analysis of chaperonin function under physiological conditions. Newly imported proteins such as the precursor of the β-subunit of the F_1-ATPase or the trimeric enzyme ornithine transcarbamylase (OTC) could no longer assemble into oligomeric complexes following the depletion of functional Hsp60 (Cheng et al., 1989). In addition, Hsp60 was found to be essential for its own assembly (Cheng et al., 1990). The primary function of the chaperonin system in these reactions is probably to mediate the folding of the imported protein subunits into a conformation competent for spontaneous oligomeric assembly. This became clear through the demonstration that the folding of the monomeric protein dihydrofolate reductase (DHFR), imported as a mitochondrial fusion protein, is dependent on Hsp60 in the matrix compartment (Ostermann et al., 1989; Martin et al., 1992). The recent *in vitro* assembly of active OTC from unfolded protein, mediated by the bacterial chaperonin GroEL confirmed this view (Zheng et al., 1993).

A. Hsp60 Activity is Regulated by the Cochaperonin Hsp10

The Hsp60 chaperonins require MgATP for their function in assisting protein folding. Apparently, ATP binding and hydrolysis are necessary to mediate conformational changes of the chaperonin complex that affect the affinity for substrate protein (Martin et al., 1991; Langer et al., 1992a,b; Saibil et al., 1993). GroEL has a weak K^+-dependent ATPase activity and contains 14 binding sites for ATP, that is, one per monomer (Viitanen et al., 1990; Martin et al., 1991; Bochkareva et al., 1992). The *in vitro* refolding of some proteins by GroEL, such as DHFR, requires only the addition of MgATP to the chaperonin/substrate complex (Martin et al., 1991; Viitanen et al., 1991). In most cases, however, an efficient GroEL-mediated folding reaction also requires the presence of the cochaperonin GroES (Viitanen et al., 1990; Buchner et al., 1991; Martin et al., 1991). GroES is an oligomer of 10 kDa subunits forming a single heptameric ring that binds to one end of the GroEL cylinder (Viitanen et al., 1990; Langer et al., 1992a; Saibil et al., 1993). The cochaperonin increases the cooperativity of ATP binding and hydrolysis by individual GroEL subunits and also their affinity for ADP (Gray and Fersht, 1991; Jackson et al., 1993; Martin et al., 1993a, 1993b; Todd et al., 1993). In the absence of substrate protein, this activity results in a 50% inhibition of the GroEL ATPase. It appears that GroES plays a crucial role in the coupling of ATP-hydrolysis by GroEL to productive release of substrate protein. A detailed model for the reaction cycle of GroEL and its regulator GroES during protein folding has recently been proposed (Martin et al., 1993a; see also Figure 1).

Under physiological conditions, GroEL forms an asymmetrical complex with GroES. Seven ADP are bound tightly to the GroES-associated ring of GroEL. Binding of substrate protein to the opposite ring accelerates ATP-hydrolysis in that ring and triggers the dissociation of the tightly bound ADP, which in turn leads to the release of GroES. Subsequent binding of ATP weakens the interaction with the substrate protein. GroES rebinds and induces the cooperative ATP-hydrolysis that allows the coordinated release of substrate protein for folding. At least partial folding is thought to occur in the central cavity of GroEL. Partially folded protein might rebind to the chaperonin, thus going through successive cycles of ATP-hydrolysis until folding is completed.

Proteins homologous to GroES (Hsp10s) have also been identified in mitochondria, indicating that mitochondrial Hsp60 functions in a similar reaction cycle as outlined previously for GroEL and GroES (Lubben et al., 1990; Hartman et al., 1992; Rospert et al., 1993a, Höhfeld and Hartl, 1994; Figure 1). The recent isolation of a yeast Hsp10 protein and the cloning of the corresponding gene made it possible to analyze the function of the cochaperonin *in vivo* (Rospert et al., 1993b; Höhfeld and Hartl, 1994). Like Hsp60, the cochaperonin is also an essential protein of yeast under all growth conditions. A temperature-sensitive *hsp10* mutant was identified in which Hsp10 could be inactivated by a treatment of isolated mitochondria at 37°C (Höhfeld and Hartl, 1994). The alteration of a proline residue at position 36 of Hsp10 to serine is responsible for the observed temperature-sensitive phenotype. The mutation lies in a region of Hsp10 that is homologous to the proposed GroEL-binding loop of GroES (Landry et al., 1993). Mutant Hsp10 was shown to have a reduced binding affinity for the chaperonin (Höhfeld and Hartl, 1994), which underlines the importance of the flexible loop region (residues 25–40 in Hsp10 and 17–32 in GroES) for the chaperonin/cochaperonin interaction. In addition, the results demonstrate the high degree of structural homology between the bacterial and the mitochondrial chaperonin systems. The reduced interaction between Hsp10 and Hsp60, observed in *hsp10* mitochondria at the nonpermissive temperature, resulted in a phenotype similar to that observed after inactivation of Hsp60 in *mif4* mitochondria (see above). The α-subunit of the matrix processing protease (α-MPP) aggregated following its translocation and newly-imported OTC did not assemble to the enzymatically active trimer (Höhfeld and Hartl, 1994). Obviously,

Hsp60 requires the regulatory activity of Hsp10 to fulfill its chaperonin function under physiological conditions. The Hsp60-mediated folding of DHFR in the mitochondrial matrix was not affected in the *hsp10* mutant strain (Höhfeld and Hartl, 1994), consistent with the observation that denatured DHFR can be refolded by GroEL in the absence of GroES *in vitro* (Martin et al., 1991). Small proteins such as DHFR, which spontaneously refold *in vitro*, may not require the co-chaperonin for their folding *in vivo*.

B. Role of Hsp60/Hsp10 in Intramitochondrial Protein Sorting

The analysis of yeast mutants defective in Hsp60 and Hsp10 function demonstrated that the chaperonin system is also involved in intramitochondrial protein sorting (Cheng et al., 1989; Hartl and Neupert, 1990; Höhfeld and Hartl, 1994). Some proteins of the intermembrane space, such as the Rieske iron sulfur protein (Fe/S), are first translocated into the matrix compartment before they are re-exported across the inner mitochondrial membrane (Hartl et al., 1986; Isaya et al., 1992). During its sorting, the Fe/S protein is proteolytically processed in two steps (see the Chapter by W. Fenton and F. Kalousek). The inactivation of the Hsp60/Hsp10 system resulted in the accumulation of intermediate-sized Fe/S protein in the matrix compartment in an aggregated form (Cheng et al., 1989; Höhfeld and Hartl, 1994). The chaperonin appears to participate in maintaining the newly-imported protein in a loosely-folded conformation that allows its subsequent re-translocation across the inner membrane. It was suggested that the chaperonin has an "antifolding" activity in this process (see also the chapter by R. Stuart et al.).

The experimental data indicate that the Fe/S protein binds to the Hsp60 complex but cannot be folded by the chaperonin, which could be due to the presence of the re-export signal. The N-terminal extrasequence may inhibit the chaperone-mediated folding of the Fe/S-protein which may then merely cycle between Hsp60-bound and free states. The chaperonin may thus stabilize the loosely-folded conformation of intermediate-sized Fe/S until contact with the reexport machinery in the inner membrane is made.

An antifolding activity of Hsp60/Hsp10 has also been proposed for the sorting of cytochrome b_2 to the intermembrane space (Cheng et al., 1989; Koll et al., 1992). A participation of the chaperonin system on this pathway has been disputed based on the finding that sorting of cytochrome b_2 can occur in a Hsp60-deficient mutant strain (Glick et al., 1992; Hallberg et al., 1993). Apparently, Hsp60 is not essential for the sorting of cytochrome b_2. However, a reduction in the kinetics of cytochrome b_2 maturation was recently observed in a *hsp10* mutant strain, suggesting that, in the presence of functional Hsp60, at least part of the imported cytochrome b_2 interacts with the chaperonin system (Höhfeld and Hartl, 1994). Antifolding activity in protein export has also been demonstrated for GroEL/GroES during the export of pre-β-lactamase in *E. coli* (Kusukawa et al., 1989).

C. Role of Hsp60/Hsp10 during Heat Stress

Hsp60 and Hsp10 are expressed at elevated rates under heat shock conditions (Cheng et al., 1989; Höhfeld and Hartl, 1994), indicating an important role in the stabilization of pre-existing proteins against denaturation. In fact, Hsp60 binds to a variety of polypeptides in organelles exposed to heat stress, and prevents the thermal inactivation of imported DHFR in the mitochondrial matrix (Martin et al., 1992). At increased temperatures, partially unfolded matrix proteins are recognized as substrates by the Hsp60/Hsp10 system. Binding to the chaperonin may then stabilize the substrate protein until conditions more favorable for folding are established.

The essential function of Mdj1p, the mitochondrial DnaJ homolog, in protein folding at 37°C indicates that the Hsp70 system also participates in the folding of matrix proteins (Rowley et al., 1994). Hence, it seems plausible that the Hsp70 and Hsp60 chaperone systems cooperate in preventing protein denaturation under stress conditions.

D. Sequential Action of Hsp70 and Hsp60 in Protein Folding

The sequential action of the Hsp70 and Hsp60 family members in protein folding was demonstrated in a reconstituted system using the homologous chaperone proteins of *E. coli* (Langer et al., 1992a). Upon dilution from denaturant, aggregation of unfolded rhodanese is inhibited by the combined activity of DnaK and DnaJ. At a molar ratio of 1:5:2 of rhodanese:DnaK:DnaJ and in the presence of MgATP, about 80% of the rhodanese molecules were found in a high molecular weight complex of 230 kDa. The dissociation of this complex required GrpE. However, addition of GrpE did not result in efficient folding of the DnaK/DnaJ-stabilized rhodanese. The presence of GroEL/GroES in the reconstituted system was necessary in order to achieve folding to the native state. Based on these observations, it was proposed that both chaperone systems cooperate in a sequential pathway of protein folding, with the Hsp70 system acting at an early stage in the stabilization of the unfolded polypeptide, followed by the Hsp60-mediated folding to the native state.

The molecular basis of this pathway appears to be the differential specificity of the two chaperone families for structural features of the polypeptide chain exposed at different stages of the folding process. Hsp70 is known to recognize extended peptide segments enriched in hydrophobic residues (Flynn et al., 1991; Landry et al., 1992; Blond-Elguindi et al., 1993). DnaK/DnaJ-stabilized rhodanese has physical properties characteristic of an early folding intermediate with respect to its intrinsic fluorescence, anilino-naphthalene sulfonate (ANS)-binding, and protease sensitivity (Langer et al., 1992a). Reduced carboxy-methylated α-lactalbumin, which maintains an extended conformation lacking stable secondary structure in the absence of denaturant (Ewbank and Creighton, 1991, 1993a, 1993b), efficiently binds to DnaK, but is not recognized as a substrate by GroEL (Palleros et al., 1991; Langer et al., 1992a). In contrast, GroEL stabilizes substrate proteins in a conformation resembling the "molten globule" (Martin et al., 1991), which is characterized by a compact shape containing all or part of the secondary structure but lacking stable tertiary structure. The binding specificity of GroEL was recently analyzed in detail using a set of trapped conformers of α-lactalbumin (αLA) representing structurally different folding intermediates (Hayer-Hartl et al., 1994). Different conformers of αLA can be generated by selective reduction of the protein's four disulfide bonds (Kuwajima et al., 1990, Ewbank and Creighton, 1991, 1993a, 1993b). GroEL was found to have a high affinity for αLA conformers that contain collapsed secondary structure and expose hydrophobic surfaces, whereas the native and fully-unfolded conformers were not recognized (Hayer-Hartl et al., 1994). Together, these results suggest that the subsequent interaction of different chaperone families with a substrate protein is based on structural characteristics transiently exposed in the course of the folding reaction.

The physiological significance of such a pathway of chaperone-mediated folding has been demonstrated by analyzing the import of the precursor of α-MPP into isolated mitochondria (Manning-Krieg et al., 1991). Co-immunoprecipitation studies revealed that newly-imported α-MPP was first associated with mt-hsp70, consistent with its role during the translocation reaction. Only after subsequent interaction with Hsp60 was α-MPP able to assemble with the second subunit of the matrix processing peptidase.

A requirement for the Hsp60/Hsp10 chaperonin system has been observed for the folding and assembly of a wide variety of matrix proteins (Cheng et al., 1989; Hallberg et al., 1993; Höhfeld and Hartl, 1994). It is possible that the newly-translocated precursor proteins are directly channeled from mt-hsp70 to Hsp60/Hsp10 for completion of folding (Figure 1). Whether the functional cooperation of Hsp70 and Hsp60 requires a physical interaction between both chaperone teams remains to be determined.

IV. MITOCHONDRIAL PROTEIN FOLDING—A MODEL FOR CHAPERONE-MEDIATED FOLDING IN THE CYTOSOL

The sequential interaction of two functionally distinct chaperone systems during intracellular protein folding appears to be a general principle not limited to mitochondria. Successive binding of a precursor protein to Hsp70- and Hsp60-homologs was observed in chloroplasts following translocation into the stroma (Tsugeki and Nishimura, 1993). Members of the Hsp70 family of molecular chaperones also participate in protein folding in the eukaryotic cytosol, where they probably cooperate with recently identified DnaJ homologs (reviewed in Hendrick and Hartl, 1993; Cyr et al., 1994). The eukaryotic cytosol also contains a functional homolog of Hsp60, the large hetero-oligomeric TCP-1 ring complex (TRiC), which mediates protein folding *in vitro* and *in vivo* (Gao et al., 1992, 1993; Frydman et al., 1992).

Do the cytosolic chaperone systems cooperate in the folding of a polypetide chain during and after its translation? According to the mitochondrial model reaction, cytosolic Hsp70s may bind to the nascent chain as it emerges from the ribosome in order to stabilize the unfolded polypeptide and to prevent off-pathway reactions. Hsp70 activity seems to be required at least until a chain length suitable for productive folding has been synthesized. Productive folding may subsequently involve the transfer to a chaperonin system such as TRiC. Evidence for the existence of such a pathway is accumulating. Constitutively expressed, cytosolic Hsp70s were found to be associated with ribosome-bound polypeptide chains, supporting a cotranslational function of the chaperone proteins (Beckmann et al., 1990; Nelson et al., 1993; Frydman et al., 1994). Using firefly luciferase as a model protein, the essential role of molecular chaperones for protein folding in the context of translation has recently been revealed (Frydman et al., 1994). Upon translation of luciferase in reticulocyte lysate, the growing polypeptide interacts with a specific set of chaperone proteins including Hsp70, the DnaJ homolog Hsp40, and TRiC. A chaperone assembly containing nascent chains was purified as a high molecular mass complex of ~1,200 kDa. When the nascent chain was released from the ribosomes, the chaperone complex was able to mediate subsequent ATP-dependent folding. TRiC did not bind to a truncated form of luciferase, which exposes only ~40 amino acids outside the ribosome, whereas Hsp70, in cooperation with Hsp40, stabilized such short ribosome-bound chains. In addition, partial depletion of Hsp70 from the lysate resulted in a reduction of TRiC-bound nascent chains (Frydman et al., 1994), further indicating a defined order of chaperone interactions. Parallel folding pathways might exist in the eukaryotic cytosol, comprising additional assisting components. However, the data described earlier indicate that upon translation, polypeptides are guided to their final conformation by cooperating chaperone systems in a fashion similar to the chaperone-mediated folding in mitochondria.

V. PERSPECTIVES

The analysis of mitochondrial protein import has contributed enormously to our current understanding of the function of molecular chaperone proteins in intracellular protein folding. It appears that the key components participating in the folding of imported proteins in the mitochondrial matrix have been identified and a detailed model for their interaction has been proposed. In the future, the application of biophysical and structural methodology should reveal further insights into the mechanistic details of chaperone-mediated folding reactions.

ACKNOWLEDGMENTS

W. Neupert and N. Pfanner are acknowledged for sharing unpublished data, and F. U. Hartl and D. McColl for critically reading the manuscript. Jörg Höhfeld is supported by a postdoctoral fellowship of the Deutsche Forschungsgemeinschaft.

REFERENCES

Anfinsen, C. B. (1973). Principles that govern the folding of protein chains. Science 181, 223–230.

Beckmann, R. P., Mizzen, L. A., & Welch, W. J. (1990). Interaction of hsp70 with newly synthesized proteins: Implications for protein folding and assembly. Science 248, 850–854.

Blond-Elguindi, S., Cwirla, S. E., Dower, W. J., Lipshutz, R. J., Sprang, S. R., Sambrook, J. F., & Gething, M-J. (1993). Affinity panning of a library of peptides displayed on bateriophages reveals the binding specificity of BiP. Cell 75, 717–729.

Bochkareva, E. S., Lissin, N. M., & Girshovich, A. S. (1988). Transient association of newly synthesized unfolded proteins with the heat-shock GroEL protein. Nature (London) 336, 254–257.

Bochkareva, E. S., Lissin, N. M., Flynn, G. C., Rothman, J. E., & Girshovich, A. S. (1992). Positive cooperativity in the functioning of molecular chaperone GroEL. J. Biol. Chem. 267, 6796–6800.

Bolliger, L., Deloche, O., Glick, B. S., Georgopoulos, C., Jenö, P., Kronidou, N., Horst, M., Morishima, N., & Schatz, G. (1994). A mitochondrial homolog of bacterial GrpE interacts with mitochondrial hsp70 and is essential for viability. EMBO J. 13, 1998–2006.

Braig, K., Otwinowski, Z., Hegde, R., Boisvert, D. C., Joachimiak, A., Horwich, A. L., & Sigler, P. B. (1994). The crystal structure of the bacterial chaperonin GroEL at 2.8 Å. Nature (London) 371, 578–586.

Brodsky, J. L., & Schekman, R. (1993). A Sec63-BiP complex form yeast is required for protein translocation in a reconstituted proteoliposome. J. Cell Biol. 123, 1355–1363.

Buchberger, A., Schröder, H., Büttner, M., Valencia, A., & Bukau, B. (1994). A conserved loop in the ATPase domain of the DnaK chaperone is essential for stable binding of GrpE. Nature Struct. Biol. 1, 95–101.

Buchner, J., Schmidt, M., Fuchs, M., Jaenicke, R., Rudolph, R., Schmid, F. X., & Kiefhaber, T. (1991). GroE facilitates refolding of citrate synthase by suppressing aggregation. Biochemistry 30, 1586–1591.

Chen, S., Roseman, A. M., Hunter, A. S., Wood, S. P., Burston, S. G., Ranson, N. A., Clarke, A. R., & Saibil, H. R. (1994). Location of a folding protein and shape changes in GroEL-GroES complexes imaged by cryo-electron microscopy. Nature (London) 371, 261–264.

Cheng, M. Y., Hartl, F.-U., Martin, J., Pollock, R. A., Kalousek, F., Neupert, W., Hallberg, E. M., Hallberg, R. L., & Horwich A. L. (1989). Mitochondrial heat-shock protein hsp60 is essential for assembly of proteins imported into yeast mitochondria. Nature (London) 337, 620–625.

Cheng, M. Y., Hartl, F-U., & Horwich, A. L. (1990). The mitochondrial chaperonin hsp60 is required for its own assembly. Nature (London) 348, 455–458.

Cyr, D. M., Langer, T., & Douglas, M. G. (1994). DnaJ-like proteins: Molecular chaperones and specific regulators of hsp70. Trends Biochem. Sci. 19, 176–181.

Ewbank, J. J., & Creighton, T. E. (1991). The molten globule protein conformation probed by disulphide bonds. Nature (London) 350, 518–520.

Ewbank, J. J., & Creighton, T. E. (1993a). Structural characterization of the disulfide folding intermediates of bovine α-lactalbumin. Biochemistry 32, 3694–3707.

Ewbank, J. J., & Creighton, T. E. (1993b). Pathway of disulfide-coupled unfolding and refolding of bovine apha-lactalbumin. Biochemistry 32, 3677–3693.

Flynn, G. C., Rohl, J., Flocco, M. T., & Rothman, J. E. (1991). Peptide-binding specificity of the molecular chaperones BiP. Nature (London) 353, 726–730.

Frydman, J., Nimmesgern, E., Erdjument-Bromage, H., Wall, J. S., Tempst, P., & Hartl, F-U. (1992). Function in protein folding of TRiC, a cytosolic ring-complex containing TCP1 and structurally related subunits. EMBO J. 11, 4767–4778.

Frydman, J., Nimmesgern, E., Ohtsuka, K., & Hartl, F-U. (1994). Folding of nascent polypeptide chains in a high molecular mass assembly with molecular chaperones. Nature (London) 370, 111–117.

Gambill, B. D., Voos, W., Kang, P. J., Miao, B., Langer, T., Craig, E. A., & Pfanner, N. (1993). A dual role for mitochondrial heat shock protein 70 in membrane translocation of preproteins. J. Cell Biol. 123, 109–117.

Gao, Y., Thomas, J. O., Chow, R. L., Lee, G. H., & Cowan, N. J. (1992). A cytoplasmic chaperonin that catalyzes beta-actin folding. Cell 69, 1043–1050.

Gao, Y., Vainberg, I. E., Chow, R. L., Lee, G. H., & Cowan, N. J. (1993). Two cofactors and cytoplasmic chaperonin are required for the folding of alpha- and beta-tubulin. Mol. Cell. Biol. 13, 2478–2485.

Glick, B. S., Brandt, A., Cunningham, K., Müller, S., Hallber, R. L., & Schatz, G. (1992). Cytochrome c_1 and b_2 are sorted to the intermembrane space of yeast mitochondria by a stop-transfer mechanism. Cell 69, 809–822.

Goloubinoff, P., Christeller, J. T., Gatenby, A. A., & Lorimer, G. H. (1989). Reconstitution of active dimeric ribulose bisphosphate carboxylase from an unfolded state depends on two chaperonin proteins and MgATP. Nature (London) 342, 884–889.

Gray, T. E., & Fersht, A. R. (1991). Cooperativity in ATP hydrolysis by GroEL is increased by GroES. FEBS Lett. 292, 254–258.

Hallberg, E. M., Shu, Y., & Hallberg, R. L. (1993). Loss of mitochondrial hsp60 function: Nonequivalent effects on matrix-targeted and intermembrane-targeted proteins. Mol. Cell. Biol. 13, 3050–3057.

Hartl, F-U., Schmidt, B., Wachter, E., Weiss, H., & Neupert, W. (1986). Transport into mitochondria and intramitochondrial sorting of the Fe/S protein of ubiquinol-cytochrome c reductase. Cell 47, 939–951.

Hartl, F-U., & Neupert, W. (1990). Protein sorting to mitochondria: Evolutionary conservations of folding and assembly. Science 247, 930–938.

Hartl, F-U., Hlodan, R., & Langer, T. (1994). Molecular chaperones in protein folding: The art of avoiding sticky situations. Trends Biochem. Sci. 19, 20–25.

Hartman, D. J., Hoogenraad, N. J., Condron, R., & Hoj, P. B. (1992). Identification of a mammalian 10-kDa heat shock protein, a mitochondrial chaperonin 10 homologue essential for assisted folding of trimeric ornithine transcarbamoylase *in vitro*. Proc. Natl. Acad. Sci. USA 89, 3394–3398.

Hayer-Hartl, M. K., Ewbank, J. J., Creighton, T. E., & Hartl, F-U. (1994). Conformational specificity of the chaperonin GroEL for the compact folding intermediates of α-lactalbumin. EMBO J. 13, 3192–3202.

Hemmingsen, S. M., Woolford, C., Van der Vies, S. M., Tilly, K., Dennis, D. T., Georgopoulos, C. P., Hendrix, R. W., & Ellis, R. J. (1988). Homologous plant and bacterial proteins chaperone oligomeric protein assembly. Nature (London) 333, 330–334.

Hendrick, J. P., & Hartl, F-U. (1993). Molecular chaperone functions of heat shock proteins. Annu. Rev. Biochem. 62, 349–384.

Hendrick, J. P., Langer, T., Davis, T. A., Hartl, F-U., & Wiedmann, M. (1993). Control of folding and membrane translocation by binding of the chaperone DnaJ to nascent polypeptides. Proc. Natl. Acad. Sci. USA 90, 10216–10220.

Höhfeld, J., & Hartl., F.-U. (1994). Role of the chaperonin cofactor hsp10 in protein folding and sorting in yeast mitochondria. J. Cell Biol. 126, 305–315.

Horwich, A. L., & Willison, K. R. (1993). Protein folding in the cell: Functions of two families of molecular chaperone, hsp60 and TF55-TCP1. Phil. Trans. R. Soc. B339, 313–326.

Hutchinson, E. G., Tichelaar, W., Hofhaus, G., Weiss, W., & Leonard, K. R. (1989). Identification and electron microscopic analysis of a chaperonin oligomer from *Neurospora crassa*. EMBO J. 8, 1485–1490.

Ikeda, E., Yoshida, S., Mitsuzawa, H., Uno, I., & Toh-E, A. (1994). YGE1 is a yeast homologue of *Escherichia coli* GrpE and required for mitochondrial functions. FEBS Lett. 339, 265–268.

Isaya, G., Kalousek, F., & Rosenberg, L. E. (1992). Amino-terminal octapeptides function as recognition signals for the mitochondrial intermediate peptidase. J. Biol. Chem. 267, 7904–7910.

Jackson, G. S., Staniforth, R. A., Halsall, D. J., Atkinson, T., Holbrook, J. J., Clarke, A. R., & Burston S. G. (1993). Binding and hydrolysis of nucleotides in the chaperonin catalytic cycle: Implications for the mechanism of assisted protein folding. Biochemistry 32, 2554–2563.

Kang, P.-J., Ostermann, J., Shilling, J., Neupert, W., Craig, E. A., & Pfanner, N. (1990). Requirement of hsp70 in the mitochondrial matrix for translocation and folding of precursor proteins. Nature (London) 348, 137–143.

Koll, H., Guiard, B., Rassow, J., Ostermann, J., Horwich, A. L., Neupert, W., & Hartl, F-U. (1992). Antifolding activity of hsp60 couples protein import into the mitochondrial matrix with export to the intermembrane space. Cell 89, 1163–1175.

Kusukawa, N., Yura, T., Ueguchi, C., Akiyama, Y., & Ito, K. (1989). Effects of mutations in heat-shock genes *groES* and *groEL* on protein export in *Escherichia coli.* EMBO J. 8, 3517–3521.

Kuwajima, K., Ikeguchi, M., Sugawara, T., Hiraoka, Y., & Sugai, S. (1990). Kinetics of disulfide bond reduction in alpha-lactalbumin by dithiothreitol and molecular basis of superreactivity of the Cys6-Cys120 disulfide bond. Biochemistry 29, 8240–8249.

Laloraya, S., Gambill, B. D., & Craig, E. A. (1994). A role for a eukaryotic GrpE-related protein, Mge1p, in protein translocation. Proc. Natl. Acad. Sci. USA 91, 6481–6485.

Landry, S. J., Zeilstra-Ryalls, J., Fayet, O., Georgopoulos, C., & Gierasch, L. M. (1992). Different conformations for the same polypeptide bound to chaperones DnaK and GroEL. Nature (London) 355, 455–457.

Landry, S. J., Zeilstra-Ryalls, J., Fayet, O., Georgopoulos, C., & Gierasch, L. M. (1993). Characterization of a functionally important mobile domain of GroES. Nature (London) 364, 255–258.

Langer, T., Lu, C., Echols, H., Flanagan, J., Hayer, M. K., & Hartl, F-U. (1992a). Successive action of molecular chaperones DnaK, DnaJ, and GroEL along the pathway of assisted protein folding. Nature (London) 356, 683–689.

Langer, T., Pfeifer, G., Martin, J., Baumeister, W., & Hartl, F-U. (1992b). Chaperonin-mediated protein folding: GroES binds to one end of the GroEL cylinder which accommodates the protein substrate within its central cavity. EMBO J. 11, 4757–4765.

Liberek, K., Marszalek, J., Ang, D., Georgopoulos, C., & Zylic, M. (1991). *Escherichia coli* DnaJ and GrpE heat shock proteins jointly stimulate ATPase activity of DnaK. Proc. Natl. Acad. Sci. USA 88, 2874–2878.

Lubben, T. H., Gatenby, A. A., Donaldson, G. K., Lorimer, G. H., & Viitanen, P. V. (1990). Identification of a groES-like chaperonin in mitochondria that facilitates protein folding. Proc. Natl. Acad. Sci. USA 87, 7683–7687.

Manning-Krieg, U. C., Scherer, P. E., & Schatz, G. (1991). Sequential action of mitochondrial chaperonines in protein import into the matrix. EMBO J. 10, 3273–3280.

Martin, J., Langer, T., Boteva, R., Schramel, A., Horwich A. L., & Hartl, F-U. (1991). Chaperonin-mediated protein folding at the surface of groEL through a "molten globule"-like intermediate. Nature (London) 352, 36–42.

Martin, J., Horwich, A. L., & Hartl, F-U. (1992). Prevention of protein denaturation under heat stress by the chaperonin hsp60. Science 258, 995–998.

Martin, J., Mayhew, M., Langer, T., & Hartl, F-U. (1993a). The reaction cycle of GroEL and GroES in chaperonin-assisted refolding. Nature (London) 366, 228–233.

Martin, J., Geromanos, S., Tempst, P., & Hartl, F-U. (1993b). Identification of nucleotide-binding regions in the chaperonin proteins GroEL and GroES. Nature (London) 366, 279–282.

Nelson, R. J., Ziegelhoffer, T., Nicolet, C., Werner-Washburne, M., & Craig, E. A. (1993). The translation machinery and 70 kd heat shock protein cooperate in protein synthesis. Cell 71, 97–105.

Ostermann, J., Horwich, A. L., Neupert, W., & Hartl, F-U. (1989). Protein folding in mitochondria requires complex formation with hsp60 and ATP hydrolysis. Nature (London) 341, 125–130.

Ostermann, J., Voos, W., Kang, P. J., Craig, E. A., Neupert, W., & Pfanner, N. (1990). Precursor proteins in transit through mitochondrial contact sites interact with hsp70 in the matrix. FEBS Lett. 277, 281–284.

Palleros, D. R., Welch, W. J., & Fink, A. L. (1991). Interaction of hsp70 with unfolded proteins: Effects of temperature and nucleotides on the kinetics of binding. Proc. Natl. Acad. Sci. USA 88, 5719–5723.

Palleros, D. R., Reid, K. L., Shi, L., Welch, W. J., & Fink, A. L. (1993). ATP-induced protein-HSP70 complex dissociation requires K^+ but not ATP hydrolysis. Nature (London) 365, 664–666.

Rassow, J., Maarse, A. C., Krainer, E., Kübrich, M., Müller, H., Meijer, M., Craig, E. A., & Pfanner, N. (1994). Mitochondrial protein import: Biochemical and genetic evidence for interaction of matrix hsp70 and the inner membrane protein MIM44. J. Cell Biol. 127, 1547–1556.

Rospert, S., Glick, B. S., Jenö, P., Schatz, G., Todd, M. J., Lorimer, G. H., & Viitanen, P. V. (1993a). Identification and functional analysis of chaperonin 10 the groES homolog form yeast mitochondria. Proc. Natl. Acad. Sci. USA 90, 10967–10971.

Rospert, S., Junne, T., Glick, B. S., & Schatz, G. (1993b). Cloning and disruption of the gene encoding yeast mitochondrial chaperonin 10, the homolog of *E. coli* groES. FEBS Lett. 335, 358–360.

Rowley, N., Prip-Buus, C., Westermann, B., Brown, C., Schwarz, E., Barrell, B., & Neupert, W. (1994). Mdj1p, a novel chaperone of the DnaJ family, is involved in mitochondrial biogenesis and protein folding. Cell 77, 249–259.

Saibil, H. R., Zheng, D., Roseman, A. M., Hunter, A. S., Watson, G. M. F., Chen, S., auf der Mauer, A., O'Hara, B. P., Wood, S. P., Mann, N. H., Barnett, L. K., & Ellis, R. J. (1993). ATP induces large quaternary rearrangements in a cage-like chaperonin structure. Curr. Biol. 3, 265–273.

Scherer, P. E., Krieg, U. C., Hwang, S. T., Vestweber, D., & Schatz, G. (1990). A precursor protein partly translocated into yeast mitochondria is bound to a 70 kd mitochondrial stress protein. EMBO J. 9, 4315–4322.

Schneider, H.-C., Berthold, J., Bauer, M. F., Dietmeier, K., Guiard, B., Brunner, M., & Neupert, W. (1994). Mitochondrial hsp70/MIM44 complex facilitates protein import. Nature (London) 371, 768–774.

Schröder, H., Langer, T., Hartl, F-U., & Bukau, B. (1993). DnaK, DnaJ, GrpE form a cellular chaperone machinery capable of repairing heat-induced protein damage. EMBO J. 12, 4137–4144.

Stuart, R. A., Cyr, D. M., Craig, E. A., & Neupert, W. (1994). Mitochondrial molecular chaperones: Their role in protein translocation. Trends Biochem. Sci. 19, 87–92.

Szabo, A., Langer, T., Schröder, H., Flanagan, J., Bukau, B., & Hartl, F. U. (1994). The ATP hydrolysis-dependent reaction cycle of the *E. coli* hsp70 system, DnaK, DnaJ and GrpE. Proc. Natl. Acad. Sci. USA 91, 10345–10349.

Todd, M. J., Viitanen, P. V., & Lorimer, G. H. (1993). Hydrolysis of adenosine-5'-triphosphate by *Escherichia coli* GroEL: Effects of GroES and potassium ion. Biochemistry 32, 8560–8567.

Tsugeki, R., & Nishimura, M. (1993). Interaction of homologues of hsp70 and Cpn60 with ferredoxin-$NADP^+$ reductase upon import into chloroplasts. FEBS Lett. 320, 198–202.

Viitanen, P. V., Lubben, T. H., Reed, J., Goloubinoff, P., O'Keefe, D. P., & Lorimer, G.H. (1990). Chaperonin-facilitated refolding of ribulosebisphosphate carboxylase and ATP hydrolysis by chaperonin 60 (groEL) are K^+ dependent. Biochemistry 29, 5665–5671.

Viitanen, P. V., Donaldson, G. K., Lorimer, G. H., Lubben, T. H., & Gatenby, A. A. (1991). Complex interactions between the chaperonin 60 molecular chaperone and dihydrofolate reductase. Biochemistry 30, 9716–9723.

Voos, W., Gambill, D., Guiard, B., Pfanner, N., & Craig, E. A. (1993). Presequence and mature part of preproteins strongly influence the dependence of mitochondrial protein import on heat shock protein 70 in the matrix. J. Cell Biol. 123, 119–126.

Wachter, C., Schatz, G., & Glick, B. S. (1994). Protein import into mitochondria: The requirement for external ATP is precursor-specific whereas intramitochondrial ATP is universally needed for translocation into the matrix. Mol. Biol. Cell 5, 465–474.

Zheng, X., Rosenberg, L. E., Kalousek, F., & Fenton W. A. (1993). GroEL, GroES, and ATP-dependent folding and spontaneous assembly of ornithine transcarbamylase. J. Biol. Chem. 268, 7489–7493.

ASSEMBLY OF MULTISUBUNIT COMPLEXES IN MITOCHONDRIA

Mark Prescott, Rodney J. Devenish, and Phillip Nagley

I. Introduction 300
II. Assembly of Complexes Exclusively Containing Nuclearly Encoded Proteins 303
A. Assembly of Citrate Synthase and Ornithine Transcarbamoylase 303
B. Assembly of 2-Oxoacid Dehydrogenase Complexes 304
C. Complex II of the Respiratory Chain 306
III. Assembly of Complexes Containing Nuclearly and Mitochondrially Encoded Proteins 307
A. Complex I of the Respiratory Chain 307
B. ATP Synthase 311
C. Complex III of the Respiratory Chain 317
D. Complex IV of the Respiratory Chain 320
IV. Future Directions and Perspectives 326
A. Pleiotropic Effects and Supramolecular Assembly 326
B. Factors Controlling Assembly 329
C. Disassembly 330
Acknowledgments 330
References 331

Advances in Molecular and Cell Biology,
Volume 17, pages 299–339.

ISBN: 0-7623-0144-9

I. INTRODUCTION

Many proteins imported into mitochondria become assembled into enzyme complexes. The assembly process, the focus of this chapter, involves a series of events and interactions subsequent to those that take place specifically to deliver a protein into the interior of the organelle.

Consideration is made in detail elsewhere in this volume concerning the key steps of the import process. An important feature of this process is the need to maintain the imported protein in an unfolded configuration during import, while allowing the subsequent folding of the protein at the appropriate time and place within the organelle. The machinery for this aspect involves cytoplasmic Hsp70 proteins and, within the mitochondrial matrix, mitochondrial Hsp70 proteins and the mitochondrial homologs of the prokaryotic heat-shock proteins GrpE and DnaJ (Pfanner et al., 1994). It is known that some proteins must then interact with complexes of Hsp60 and Hsp10 (chaperonins), enabling proper folding of imported proteins.

Is folding synonymous with assembly? We take assembly to mean, for the purposes of this essay, the interaction of a protein with one or more other proteins (either the same or different), as part of the process of building up some functional or organized structural unit (a complex). It is highly likely that the final shape adopted by a protein that has become assembled, in this sense, will be different from that which it adopts as a monomer in free solution, for example, having just been released from its Hsp60/Hsp10 chaperonin complex. The proteins and other molecules with which an assembling subunit interacts to generate the assembled complex will clearly influence the structure such a given protein adopts. Such interactions may be expected to have a mutual effect on many, if not all, the subunits in a particular given assembled complex. While folding events are important steps leading towards assembly, in this article we will focus on the interactive aspects of assembly leading to defined macromolecular associations.

In our discussion of particular mitochondrial enzyme complexes (see following), we will touch on the experimental techniques used in studying the assembly processes. These include biochemical approaches for identifying the composition and biogenesis of mature complexes and, where relevant, sub-assemblies or intermediates that can be defined as part of the assembly process. Immunochemical approaches are also used, involving the use of antibodies, specifically interacting with one or more subunits of a given complex. In the systems we survey, there has been a prominent and productive use of molecular genetic approaches, particularly in the yeast *Saccharomyces cerevisiae*. This has enabled identification of genes encoding not only subunits of enzyme complexes, but also other assembly factors. The molecular genetic approach has led further to the analysis of both the expression and aspects of the structure of these proteins.

A salient feature of mitochondrial molecular biology is the presence of mitochondrial DNA (mtDNA) and the mitochondrial protein synthesizing system. The mtDNA genome encodes essential components of several key enzyme complexes

Table 1. Oligomeric Enzyme Complexes from Mitochondria Considered in this Review[a]

Complex[b]	*Source Organism*	*Location*[c]	*Estimated Number of Non-identical Subunits*	
			Nuclearly Encoded	*Mitochondrially Encoded*
1. Citrate synthase	Mammalian	M	1	NIL
2. Ornithine transcarbamoylase	Mammalian	M	1	NIL
3. α-ketoglutarate dehydrogenase	*S. cerevisiae*/ mammalian	M	4	NIL
4. Pyruvate dehydrogenase	*S. cerevisiae*/ mammalian	M	5	NIL
5. Complex II (Succinate - CoQ reductase)	*S. cerevisiae*	M, IM	4	NIL
6. Complex I (NADH - CoQ reductase)	Mammalian (bovine) *Neurospora crassa*	IM (M, IMS)	41 (Mammalian) 32 (*N. crassa*)	7
7. Complex V (ATP synthase)	*S. cerevisiae*	IM (M)	11	3
8. Complex III ($CoQH_2$ - cytochrome *c* reductase)	*S. cerevisiae*	M (M, IMS)	9	1
9. Complex IV (cytochrome *c* oxidase)	*S. cerevisiae*	IM (M, IMS)	10	3

Notes: [a]References are found in the section of this chapter dealing with each enzyme complex.

[b]Name used in review is listed first (other names in parenthesis). CoQ indicates coenzyme Q (ubiquinone).

[c]OM, outer membrane; IMS, inner membrane space; IM, inner membrane; M, Matrix. For some complexes the primary location is given together with the location (in parenthesis) of other defined domains of the complex.

involved in the electron transport chain and ATP synthase. In order to understand the assembly of imported subunits into such enzyme complexes it is thus necessary to take into consideration the role of mitochondrial gene products in the biogenesis and functional assembly of these enzyme complexes.

Our survey of the assembly of multisubunit complexes in mitochondria encompasses the nine enzyme complexes listed in Table 1. These have been selected on the basis of their illustrating particular principles or features, which are representative of more general themes, or which have entailed the application of particularly insightful experimental approaches in their analysis. Space does not permit consideration of all information published on the topic of protein assembly in mitochondria and, therefore, omission of some aspects is inevitable.

Table 2. Some Nuclear Genes that Do Not Encode Structural Components But Are Required at the Post-translational Level for Assembly of Mitochondrial Complexes

Nuclear Gene[a]	*Defective Complex*	*Assembly Defect in Mutant Phenotype*	*Putative Function of Gene Product*
"Unnamed"	α-ketoglutarate dehydrogenase	Lower affinity of E_1 for E_2 core	Post-translational modification of E_1 or E_2; promotes correct subunit interaction
ATP 10	ATP synthase	Loose physical coupling of F_1 and F_0 sectors	Chaperone induced assembly of F_0 subunits
ATP11, ATP12	ATP synthase	Accumulation of unassembled F_1 subunits α and β	Promotion of correct interaction of subunits
BCS1	Complex III	Failure to assembly Rieske protein into holoenzyme	Insertion of Fe-S cluster into Rieske apoprotein; chaperone function
CBP3	Complex III	Reduced assembly of apocytochrome b, Rieske protein, subunits IV and V	Insertion of prosthetic groups into apocytochrome *b* or Rieske apoprotein; promotes assembly of structural subunits
COX10, COX11	Complex IV	Failure to assemble subunits IV and V; absence of subunit I	Post-translational modification of subunits; delivery of prosthetic groups to apoprotein; chaperone function
PET117, PET191	Complex IV	Failure to assemble an active cytochrome *c* oxidase complex	Post-translational modification of polypeptides; delivery of prosthetic groups to apoprotein; chaperone function

Note: [a]References are found in the sections of this chapter dealing with the relevant enzyme complexes.

Those enzyme complexes (Table 1) we have chosen for review highlight themes that deal, in general, with various aspects of the assembly process. Thus, we consider complexes assembled from imported subunits, and those that involve interaction and assembly of both mitochondrial and nuclear gene products. The topology of assembly is also addressed in relation to the locations of complexes associated with one or more spatial domains within the mitochondria, such as the matrix space or inner membrane (see Table 1). We also place special emphasis on nuclear gene products that participate in the assembly reactions, but which are not themselves retained in the final assembled complex. Some of these are listed in Table 2. These assembly factors are a subset of a more general group of proteins, thus far identified as nuclear gene products, which are variously involved in the processing, maturation, and assembly of both imported and mitochondrial gene products. It is important to distinguish, on the one hand, general chaperones such as Hsp70 and Hsp60 which apparently have a wide range of different substrate proteins for unfolding/refolding reactions and, on the other, the assembly factors (Table 2) which appear to have a specifically restricted range of substrate proteins that are associated with the assembly of individual enzyme complexes in mitochondria. Other aspects we touch on in this review include the chemical modification of particular subunits (such as heme in cytochromes, or other prosthetic groups) and the coordination of the assembly of the various subunits within a key enzyme complex. Understanding the regulation of assembly presents one of the major challenges for the future.

II. ASSEMBLY OF COMPLEXES EXCLUSIVELY CONTAINING NUCLEARLY ENCODED PROTEINS

A. Assembly of Citrate Synthase and Ornithine Transcarbamoylase

We start by considering the relatively simple situations that serve to highlight the importance of chaperones for the assembly of enzyme complexes. In these two cases, citrate synthase and ornithine transcarbamoylase, homo-oligomers of imported proteins are assembled. Assembly reactions have been carried out *in vitro* to study the role of chaperones. The first complex, citrate synthase, is a dimeric enzyme of two identical subunits. The dissociation of the complex and unfolding of the subunits by denaturants such as guanidinium hydrochloride are essentially irreversible, when restoration of function is used as a criterion of reassembly. However, functional reassembly of the enzyme complex can be achieved in the presence of both bacterial chaperones cpn60 and cpn10, and hydrolysable ATP (Buchner et al., 1991). Similarly, chemically denatured ornithine transcarbamoylase could be successfully reconstituted into its trimeric active form in a process requiring bacterial cpn60, rat mitochondrial Hsp10, and hydrolysable ATP. The time course of reconstitution was consistent with a relatively fast Hsp-dependent folding of monomers followed by a rate-limiting and Hsp-dependent assembly of folded monomers into catalytically active trimers (Hartman et al., 1992).

Note that in these studies the assembly reactions analyzed represent the fate of imported proteins after processing to remove the cleavable N-terminal leader. It is likely, although by no means explicitly demonstrated in every case, that such chaperonin-mediated folding and assembly events take place for all polypeptides imported into mitochondria that are subsequently assembled into complexes.

B. Assembly of 2-Oxoacid Dehydrogenase Complexes

The level of complexity increases when considering those enzymes containing both sets of nonidentical subunits and a multiplicity of identical subunits. We now examine aspects of assembly of 2-oxoacid dehydrogenases, which are the multi-enzyme systems including pyruvate dehydrogenase complex (PDHC) and α-keto-glutarate dehydrogenase (KGDHC). These complexes are responsible for the oxidative decarboxylation of their substrates through the concerted action of multiple copies of three associated enzyme complexes (E_1, a substrate-specific decarboxylase dehydrogenase; E_2, a dihydrolipoamide acyltransferase; and E_3, a dihydrolipoamide dehydrogenase). Each multi-enzyme system has a molecular mass in excess of two million. Architecturally, each has a quartenary structure of a polyhedral oligomeric E_2 core having an "open" structure to which E_1 and E_3 are bound by noncovalent bonds. The E_2 core is composed of 24 (KGDHC) or 60 (PDHC) identical subunits. A morphological unit of three E_2 subunits may be important in the assembly of the core structure (for review on structure-function see Reed and Hackert, 1990). By way of example, PDHC contains 30 tetrameric copies of E_1 and six copies of the dimeric E_3 (Barrera et al., 1972).

Eukaryotic PDHC complexes, but not other 2-oxoacid dehydrogenases, contain in addition small amounts of another polypeptide, tightly associated with the E_2 core, designated protein X or E_3 binding protein (De Marcucci and Lindsay, 1985; Jilka et al., 1986).

In the case of 2-oxoacid dehydrogenases it is likely that assembly proceeds through a series of discrete assembly intermediates *in vivo*. As with other complexes discussed ahead, the identification or isolation of these intermediates, or the resolution of the complexes into smaller related structures, has become central to defining the assembly process of a particular complex. Insight has been obtained by studying yeast mutants that lack one of the subunits, or by over-expressing individual subunits in the presence of others.

The assembly of yeast KGDHC has been studied in *S. cerevisiae* strains carrying mutations in *KGD1* and *KGD2*, the structural genes for E_1 and E_2 (Repetto and Tzagoloff, 1991). Since partner subunits of the complex in these mutants were stable in the absence of either E_1 or E_2, the assembly intermediates generated *in vivo* could be readily isolated and studied. Mitochondrial extracts from a strain lacking E_1 contained the oligomeric E_2 core structure. Further, E_2 assembled with

E_3, thus indicating that E_1 is required neither for the assembly of the E_2 core, the E_3 dimers, nor the formation of the E_2-E_3 sub-assembly. On the other hand, mitochondrial extracts from a yeast strain lacking E_2 demonstrated the presence of E_1 and E_3 dimers. However, such dimers were not associated with one another, suggesting that E_1 and E_3 predominantly interact with E_2 during assembly. Significantly, correctly assembled KGDHC activity could be restored by mixing appropriate ratios of mitochondrial extracts from mutants lacking in E_1 or E_2; this suggests that the sub-assemblies are in a native configuration and, therefore, may represent true intermediates in the assembly pathway. Similar complexes have been isolated in resolution and reconstitution studies of KGDHC from pig heart (Tanaka et al., 1972).

One may consider that the assembly of complexes consisting of nonidentical subunits and distinct assembly intermediates should require the coordination of those events leading to the production of intermediates or the fully assembled complex. Nevertheless, aspects of KGDHC assembly display elements of a random process, as opposed to a strictly cooperative process that relies upon the presence of the correct stoichiometric amounts of subunits provided by regulated and coordinated gene expression. Thus, the over-expression in yeast of subunits of the E_2 core, relative to E_1 and E_3, led to a preponderance of incompletely assembled complexes with sub-stoichiometric contents of E_1. By contrast, over-expression of E_1 resulted in the appearance of mature intact complexes, the excess E_1 appearing as physically separate dimers (Repetto and Tzagoloff, 1991).

The correct assembly of KGDHC depends on the expression of nuclear genes that encode proteins that are not components of the fully assembled complex. A biochemical screen of mutant strains of *S. cerevisiae* unable to grow on respiratory substrates (referred to as *pet* mutants) resulted in the identification a gene that appears to be responsible for ensuring the correct interaction of E_1 with the E_2-E_3 subcomplex. The product of this gene appears to be complex-specific as mutations in this gene do not appear to affect the assembly of the related complex PDHC (Repetto and Tzagoloff, 1991). It is suggested that this gene may encode for an enzyme involved in post-translational modification of the components, or for a factor that promotes correct subunit contacts.

In view of their structural similarity, it is reasonable to assume that the assembly of 2-oxo acid dehydrogenases is similar. The role of protein X in the assembly of PDHC has been studied with yeast strains in which the nuclear gene (*PDX1*) has been disrupted. The PDH complex isolated from such a mutant lacked both protein X and E_3 subunits, but contained an assembled E_2/E_1 sub-complex (Lawson et al., 1991), thus indicating that X is required for promoting the assembly of the E_2-E_3 complex but not the E_2 core which self-assembles in its absence. Deletion mutants showed that the putative E_2 subunit-binding domain of protein X plays an important role in binding E_3 to the E_2 core. Dissociation and reassembly studies of the bovine heart PDH complex confirm that E_3 binding protein is required for the binding of E_3 to a subcomplex comprising E_2 and E_3 binding protein (De Marcucci et al.,

1995). In other 2-oxoacid dehydrogenase complexes that lack protein X, the E_2 core binds directly to E_3 via a binding domain analogous to that found in protein X.

C. Complex II of the Respiratory Chain

Complex II, succinate-CoQ reductase, is a membrane-associated enzyme complex linked to the respiratory chain (for review see Ohnishi, 1987). The mitochondrial complex is comprised of four subunits: two integral membrane subunits and the two subunits of the succinate dehydrogenase (SDH) subcomplex. The latter two are referred to as the FP (flavoprotein) and IP (iron-sulphur protein), respectively, and may be isolated as a heterodimer (Girdlestone et al., 1981). All four polypeptides constituting complex II are nuclearly encoded and are imported into mitochondria. FP carries a covalently bound FAD moiety. IP contains three non-heme iron (Fe-S) clusters (Hederstedt and Ohnishi, 1992). The integral proteins are associated with cytochrome *b*. There is limited information concerning the assembly of these prosthetic groups. The biogenesis of a functional complex requires coordination of a heme group to one of the two integral membrane subunits, the covalent attachment of the flavin cofactor to FP, and the formation and insertion of three Fe-S clusters with the IP subunit (Ackrell et al., 1992). Although little is known about the mechanism and timing of these events in relation to assembly, the Fe-S cluster of center 3 in IP is known to be required for functional membrane building of mammalian SDH (Ackrell et al., 1992). Close interaction between the heme and the SDH heterodimer is indicated by the observation that the cytochrome *b* in mammalian complexes shows different light-absorption and EPR spectra depending on whether SDH is bound or not (Yu et al., 1987). These studies were carried out by using classical biochemical approaches involving subfractionation and reconstitution.

The recent isolation of the IP and FP genes from the yeast *S. cerevisiae* has made it possible for questions of assembly *in vivo* to be addressed (Lombardo et al.,1990; Robinson and Lemire, 1992). Targeted gene disruption has been used to create individual strains lacking either IP or FP (Lombardo and Scheffler, 1989; Schmidt et al., 1992a). The mitochondrial steady state levels of their respective partner subunits, IP in the FP-deficient strain and FP in the IP-deficient strain, were determined and found to be significantly reduced. These observations were taken to indicate that the assembly of IP and FP into an intact complex is necessary for their stability. Moreover, interaction of IP or FP individually with the integral membrane proteins of complex II may be considered insufficient to protect against degradation of each of these soluble subunits. Therefore, the interaction of soluble subunits with the membrane sector may not occur as an independent step in the assembly process. Subunits FP and IP may assemble in the matrix prior to membrane attachment. That the heterodimer can be isolated intact from assembled complexes by treatment with chaotropic ions provides support to this latter view (Girdlestone et al., 1981). Interestingly, significant excesses of the FP subunit can be detected in the matrix of mammalian mitochondria in the presence of assembled complexes, suggesting that this subunit may be drawn for assembly from a free pool existing in the mitochondrial matrix (Lombardo and Scheffler, 1989). It is not known whether FP is turned over in these cells.

There are two evolutionarily conserved consecutive lysine residues situated near the C-terminus of the IP protein (Lombardo et al., 1990). These were earlier suggested to play a role in the interaction of SDH with the integral membrane subunits of complex II (Yu and Yu, 1981), but more recent experiments on the C-terminal region involving site-directed mutagenesis (Schmidt et al., 1992a) do not support this suggestion. Substitution of one or both of the conserved lysines for threonine failed to affect the stability or assembly of the complex. On the other hand, the removal of seven or more amino acids from the C-terminus, including the two lysine residues, resulted in the lack of assembly as indicated by the abolition of SDH activity and much reduced steady state levels of the truncated polypeptides in the mitochondria. The last four amino acids of IP, which do not include the two lysine residues, are not essential for assembly as their removal was without effect.

The assembly of complex II has been studied using chimeric human/yeast genes encoding IP, expressed in a yeast strain lacking endogenous IP. This novel approach has allowed the identification of regions of IP that are important for assembly. Thus, substitution of a stretch of 50 amino acids located between the sites of the first and second Fe-S clusters and encompassing a 15 amino acid segment displaying the greatest variability between species, prevents the formation of an active complex. In cells expressing chimeric IP, both it and FP are degraded upon entry into the mitochondria (Saghbini et al., 1994).

III. ASSEMBLY OF COMPLEXES CONTAINING NUCLEARLY AND MITOCHONDRIALLY ENCODED PROTEINS

A. Complex I of the Respiratory Chain

Our discussions now turn to those complexes that contain mitochondrial gene products as well as imported proteins. Complex I (NADH-CoQ reductase) of the mitochondrial electron transport chain transfers electrons from NADH via FMN and a series of Fe-S clusters to ubiquinone, linking this process with the translocation of protons across the inner membrane. The enzyme is the largest and the most structurally complex of all the membrane-associated electron transport complexes. The complex from bovine heart mitochondria has an estimated molecular mass of 880 kDa; that from the fungus *Neurospora crassa*, approximately 700 kDa. This complex has recently been reported to contain at least 41 different polypeptides in bovine heart (Walker, 1992) and 32 different polypeptides in *N. crassa* (for a review, see Weiss et al., 1991); in both cases seven subunits are mitochondrially encoded.

We discuss here those studies addressing the assembly of complex I from mammals and the fungus *N. crassa*. In contrast, *S. cerevisiae* does not express

NADH-CoQ reductase activity of the proton-pumping complex I type (de Vries and Grivell, 1988) and is not considered further here.

A low resolution three-dimensional structure has been reconstructed by electron microscopic analysis of two-dimensional membrane crystals of complex I (Hofhaus et al., 1991). The complex appears as an 'L'-shaped structure with two arms perpendicular to each other such that the peripheral component forms the arm protruding into the matrix space. Figure 1 (panel A) illustrates the overall organization of the complex. Smaller related structures were generated and visualized by electron microscopy and related to the intact complex. The hydrophobic segment (350 kDa) containing all seven of the mitochondrially encoded subunits and some of the nuclearly encoded subunits, which can be isolated upon dissection of the complex with chaotropic ions, was identified as the arm embedded in the membrane (Figure 1, panel B). Further studies (Friedrich et al., 1989; Hofhaus et al., 1991) indicate that the peripheral arm corresponds to a small sub-complex of 350 kDa, containing only nuclearly encoded subunits that may be isolated after inhibition of mitochondrial protein synthesis (Figure 1, panel C).

While the crystallographic data indicate that the enzyme is monomeric (Leonard et al., 1987) kinetic evidence is accumulating that suggests the complex in mammalian mitochondria may exist as a functional dimer when fully assembled and associated within the membrane (Van Belzen et al., 1992).

Despite the complicated nature of complex I, significant steps have been made towards understanding aspects of its assembly. Success has, in part, been due to the resolution of the complex into related sub-complexes of smaller size. This has been achieved through manipulating the growth conditions of *N. crassa*. Using this approach, evidence has been obtained suggesting that the two arms of the complex are assembled along separate pathways before associating to form the final complex. Thus, inhibition of organellar protein synthesis with chloramphenicol results in the synthesis of a small sub-complex (350 kDa; Figure 1, panel C) in place of the intact complex I normally found (770 kDa; Friedrich et al., 1989). Using immunological and enzymatic methods it was determined that this sub-complex contained at least 13 nuclearly encoded subunits of the larger complex and includes the FMN moiety, Fe-S centres and the high affinity binding site. Conversely, growth of *N. crassa* under manganese limitation results in the disturbed assembly of complex I such that a sub-assembly (350 kDa; Figure 1, panel D) accumulates that contains those subunits characteristic of the membrane arm of complex I but has no NADH-reductase activity (Schmidt et al., 1992b).

The pre-assembly of the membrane arm of the complex was demonstrated by following the flux of radioactivity in pulse-labeled cells of *N. crassa* (Tuschen et al., 1990). Labeled polypeptides were found to accumulate transiently in a 350 kDa intermediate complex (Figure 1, panel F) that could be chased into the intact complete complex. This intermediate, together with the small sub-assembly of the complex formed in the presence of chloramphenicol, can account for almost all the subunits of a fully assembled complex I.

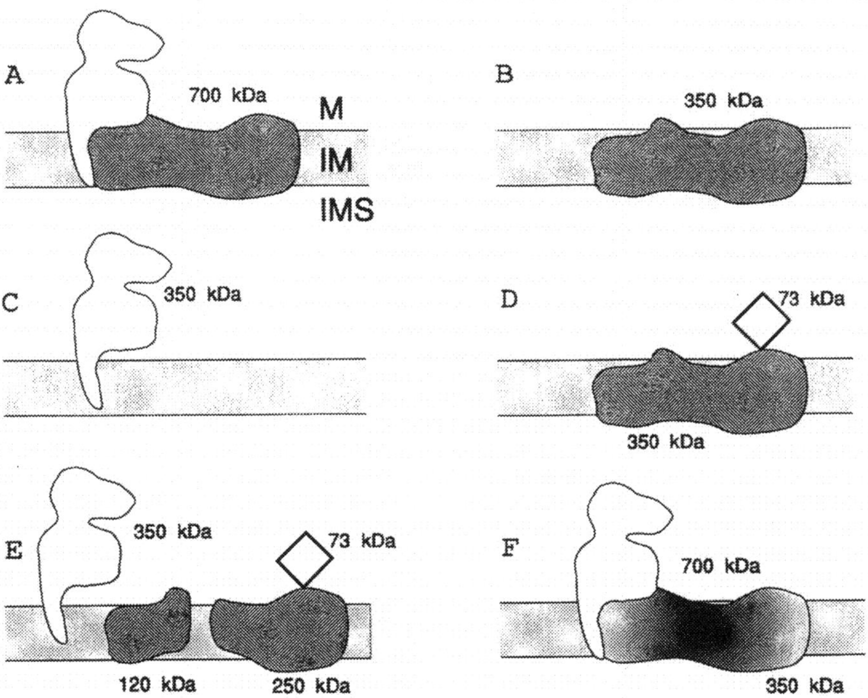

Figure 1. Representation of complex I and its experimentally generated sub-assemblies. Overall organization of complex I from *N. crassa* derived from electron microscopic analysis of membrane crystals is shown in A. The membrane arm (dark stippled) contains all the mitochondrially encoded subunits and some nuclearly encoded subunits. Mitochondrial membranes treated with a chaotropic agent, NaBr, lack the peripheral arm (B). Mitochondria from cells grown in the presence of an inhibitor of mitochondrial protein synthesis, chloramphenicol, lack the membrane arm but retain the peripheral arm composed of 13 nuclearly encoded subunits (C). The membrane arm alone may be isolated from mitochondria of cells grown in the presence of limiting Mn^{++} (D). Two sub-assembly intermediates of the membrane arm and a pre-assembled peripheral arm accumulate in a strain carrying a disrupted nuclear gene encoding a 21 kDa subunit of the membrane arm (E). A 73 kDa polypeptide presumed to have a chaperone function is shown associated with the assembly intermediates accumulated in D and E. This polypeptide is arbitarily shown with a matrix location. Pulse-labeling studies of mitochondrial proteins showed a transient labeling of subunits located in the membrane arm indicated by the graded stippling (F). M, matrix; IM, inner membrane; IMS, inter-membrane space. The inner membrane is shown lightly stippled. Sizes of subcomplexes are indicated (in kDa). References are provided in the text.

Further insight into the assembly of the membrane arm of the complex has been gained from studies using a molecular genetic approach in *N. crassa*. Gene disruption in *N. crassa* of a nuclearly encoded subunit of the membrane arm (21.3 kDa) generated a mutant strain (*nuo*21) that could not form a functional complex I (Nehls et al., 1992). The assembly of complex I in the mutant (Figure 1, panel E) was determined by analysis of radiolabelled mitochondrial membrane extracts using density gradient centrifugation and immunoprecipitation of sub-assemblies using antibodies directed toward one or all of the subunits. All subunits, with the exception of the 21.3 kDa subunit, could be accounted for in one of three sub-assemblies of complementary subunit composition. An intact peripheral arm (350 kDa) containing all the prosthetic groups with redox function, and able to catalyze NADH oxidation by artificial electron acceptors, was seen to accumulate in the mutant mitochondria. These data provide further proof of independent assembly pathways for the two arms of complex I. The final assembly of the membrane arm is blocked, leading to an accumulation of two membrane arm sub-assemblies believed to be intermediates in the assembly pathway. The larger intermediate (250 kDa) contained four of the seven mitochondrially encoded subunits and six nuclear encoded subunits. The smaller intermediate (120 kDa) contained two mitochondrially and two nuclearly encoded subunits.

One of the four Fe-S clusters, N2, has been localized to the preassembled membrane arm (Schmidt et al., 1992b). Its absence from both membrane arm sub-assemblies of the *nuo*21 strain lacking the 21.3 kDa subunit has been taken to indicate that its insertion into the complex may occur after association of the membrane arm sub-assemblies but before assembly of the membrane and peripheral arms. Alternatively, subunits from both of these sub-assemblies may be required for its insertion.

A polypeptide (73 kDa), not found in the fully assembled complex, was found to be associated with the 250 kDa membrane arm intermediate identified in the *nuo*21 mutant (Nehls et al., 1992; Figure 1, panel E) pre-assembled arm generated by Mn^{++}-limited growth (Schmidt et al., 1992b; Figure 1, panel D). It was presumed that this polypeptide may serve as a chaperone to keep these intermediates in an assembly-competent state (Schmidt et al., 1992b; Nehls et al., 1992). Thus, it may be categorized as an assembly factor as represented by products of those nuclear genes that are required for complex assembly (see following, Table 4).

More recently other phenotypes resulting from the absence of particular subunits of complex I in *N. crassa* have been described. Mutant strains of disrupted complex I for the 51 kDa (Fecke et al., 1994) or the 21.3 kDa subunit (Alves and Videira, 1994) of the peripheral arm assemble almost complete complexes. An accumulation of intermediates is seen in the strain lacking the 21.3 kDa subunit. Similarly, a mutant lacking the 12.3 kDa subunit of the membrane arm separately assembles the peripheral arm and most of the membrane arm of the complex (Duarte et al., 1995). By contrast, a more severe defect is apparent in a strain lacking the 29.9 kDa subunit of the peripheral arm. The membrane arm of the complex accumulates but the peripheral arm is lacking suggesting the 29.9 kDa subunit is involved in an early step of complex I assembly (Duarte et al., 1995).

Some observations are beginning to be made in human cell lines lacking individual mitochondrially encoded subunits of complex I. A human cell line lacking the ND4 subunit exhibited a total loss of complex I activity and lack of assembly of the mitochondrially encoded subunits (Hofhaus and Attardi, 1993). By contrast, in a mutant cell line in which ND5 subunit was not detectable and that was nearly totally deficient in complex I activity, the mitochondrially encoded subunits were assembled, although with either decreased efficiency, or a reduced stability of the assembled complex (Hofhaus and Attardi, 1995).

Hall and Hare (1990) have studied the assembly of mitochondrially and nuclearly encoded proteins into the mitochondria of rat hepatoma cells. They proposed a model for the assembly of complex I in which a scaffold of imported proteins forms the core of the enzyme and later arriving, mitochondrially encoded, proteins attach to the scaffold in a time-dependent manner. Similarly, studies of complex I in *N. crassa* show that newly synthesized polypeptides are incorporated into the holoenzyme at markedly different rates (Videira and Werner, 1989). Two mitochondrially encoded proteins (29 kDa and 12 kDa) were the first to appear in the enzyme. These authors estimated the pool sizes of a number of the components of the complex as being small, in agreement with Hall and Hare (1990). The rapidly assembled proteins were suggested to serve as recognition sites for the integration of partner polypeptides. The polypeptides most likely to form these keystones of assembly are those present in the membrane in larger amounts and should correspond to those with the larger precursor pools. In the studies of Videira and Werner (1989) they were represented by those subunits that appeared to assemble late, their slower assembly depending on their dilution by large pool sizes. This scenario implied that some of the mitochondrial proteins would be the last to assemble, in agreement with the model of Hall and Hare (1990). Thus, similar conclusions have been reached by interpretations of data derived from the two different systems. A key observation is that assembly occurs by an ordered sequence of events.

B. ATP Synthase

Mitochondrial ATP synthase (Figure 2) catalyzes the formation of ATP from ADP and inorganic phosphate, driven by the proton-motive force generated by respiratory electron transport. The complex comprises two major sectors F_1 and F_0. F_1 catalyses ATP hydrolysis and is water soluble; F_0 is located in the inner mitochondrial membrane and constitutes the proton channel. There is a further panel of subunits whose classification is less well defined. According to Nagley (1988) these subunits are designated F_A (associated components), being part of neither the clearly definable soluble F_1 sector nor the membrane-integral F_0 sector. It is likely that the F_A subunits make physical and functional links between F_1 and F_0, lying within the stalk portion of the complex, which in electron micrographs is visualized as linking the bulbous F_1 section to the membrane. A schematic representation of the distribution of subunits of the F_1, F_A, and F_0 sectors of ATP synthase from the yeast *S. cerevisiae* is shown in Figure 2.

The biogenesis of mitochondrial ATP synthase requires contributions from both the mitochondrial and nuclear genomes (see review by Cox et al., 1992). In *S. cerevisiae*, three F_0 subunits (6, 8, and 9) are specified by mitochondrial genes (see Figure 2). At least two of these subunits, 6 and 8 (known as A6L in mammalian mitochondria), are encoded in mitochondrial DNA (mtDNA) of all organisms so far examined. In some yeasts and most plants subunit 9 is encoded in mtDNA. By contrast, genes encoding subunit 9 in metazoa and the filamentous fungi including

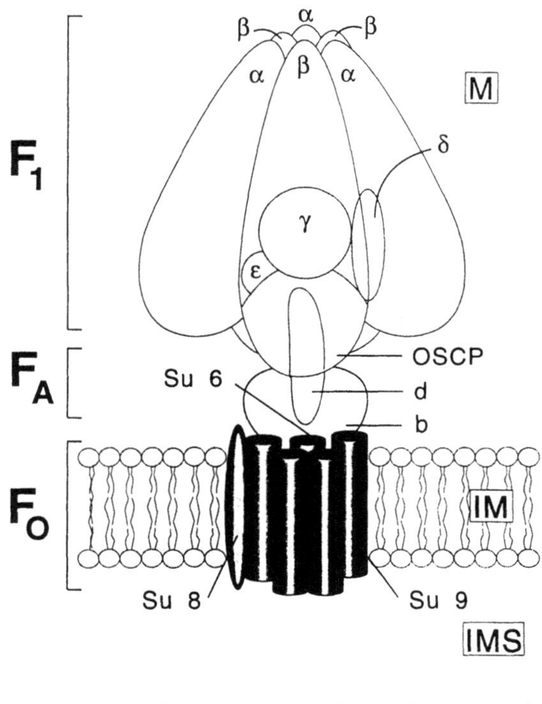

Figure 2. Subunit arrangement of yeast mtATP synthase (after Cox et al., 1992). Those subunits of mitochondrial genetic origin are shown stippled. Subunit stoichiometry in F_1, the nucleotide binding and catalytic domain, is $\alpha_3\beta_3\gamma\delta\varepsilon$. F_A, drawn to represent the "stalk," is comprised of OSCP, d, and b (stoichiometry probably 1 or 2 of each per complex). In F_O, which encompasses the proton channel across the membrane, there are probably 6–12 subunit 9 proteins and one subunit 6 per complex. The stoichiometry of subunit 8 in F_O is not well defined (perhaps 1 or 2 per complex). Yeast subunit has a homolog in mammalian mtATPase denoted subunit (see Cox et al., 1992). M, matrix; IM, inner membrane; IMS, inter-membrane space. Inner membrane is represented as a stylised phospholipid bilayer.

N. crassa are located in the nucleus. The mitochondrially encoded subunit 9 proteins are made without a leader sequence, their N-terminal amino acid being N-formylmethionine. However, the nuclearly coded subunit 9 proteins are first biosynthesized as precursors bearing an extended N-terminal leader that targets subunit 9 to the mitochondria, at the same time keeping the very hydrophobic passenger protein soluble in the aqueous cytosol (see Cox et al., 1992).

A knowledge of the relative arrangement of subunits in a complex will give important clues about assembly pathways. Recently, the structure of the F_1 sector from bovine heart mitochondria has been solved by X-ray crystallography at 2.8 Å resolution (Abrahams et al., 1994). The spherical part of F_1 made up of 3α and 3β subunits, is traversed by a single α-helix made from the C-terminal region of the γ subunit. This helix extends out into a region that forms the stalk in the intact complex. A second α-helical segment of the γ subunit is wrapped around the lower part of the first α-helix to form an anti-parallel coiled-coil structure. The solved structure does not include the δ and ε subunits, although they were present in the crystals. Structural data for the F_A and F_o sectors are now awaited to enhance our overall understanding of the complex, particularly the proton-conducting structures.

In general, following their import, processing and handling by molecular chaperones the nuclearly encoded subunits are assembled into an oligomeric array. The assembly of F_1 has not received the attention it perhaps deserves. Burns and Lewin (1986) studied the rate of import of subunits and assembly of F_1 *in vitro*, employing a polyclonal anti-F_1-β antibody. They showed that newly imported subunits associate with each other almost immediately following their import and that intra-mitochondrial pools of unassembled subunits are apparently very small. Evidence was obtained that F_1-β subunits were imported and assembled at a lower rate than two other subunits, F_1-α and F_1-γ, suggesting that F_1 assembly may be limited by the availability of β-subunits.

F_1 assembly is, to a certain extent, independent of F_0. Thus, the F_1 sector is able to assemble in the mitochondrial matrix of cytoplasmic petite mutants of yeast, which completely lack mtDNA (*rho*^{0}) and therefore produce no F_0 subunits. The assembled F_1 has a cold-labile ATPase activity as a consequence of its lack of integration with F_0 (Schatz, 1968). Nevertheless, on the basis of temperature-dependence of enzyme kinetics of F_1 in *rho*^{0} mitochondria, Orian and colleagues (1984) concluded there is a loose association of F_1 with the mitochondrial inner membrane, in spite of the lack of an assembled membrane anchor F_0 sector. It is considered that in wild-type cells association of the already assembled F_1 with the inner mitochondrial membrane occurs via interaction with F_0 and F_A subunits, inserted directly into or associated peripherally with the membrane.

An assembly pathway for the individual F_0 sector subunits of the yeast complex has been deduced by analysis of a series of mutants bearing early frameshift mutations in each of the mitochondrial genes encoding subunits 9, 8, and 6 (Linnane et al., 1985). In this study the proteins of mutant cells were radiolabelled, grown in a chemostat to minimise glucose repression affects, then analyzed for the nature of ATP synthase subunits in mitochondrial lysates immunoabsorbed to an immobilized anti-F_1-β monoclonal antibody (Hadikusumo et al., 1984, 1988; Marzuki et al., 1989). The assembly pathway involves the sequential addition of subunits 9, 8, and 6 to an independently formed F_1 sector, itself assembled from imported subunits. The suggestion has been made that there is dependence on F_1 for proper F_0 assembly.

Analysis of the respiratory properties of a mutant null for the synthesis of the F_1-β subunit indicated that a functional proton channel formed by F_0 proteins is not detected in the inner mitochondrial membrane (Takeda et al., 1985). It was not established whether a functional proton-transporting channel is not assembled as such, or is present but blocked by another protein prior to the assembly step involving F_1-β subunit.

Further studies of the assembly properties of F_0 components, requiring the manipulation of mitochondrial gene products in a more controlled manner, was achieved by the functional relocation of specific mitochondrial genes to the nucleus. This strategy is denoted allotopic expression (Nagley and Devenish, 1989). Effective targeting of subunit 8 back to the mitochondria was achieved by inclusion in a chimaeric precursor protein of a mitochondrial import leader (N9L) retrieved from the nuclearly encoded subunit 9 from *N. crassa* (Nagley et al., 1988). Although imported efficiently into isolated mitochondria, detectable assembly of this retargetted subunit 8 into ATP synthase *in vitro* was possible only in mitochondria containing a mosaic of partially and fully assembled ATP synthase complexes. Such a mosaic was generated by the controlled depletion of the cellular subunit 8 (Law et al., 1990) by manipulating expression of the relocated gene encoding the chimaeric subunit 8 precursor. Greatly enhanced import of chimaeric precursors bearing hydrophobic passenger proteins such as yeast subunit 8 or 9 is achieved by duplication of the N9L leader (Galanis et al., 1991; Papakonstantinou et al., 1993). Using these approaches in conjunction with a program of site-directed mutagenesis, it has been established that the evolutionarily conserved positively charged C-terminal region of subunit 8 is required for assembly of this subunit into mtATPase (Nero et al., 1990; Grasso et al., 1991; Papakonstantinou et al., 1993). It remains to be established with which subunit or subunits this assembly domain of subunit 8 interacts.

Does the observation that assembly of subunit 8 imported into isolated mitochondria requires prior depletion of cellular subunit 8 hold for other ATP synthase subunits imported *in vitro*? Early studies on the import and assembly of *N. crassa* subunit 9 into isolated mitochondria of this filamentous fungus suggests that depletion of this subunit is not required (Schmidt et al., 1983). More recent studies showed that detectable assembly into ATP synthase of OSCP and subunit b imported into isolated mitochondria does not require prior cellular depletion of the cognate subunit, although such depletion does lead to enhanced assembly *in vitro* (Prescott, Law, Nagley, and Devenish, unpublished data).

The assembly of F_A components is dependent upon F_0 assembly. In the absence of a structurally intact F_0 sector the proteins encoded by the nuclear genes *ATP7* (Norais et al., 1991) and *ATP4* (Velours et al., 1988), denoted d and b, respectively (Figure 2), were not observed to be associated with immunoabsorbed mitochondrial ATP synthase complexes (Hadikusumo et al., 1988). In particular, the absence of subunit 6 leads to a failure of d and b to assemble, despite the presence of assembled subunits 9 and 8. In these experiments, which involved radiolabelling proteins with ^{35}S-sulphate, the *ATP5* gene product (OSCP) could not be monitored because of the paucity of sulphur-containing amino acids in this protein (Uh et al., 1990). Further studies of the assembly of OSCP into ATP synthase *in vitro* and *in vivo* have been facilitated by the construction of an OSCP variant substituted with two C-terminal methionine residues for the purposes of enhanced radiolabelling, and which has functional and assembly characteristics comparable to those of wild-type OSCP (Bateson et al., 1996).

Complementary experiments on the influence of F_A subunits on F_0 assembly utilize null mutants in the relevant yeast nuclear genes. Disruption of each of the cloned genes for each of the F_A subunits b, OSCP and d leads to nonassembly not only of the other F_A subunits but also of subunit 6. This indicates the mutual inter-dependence for assembly of subunit 6 and each of the F_A subunits (Norais et al., 1991; Paul et al., 1989, 1992; Devenish et al, 1992; Prescott et al., 1994; Prescott, Straffon, Andrews, Nagley, and Devenish, unpublished data). The lack of subunit d from mitochondria also affects biosynthesis of subunit 6 (Norais et al., 1991).

Characterization of sub-assemblies of multisubunit complexes generated *in vivo* as a result of genetic or physiological manipulation, or *in vitro* as a result of chemical means, is a useful approach for understanding assembly processes. In order to facilitate the isolation of sub-assemblies or intact mtATPase from *S. cerevisiae* individual subunits of the F_A sector of the complex have been tagged at the C-terminus with hexahistidine (Bateson et al., 1996). Assembled complexes can be adsorbed to nickel charged affinity chromatography media and the constituent subunits resolved by subsequent gel electrophoresis.

Walker and colleagues have used an *in vitro* reconstitution approach to study the formation of an assembly of bovine F_A stalk subunits with bovine F_1-ATPase (F_1) (Collinson et al., 1994; Walker and Collinson, 1994). Subunits F_6, b, d, and defined fragments of both subunits b and d were overexpressed in *Escherichia coli*. Subunits b and d of bovine mtATPase have homologs in yeast; F_6 has no known homolog in yeast. The binding to isolated F_1 of the various proteins and fragments, both singly and in combination, was examined using gel filtration chromatography. A stable pentametric complex of F_1, OSCP, b, F_6, and d was isolated. A quartenary complex of the same F_A subunits was also isolated in the absence of F_1. Subunits b, F_6, and d interact with F_1 through OSCP as these subunits were unable to assemble with F_1 in the absence of OSCP. Assemblies obtained containing fragments of subunits b and d suggest an arrangement in which F_6 and d interact primarily with b and to a lesser extent with each other.

The role of proteins of the F_A sector in assembly has been studied in *S. cerevisiae* by expressing a rat cDNA encoding the yeast homolog of each subunit in a strain in which the expression of the equivalent yeast gene is lacking. Rat OSCP is able to functionally replace yeast OSCP *in vivo*. However, upon isolation mtATPase is unstable (Prescott et al., 1995); apparently the F_1 detaches from the F_o sector. The C-terminus of OSCP is important to maintain stability of the complex (Prescott, Boyle, Nagley, and Devenish, unpublished data).

Assembly of the complex has been studied in cultured human cells labeled with [^{35}S] methionine in pulse-chase experiments. Two-dimensional gel electrophoresis and fluorography were used to analyze the assembly patterns. The F_1 sector of complex appeared to be assembled first followed by an intermediate that contained F_1 and F_o subunit 9 (c). It was suggested this intermediate also contained subunits b, F_6, and OSCP, but not the mitochondrially encoded F_0 subunits 6 (a) and A6L (8). Both of the sub-assemblies accumulated when mitochondrial protein synthesis was inhibited suggesting that the incorporation of the mitochondrially encoded subunits occurs late in the formation of a fully assembled mtATPase complex (Nijtmans et al., 1995).

The assembly of F_1-α and F_1-β subunits has been shown to be dependent upon the products of the *ATP11* and *ATP12* genes. Although near wild-type levels of processed F_1-α and F_1-β subunits were found in mitochondria they were present as high molecular weight aggregates. The role of the encoded products of the *ATP11* and *ATP12* genes may be to catalyze a particular chemical modification to one of the participating subunits, or to confer an appropriate conformational structure, thus mediating the folding of the F_1-α and F_1-β subunits (Ackerman and Tzagoloff, 1990a; Bowman et al., 1991).

The product of the *ATP10* gene has been found to be associated with the inner mitochondrial membrane but does not co-fractionate with the F_1 sector or the assembled mtATPase complex (Ackerman and Tzagoloff, 1990b). Mutations in the *ATP10* gene confer a phenotype suggestive of a defect in physical coupling of the F_1 sector to the F_0 sector. The biochemical evidence suggests that the *ATP10* gene product functions in the assembly of the F_0 sector. Although these mutations appeared to have no effect upon the translation of mitochondrially encoded F_0 subunits 6 and 9, an effect upon subunit 8 could not be ruled out.

Little is known about the means by which the mitochondrion coordinates the availability of components of an enzyme complex such as ATP synthase, in such a way to provide a coordinated delivery of appropriate quantities of subunits required for assembly. Presumably there is some co-ordinated regulation of nuclear gene products as may occur in mammals (Nagley, 1991). At the level of expression of individual genes, control of F_0 assembly may be exerted through nuclear regulation of synthesis of the mitochondrially encoded subunits. The biosynthesis of subunit 9 in mitochondria is controlled by the products of the nuclear genes *AEP1* and *AEP2*

(Payne et al., 1991; Finnegan et al. 1991). Other nuclearly encoded proteins are involved in controlling the biosynthesis of mitochondrial subunits 8 and 6 (Manon and Guerin, 1989; Pelissier et al., 1992, 1995). The N-terminal processing of subunit 6 (Michon et al., 1988) also probably involves an imported protease.

C. Complex III of the Respiratory Chain

Complex III ($CoQH_2$-cytochrome *c* reductase, or the cytochrome bc_1 complex) is the second of three energy-transducing respiratory chain complexes of mitochondria. The bc_1 complex from mammalian sources is composed of 11 subunits, while that from fungal sources is found to have at least nine and possibly 10 subunits (Weiss, 1987; Yang and Trumpower, 1994) that may be organized as a whole into a structural dimer (Nalecz et al., 1985; Yang and Trumpower, 1994). All the subunits, with the exception of cytochrome *b*, are nuclearly encoded. Interestingly, the equivalent bacterial bc_1 complex contains only the three catalytic subunits, cytochrome *b*, cytochrome c_1, and the Rieske Fe-S protein (Yang and Trumpower, 1986). Although the functions of the additional subunits are largely unknown, most are required for the integration of a correctly assembled and functional complex in mitochondria (Trumpower, 1990). The molecular genetics of cytochrome *b* as such has been reviewed (Colson, 1993).

The assembly of the bc_1 complex in yeast has recently been studied in several laboratories using a biochemical genetic approach. Mutant strains lacking individual subunits were constructed by gene disruption or isolated from a collection of *pet* mutants. These strains have been examined for the resultant effects on steady state levels of bc_1 complex subunits, or in some cases, assembly of the subunits of the complex (Oudshoorn et al., 1987; Crivellone et al., 1988; Maarse et al., 1988; Tzagoloff et al., 1988; Schoppinck et al., 1988, 1989a,b,c; Gonzalez-Halphen et al., 1988; Japa and Beattie, 1989; Phillips et al., 1993; Yang and Trumpower, 1994). There is insufficient evidence to enable detailed descriptions of events leading to assembly of each subunit of the complex. Nevertheless, on the basis of the available evidence, Grivell (1989) has drawn up a general scheme that is summarized in Figure 3. The central features of this scheme are the involvement of three sub-complexes, together with a series of other unspecified interactions between individual subunits. The orderly integration of such sets of subunits leads to the final assembly of the complex as a whole. For details of data supporting the scheme in Figure 3, the reader is referred to Grivell (1989) and references therein.

Assembly of the bc_1 complex has been suggested to occur through the interaction, within the lipid bilayer, of pools of various sizes of unassembled subunits and partially assembled complexes (Crivellone et al., 1988; Attardi and Schatz, 1988). The participation of the two largest subunits of the complex, I and II, in the formation of a stable intermediate is based on biochemical genetic evidence. Thus, mutations in the genes encoding each subunit of the complex causing a failure to express the mutated proteins (with the exception of the genes encoding I and II),

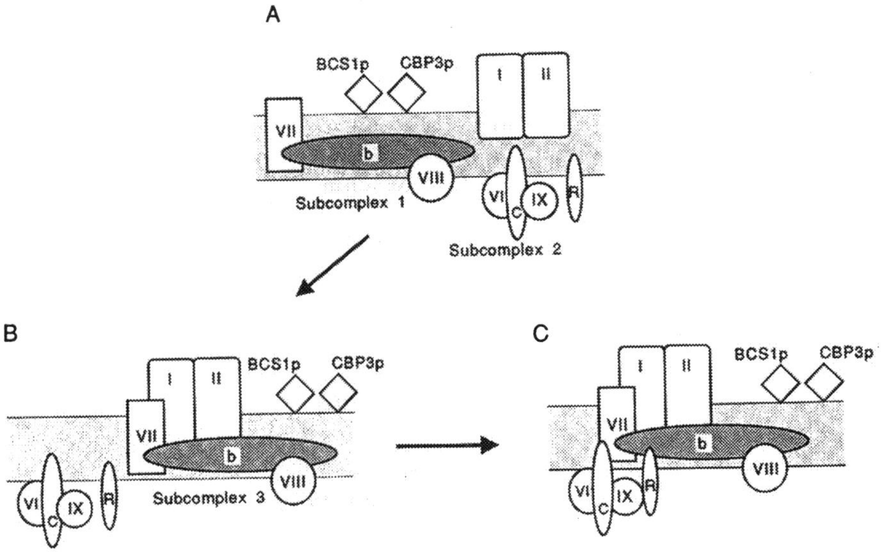

Figure 3. A model for the assembly of Complex III (the bc_1 complex) in yeast (after Grivell, 1989). Following delivery to the inner mitochondrial membrane (lightly stippled) (IM), individual subunit polypeptides assume native configurations and associate to form a number of sub-complexes (A). These sub-complexes (denoted by Arabic numerals) are envisaged subsequently to assemble, forming a functional complex (B, then C). Unless otherwise indicated, subunits of complex III are denoted by Roman numerals. The insertion of prosthetic groups into apoproteins of cytochrome *b* (b), cytochrome c_1 (C), and the Rieske protein (R) is thought to be a late event in assembly. Maturation of the Rieske protein, involving proteolytic removal of the mitochondrial target sequence, may occur after assembly into the complex. The presence of membrane-associated factors that promote assembly of the complex is indicated (CBP3p and BCS1p). Cytochrome *b* is mitochondrially encoded and is indicated by the darkly stippled appearance. M, matrix; IMS, inter-membrane space. Reference citations are provided in the text.

do not affect the assembly or steady state levels of subunits I and II (Tzagaloff et al. 1988, Crivellone et al., 1988). Sub-complex 1, comprising apocytochrome *b* and subunits VII and VIII, associates with the stable subunit I/II intermediate to form sub-complex 3 (Figure 3, panel B). The stability of sub-complex 1 appears to be enhanced by its interaction with the subunit I/II intermediate, as in the absence of subunits I or II the levels of the components of sub-complex 1 are considerably reduced (Crivellone et al., 1988). The sub-complex 3 assembly intermediate combines with sub-complex 2, comprising cytochrome c_1 and subunits VI and IX, followed by the Rieske Fe-S protein to form a fully assembled and functional complex (Figure 3, panel C). The maturation of the Rieske and cytochrome b apoproteins by the insertion of the prosthetic groups appears to be a late event and is thought to occur, as in the case of cytochrome c_1, after association with other subunits (Tzagoloff et al.,1988; Graham and Trumpower, 1991; Yang and Trumpower, 1994).

The maturation of the Rieske protein appears to occur after its assembly into the complex. Precursor polypeptides of intermediate and mature size can be isolated from the assembled bc_1 complex. Further, the insertion of the Fe-S cluster appears not to be an absolute prerequisite for correct proteolytic processing of the protein or its assembly into the complex (Graham and Trumpower, 1991; Graham et al. 1993). An interesting finding is that in mammalian mitochondria the entire targeting sequence of the Rieske protein precursor is retained by the mitochondria and appears to assemble as a component (subunit 9) of the final complex (Brandt et al., 1993). A similar association of the 22 amino acid presequence with the bc_1 complex in yeast after cleavage from the Rieske protein remains to be demonstrated.

The assembly of the Rieske protein is required for the correct maturation of cytochrome *b*, whereas assembly of the Rieske apoprotein appears not to perturb the optical properties of this cytochrome (Ljungdahl et al., 1989; Crivellone et al., 1988; Graham and Trumpower, 1991). It is therefore possible that the maturation of cytochrome *b* may occur after the assembly of the apoprotein but before insertion of the Fe-S cluster.

Surprisingly, deletion of the *QCR10* gene which encodes an 8.5 kDa subunit of the bc_1 complex (homologous with the 6.4 kDa subunit 11 of the bovine heart complex) does not affect growth of yeast on nonfermentable carbon sources. In addition, enzyme activity was reduced by 40% and the Rieske iron-sulfur protein was lost when the cytochrome bc_1 complex was purified from cells lacking subunit 10. Thus, this subunit is required during assembly for the stable association of the iron-sulfur protein with the complex (Brands et al., 1994).

A screen of respiratory deficient mutants of *S. cerevisiae* has identified two nuclear genes required for the correct assembly of the bc_1 complex. Strains bearing a lesion in the *Bcs1* gene (Nobrega et al., 1992) suffered only a modest effect on the constituent subunits of assembled bc_1 complex in comparison to the effects of lesions in genes encoding other structural subunits. Nevertheless, the Rieske Fe-S protein was almost completely absent and, further, the heme group was absent from a proportion of the assembled apocytochrome *b*. Strains carrying a lesion in the *CBP3* gene, such that a truncated form of its *CBP3p* gene product is found in mitochondria, contain reduced amounts of assembled apocytochrome *b*, subunit VII, subunit VIII, and the Rieske Fe-S protein (Wu and Tzagolff, 1989). Presumably, as a result of destabilized assembly, increased turnover of these proteins occurs. Mutants in the *CBP4* gene have a phenotype similar to that of *cbp3* mutants (Crivellone, 1994).

Immunochemical data suggest that although the protein products of both *BCS1* and *CBP3* genes are associated with the mitochondrial membrane, they do not appear to be components of the bc_1 complex. Since these proteins are mitochondrially located, they are probably not involved in the transcriptional or translational regulation of the nuclear genes encoding structural proteins of the complex. Furthermore, wild-type levels of the cytochrome b mRNA were detected in *cbp3* mutant strains. The finding that overexpression of the Rieske protein in strains of yeast carrying the mutant *BCS1* gene did not alter the mutant (*pet*) phenotype suggested that the protein may be involved with the insertion of the Fe-S cluster into the apoprotein. Alternatively, the protein might provide a specific chaperone function necessary for the assembly of the Rieske protein into the rest of the complex.

The function of Cbp3p may be to facilitate the assembly of an intermediate in the assembly pathway either in a general or specific manner, hence the reduction in the amounts of affected subunits. It may also be involved in the maturation of apocytochrome *b* or Rieske protein or postranslational modification of other structural subunits.

D. Complex IV of the Respiratory Chain

Cytochrome *c* oxidase, which catalyzes the transfer of electrons from cytochrome *c* to molecular oxygen, is the terminal member of the electron transport chain and is known as complex IV. In mitochondria of all species it consists of three mitochondrially encoded subunits, together with a species-specific number of nuclearly encoded subunits making a total number of 7–13 subunits, depending on the organism (Capaldi, 1990a, 1990b). The heme a_1 and a_3, two or three Cu atoms, and a Zn atom are located in the functional core that consists of the mitochondrially encoded subunits I and II. Subunit III, the third mitochondrial gene product, has been suggested to be involved in proton pumping (Cooper et al., 1991).

The intact complex from yeast was previously thought to contain only 9 subunits (Power et al., 1984) compared to the 13 of the mammalian complex (Capaldi et al., 1986; Kadenbach et al., 1986). With the use of mild solubilization, a complex containing three additional subunits has recently been isolated from yeast (Taanman and Capaldi, 1992). Two of the additional subunits are homologous to the mammalian subunits VIa or VIb (nomenclature of Kadenbach and Merle, 1981).

In spite of much effort made to investigate this important enzyme, the assembly of the cytochrome *c* oxidase complex is not well defined. Early studies suggested that mitochondrially encoded subunits are central to the assembly of the complex. Yeast mutant strains lacking mitochondrially encoded subunits failed to assemble the remaining subunits (Cabral and Schatz, 1978). The mitochondrially encoded subunits are synthesised on membrane-bound ribosomes. Although co-translational or post-translational insertion into the membrane has not been established, the available evidence (Capaldi, 1990b) suggests that subunit II is post-translationally translocated completely into the intermembrane space before insertion into the inner membrane. This subunit is processed N-terminally by an imported protease (Pratje et al., 1986).

A biochemical approach involving pulse-chase labelling studies in isolated mitochondria and rat hepatocytes indicates a possible sequence of early events leading to assembly of the complex. Subunits II and III assemble rapidly with each other, or with the nuclearly encoded subunit VI (VII yeast homologue), followed by a slow association with subunit I (Wielburski and Nelson, 1983; Wielburski et al., 1982). Subunits appear not to be drawn from intracellular pools of subunits (Poyton, 1980).

Attempts to understand the assembly of the complex by other biochemical means have met with little success. Approaches include the serial depletion of individual subunits (e.g., Brunori et al., 1987) or the reconstitution of constituent subunits from the isolated mammalian enzyme complex (see, for example, Planques et al. 1989). The most promising strategy to understanding assembly is through genetic approaches in yeast. We discuss primarily nuclear mutations affecting assembly of complex IV. The expression of mitochondrial genes encoding subunits of complex IV has been reviewed (Grivell, 1989; Constanzo and Fox, 1990).

Mutations in nuclear genes encoding structural subunits of the yeast complex have been isolated by screening a collection of *pet* strains, or created by targetted deletion or disruption of specific genes (Table 3). The yeast genes encoding the subunits homologous to the bovine subunits VIa and VIb have yet to be identified. The consequences of mutation in each subunit generate a complicated picture in terms of trying to draw inferences about the individual role of each subunit in complex IV function and assembly. It is not our intention to discuss the property of each mutant in detail here. We will describe selected mutants, including those affected in genes encoding structural proteins of complex IV (Table 3) or in other genes that influence the organization of this complex, in order to illustrate some general statements that can be made about assembly of cytochrome *c* oxidase.

By genetic criteria, data indicate that most nuclearly encoded subunits appear to be required for assembly of complex IV. Following inactivation of each nuclear gene by deletion or disruption, with the exception of genes encoding subunits VIII and VIb, the loss of functional assembly of the enzyme was observed. In the majority of these mutant strains, the assembly status of those remaining subunits of the complex was not determined, but was assumed to be perturbed or incomplete due to the dysfunctional nature of the enzyme complex. The absence of a detectable spectral heme in many mutants indicates the absence of the mitochondrial subunit containing the prosthetic group or the failure to insert it into the apoprotein.

In a mutant strain lacking expression of the *COX4* gene, those subunits not assembled into an intact complex did demonstrate some association with one another, as was indicated by density gradient centrifugation (Dowhan et al., 1985). It is possible that this subunit is peripheral to the complex, as a missense mutation in the *COX4* gene did not perturb the assembly of the complex itself at the permissive temperature, but the subunit IV was readily lost upon isolation of the complex. It is suggested that failure of this strain to grow at the restrictive

Table 3. The Effect Upon Assembly of Lesions in the Nuclear Genes Encoding Structural Subunits of Complex IV from *S. cerevisiae*

Nuclear Gene	*Subunit*	*Lesion Type*	*Effect Upon Assembly or Function*	*Reference*
COX4	IV	Missense mutation	Assembled but reduced stability of isolated complex. Reduced turnover number. Possible function in catalysis as distinct from assembly.	Lightowlers et al. (1991)
		Disruption	Lack of spectral aa_3 and cytochrome *c* oxidase activity. Subunit IV absent; remaining subunits present but assembly incomplete.	Dowhan et al. (1985)
		pet (*cox*4-2)	Lack of subunits IV and II. Diminished amounts of subunit III. No enzyme activity or spectral aa_3. Lack of functionalassembly.	McEwen et al. (1986)
COX4	IV	*pet* (*cox*4-1)	Diminished cytochrome *c* oxidase activity and spectral aa_3. Aberrant processing of precursor polypeptide.	Koerner et al. (1985)
COX5a	Va	*pet* (*cox*5a-*cox*5a-3 *cox*5a-4)	Perturbed assembly. Diminished amount of subunits I, II and III. Lack of enzyme activity and heme spectra. Lack of functional assembly. Subunit Va absent or extended in length.	McEwen et al. (1986)
COX5b	Vb	Disruption	Lack of respiratory function when disruption present with mutant *COX5a* gene.	Cumsky et al. (1985)
COX5a	Va	*pet* (*cox*4-2)	Truncated form of subunit Va expressed. Remaining subunits present in normal amounts. Respiration and cytochrome *c* oxidase activity 10–20% of wild-type.	Cumsky et al. (1983)

COX6	VI	Deletion	Loss of cellular respiration, cytochrome *c* oxidase activity and spectral *aa3*.	Poyton et al. (1988)
COX6	VI	*pet* (*cox*6-1)	Diminished amounts of cytochrome *c* oxidase activity and absence of spectral *aa3*. Aberrant processing of precursor polypeptide.	Koerner et al. (1985)
COX7	VII	Disruption	No spectral heme. No enzyme activity. Lack of functional assembly.	Aggelar and Capaldi (1990)
COX7	VII	Deletion	Lack of *aa3* spectra and cytochrome oxidase activity. Decreased steady state levels of subunit I. Association of subunit IV with subunit I or II, respectively, absent or reduced. Required for association between mitochondrially and nuclearly encoded subunits.	Calder and McEwen (1991)
COX8	VIII	Disruption	Functionally assembled. 20% reduction in enzyme activity and respiration. Possible role of subunit VIII in modulating activity.	Patterson and Poyton (1986)
COX9	VIIa	Deletion	Loss of respiratory function and cytochrome oxidase activity.	Wright et al. (1986)
		pet (*cox*9-1)	Loss of subunits II, VII, VIII, and altered maturation of IV. Diminished amount of II. Lack of respiratory function. Lack of spectral *aa3*.	McEwen et al. (1986)
COX12	VIb	Deletion	Temperature-dependent *pet* phenotype. Optically detectable heme but diminished enzyme activity. Grows slowly at 28°C, not at all at 37°C. Required for functional assembly but dispensable once complex assembled.	La Marche et al. (1992)

Notes: [a]*pet* indicates a mutation that results in the inability to use ethanol or glycerol as a respiratory substrate for growth of mutant cells.
[b]Designated alleles for mutant phenotypes are shown where appropriate.

temperature (37°C) is due to the dissociation of the subunit from the complex *in vivo*.

In several mutant strains the absence or reduced steady state levels of one or more subunits within the mitochondria has been demonstrated by use of antibodies with specificities directed to particular subunits. Unassembled subunits are presumed to be unprotected and therefore proteolytically degraded. Thus, diminished levels of one or more of the mitochondrially encoded subunits are found in strains carrying mutations in the *COX9* and *COX5a* genes. In the latter case, several of the nuclear subunits are also absent.

The incorporation of heme into the complex and subsequent assembly of further subunits are interdependent events. Thus, a heme-deficient mutant lacking subunits V and VII and containing diminished amounts of subunits I and IV showed no assembly of those subunits that remained (Saltzgaber-Muller and Schatz, 1978). Conversely, the insertion of heme into the complex is dependent on the assembly of certain subunits. Other post-translational events are required for the correct assembly of the complex. Thus, the maturation of subunit IV is incomplete in a strain bearing a mutant *COX9* gene and lacking assembly of several mitochondrially encoded and nuclearly encoded subunits (McEwen et al., 1986). Lack of functional assembly in mutants with aberrantly processed precursors of subunit V and VI may indicate that correct maturation is a general requirement for assembly. The maturation of the Rieske protein of complex III was found to be dependent upon the assembly of the apoprotein into the complex (Graham and Trumpower, 1991).

In certain cases, failure to synthesize subunits does not appear to affect significantly the assembly of the remaining subunits into a functional complex. It is possible that these represent subunits associated in a peripheral manner with the structure and associate late in the assembly pathway. Thus, inactivation of the *COX8* gene by disruption results in a functionally assembled complex with only a small (20%) reduction in enzyme activity compared to that of the wild-type complex. It is suggested that subunit VIII may play a role in regulating the activity of the complex.

The recent isolation of an assembled complex from yeast containing three additional subunits (VIb, and two subunits homologous to VIa and VIb from the bovine complex), but with enzyme properties indistinguishable from those of the complex previously isolated and lacking these subunits, suggests that these additional polypeptides are not required for maintaining the integrity of the smaller assembled complex (Taanman and Capaldi, 1992). However, inactivation of the gene *COX12* encoding one of these subunits (VIb from yeast) resulted in a functionally assembled complex possessing spectrally detectable heme, but with diminished enzyme activity (La Marche et al., 1992). Thus, VIb subunit appears to be required for events leading to the formation of a functional complex *in vivo*. Since a fully active complex lacking VIb can be isolated under certain conditions from nonmutant cells, it appears that loss of this subunit can be tolerated once assembled without detriment to the function or assembly of the remaining subunits. Identification and inactivation of the genes in yeast encoding those subunits homologous to the bovine homologues VIa and VIb will shed light on their role in the assembly of the complex. It has been suggested that the mammalian subunit VIb may have a role in stabilizing the dimeric membrane bound form of the enzyme *in vivo* (Capaldi, 1990a).

Disruption or deletion of the *COX7* gene (subunit VII) yields a respiratory deficient phenotype (Aggeler and Capaldi, 1990; Calder and McEwen, 1991). The absence of heme spectra and cytochrome *c* oxidase activity indicated lack of the assembled complex. It was shown that subunit I had a decreased rate of translation or there was a rapid rate of degradation of that nascent subunit. The other nuclearly encoded subunits were not substantially affected. Investigation of the assembly of subunits I and II with subunit IV revealed no association of subunit IV with subunit I and a decreased association between subunits IV and II. It was concluded that subunit VII has a role in assembly of mitochondrially and nuclearly encoded subunits.

Defining the biogenesis of cytochrome *c* oxidase in terms of assembly is complicated by the existence of isoforms of several of the nuclearly encoded subunits. These isoforms must be tolerated by other subunits of the complex during assembly. In addition to those isoforms of this complex that occur in mammals (reviewed in Lomax and Grossman, 1989), Poyton and co-workers have isolated two genes from *S. cerevisiae* encoding isoforms of subunit V, Va, and Vb (Cumsky et al., 1985).

Genetic analysis indicates that the biogenesis of complex IV in yeast is known to require the involvement of an unusually large number of nuclear genes (Tzagoloff and Diekmann, 1990). There are 18 reported to date to be involved in the synthesis of subunit I alone (Pel et al., 1992). In a separate genetic study, complementation analysis of a collection of respiratory deficient mutants having lesions in cytochrome *c* oxidase activity indicated 17 complementation groups in one class of mutants that contained near normal amounts of subunits but were defective in assembly (McEwen et al., 1986). One mutant has been characterized in detail and a novel gene, *OXA1,* identified as being required for correct assembly of complex IV (Bonnefoy et al., 1994; Altamura et al., 1996). This nuclear gene is conserved in evolution from plants to humans (Bonnefoy et al., 1994). Yeast cells lacking Oxalp have decreased levels of the three mitochondrially synthesized subunits, yet a total absence of cytochrome aa_3 and decreased cytochrome *b*. Two further nuclear genes have been characterized, *PET117* and *PET191*, whose products are not constituents of complex IV but are required in a post-translational step of biogenesis of this complex (McEwen et al., 1993).

Several genes that appear to be involved in assembly have been investigated in detail. Strains with mutations in either of the yeast nuclear genes *COX10* or *COX11* have similar phenotypes, namely lack of cytochrome *c* oxidase activity, lack of spectral heme, and failure to assemble the holoenzyme (Nobrega et al., 1990;

Tzagoloff et al., 1990). Both strains synthesize mitochondrially encoded subunits I, II, and III, as well as the nuclearly encoded subunits IV and V. These could be retrieved in unassembled or ill-defined partly assembled complexes. Subunit I was turned over rapidly, probably as a result of its failure to assemble and the consequent exposure to proteases. The products of the *COX10* and *COX11* genes are now known to be required for synthesis of heme A (Tzagoloff et al., 1993; Glerum and Tzagoloff, 1994). Strains with mutations in the *COX14* gene have normal levels of mitochondrially and nuclear encoded subunits but lack cytochrome aa_3 and have no detectable cytochrome *c* oxidase activity. The assembly-arrested phenotype of *cox14* mutants indicates that Cox14p provides an important function in the cytochrome *c* oxidase assembly pathway (Glerum and Tzagoloff, 1995).

Strains having mutations in the nuclear *SCO1* gene, depending on the particular allele, fail to accumulate subunit II, or subunits I and II, due to increased turnover of the newly synthesised subunits (Krummeck and Rödel, 1990). The stability of some of the nuclearly encoded subunits is also affected, presumably due to increased exposure to proteases in the partly assembled complex. Subunit VI is absent when subunits I and II are absent. Consistent with this finding is that subunits I and II are found to interact tightly with subunits VI and VII (Woodrow and Schatz, 1979). It is suggested that the Sco1p protein is involved in the assembly of subunits I and II. The higher stability (protease resistance) of subunit III would suggest that it does not rely on the function of the Sco1p for assembly. It is suggested that a sub-assembly of Sco1p and subunits I and II would protect against proteolysis; this may explain why cells grown in low O_2 tension synthesize subunits I and II but do not assemble them into the complex (Asson et al. 1990; Schulze and Rödel, 1989).

IV. FUTURE DIRECTIONS AND PERSPECTIVES

A. Pleiotropic Effects and Supramolecular Assembly

The assembly of higher order structures or interactions between complexes is an area of interest that remains to be explored in full. Such interactions, including those generating dimeric forms of the individual respiratory complexes, may be anticipated to play a central role in the function of the respiratory electron transport chain as a whole. The characterization of a number of respiratory deficient mutants in *S. cerevisiae* has provided an indication of the nature of these interactions and their importance in the organization of higher order structures involving neighboring complexes. Such inferred interactions between complexes may suggest some starting points for further investigation. In Table 4 are a listed a number of documented examples in yeast, where primary mutations affecting the subunits of one complex have not only a deleterious effect on that complex, but lead to pleiotropic effects on other mitochondrial complexes. Rather than detailing each pleiotropic interaction listed in Table 4 that suggests supramolecular assemblages,

Table 4. Some Pleiotropic Mitochondrial Assembly Defects in *S. cerevisiae*

Complex Containing Primary Lesion	*Primary Lesion*	*Pleiotropic Defect*	*Reference*
ATP synthase	F_0 assembly defect. Subunit 9 C-terminal truncations.	No spectral cytochrome aa_3; complex IV assembly defect	Jean-Francois et al. (1986)
Complex III	Disruption of gene encoding subunit 6. Devoid of complex III activity.	No spectral cytochrome aa_3; complex IV assembly defect	Schmitt et al. (1990)
ATP synthase	F_1 assembly defect. *ATP11* and *ATP12* mutant alleles.	Lowered concentrations of QH_2-cytochrome *c* reductase and cytochrome *c* oxidase	Ackerman and Tzagoloff (1990)
ATP synthase	Subunit 6 non-assembly or C-terminal truncation.	Lowered activity of NADH-cytochrome *c* reductase and cytochrome *c* oxidase	Choo et al. (1985)
Complex III	Non-assembly of iron sulphur protein into bc_1 complex. *BCS1* mutant allele.	Reduced spectral aa_3 and cytochrome *c* oxidase activity	Nobrega et al. (1992)
ATP synthase	Amino acid substitutions in subunit 8. Complex apparently assembled.	Reduced NADH-cytochrome *c* reductase and cytochrome *c* oxidase	Marzuki et al. (1989)
Complex IV	Probably frameshift mutation in subunit VI. Aberrantly processed precursor.	Lowered levels of complex III activity and ATPase	Koerner et al. (1985)
Complex IV	Mutant *OXA1*. Incorrect assembly of complex IV.	Incorrect assembly of ATP synthase.	Altamura et al. (1996)

we focus our attention on a small number of selected examples to make some general points.

A possible organizational role of the F_0 sector of ATP synthase in the assembly of other mitochondrial complexes is indicated by results obtained by Jean-Francois and colleagues (1986). Some strains of yeast containing lesions in the mitochondrial gene (*oli1*) encoding subunit 9 of mitochondrial ATP synthase not only fail to assemble subunit 9 but also lack spectrally detectable cytochrome aa_3 characteristic of cytochrome *c* oxidase, indicating the perturbed assembly of the latter complex. The perturbation occurs at the post-translational level, as all the constituent cytochrome c oxidase subunits were synthesized. The characteristic absorption spectrum of complex III was also diminished, indicating a possible further pleiotropic extension of the primary lesion. Mutants containing subunit 9 variants that did assemble ATP synthase showed a mild reduction in spectral cytochrome aa_3 and little modulation of enzyme activity of complexes III and IV, suggesting that a correct overall structure of ATP synthase is probably required for organization and stability of surrounding enzyme complexes. The multimeric nature of subunit 9 in the ATP synthase complex, probably forming a barrel structure in the lipid bilayer (Cox et al., 1992), may have an important role in membrane organization.

Insertional inactivation of the gene encoding subunit 6 of complex III (*QCR6*) results in a strain that is not only devoid of complex III activity but contains less than 2% of the wild-type cytochrome *c* oxidase activity (Schmitt et al., 1990). This contrasts with the effect of deletion of *QCR6*, where 50% of the wildtype complex III activity and all of the cytochrome c oxidase activity remain (Schoppinck et al., 1988). In order to explain these findings, Schmitt and colleagues (1990) suggested that complexes III and IV are assembled through a common nucleation intermediate. It is envisaged that the aberrant protein encoded by the disrupted *QCR6* gene is inserted into this nucleation complex thus interfering with the correct assembly of the subunits of these two respiratory complexes.

Gene products that are not apparently components of the assembled complexes can also have pleiotropic effects on assembly, although in some cases this may be an indirect effect as a result of a primary effect on a structural component. Thus, mutations in the nuclear genes *ATP11* and *ATP12* (Ackerman and Tzagoloff, 1990a) led to the non-assembly of F_1-α and F_1-β subunits and to reduced activities of cytochrome *c* oxidase and complex III. This result did not appear to be due to reduced mitochondrial translation. Although a direct effect of the products of the *ATP11* and *ATP12* genes on these additional complexes could not be ruled out, it would appear that the structure of the F_1 sector of ATP synthase, extrinsic to the membrane, has an organizational effect upon assembly. This conclusion would be consistent with the observation that mutations in the structural genes for F_1-α and F_1-β cause similar pleiotropic effects (Tzagoloff and Diekmann, 1975; Takeda et al., 1985, 1986).

Mutations in the *RCA1* (*YTA12*) and *AFG3* (*YTA10*) genes have been shown to inhibit F_1 assembly in yeast (Paul and Tzagoloff, 1995). Defective assembly of both cytochrome *c* oxidase and the bc_1 complex is also seen in mutant strains. The products of these two genes have been proposed to be ATP-dependent proteases, based on the presence of a sequence motif characteristic of Zn-binding proteases (Pajic et al., 1994: Tauer et al.. 1994).

The product of the gene *BCS1* is required for the expression of a functional Rieske Fe-S subunit of complex III. In a strain lacking functional protein encoded by the *BCS1* gene, the steady state levels of other subunits of complex III are not affected and they are probably assembled. However, a pleiotropic effect on cytochrome *c* oxidase, resulting in a 30–40% reduction in activity, was observed (Nobrega et al., 1992). It has been suggested that these effects are characteristic of pet mutants in general. However, mutation of another nuclear gene, *CBP3*, such that its product was rendered nonfunctional, resulted in a strain having reduced levels of a number of complex III subunits including the Rieske Fe-S protein and subunits IV and V, but wild-type levels of cytochrome c oxidase and ATP synthase activities were found (Wu and Tzagoloff, 1989).

B. Factors Controlling Assembly

In addition to defining assembly in terms of sequential events, ultimately one would wish to define those factors involved in controlling the order and timing of events leading to the production of sub-assemblies and fully assembled complexes. Conceivably, this may be achieved at the level of transcription, translation, or through post-translational events, thus regulating the appearance and supply of assembly-competent structural subunits or assembly factors. The importance of prosthetic groups to the assembly and function of complex III and IV that are part of the final complex has already been described. However, there are reports that proteins of some complexes that undergo chemical modification in a transient manner such that the modifications are not detectable in the assembled complex.

Three imported mitochondrial proteins in *N. crassa*, subunit VI of cytochrome *c* oxidase, and two subunits of ATP synthase have been shown to contain a covalently attached derivative of pantothenic acid, 4′-phosphopantotheine (Plesofsky-Vig and Brambl, 1984; Lakin-Thomas and Brody, 1985; Brambl and Plesofsky-Vig, 1986). Deprivation of pantothenic acid from the developing spores of a pantothenate auxotroph of *N. crassa* leads to the partial assembly of the cytochrome *c* oxidase and ATP synthase complexes. Subunits of these complexes were specifically labelled when spores were grown in the presence of radiolabelled pantothenate. However, 4′-phosphopantetheine was absent in fully assembled complexes and was presumably removed during their assembly. These results suggest that covalent modification of specific subunits is required for the complete assembly of their respective enzymes. Further, removal of the pantothenate derivative from the modified subunits may occur before assembly of the enzyme complexes is completed, or soon after completion, since most of the protein-associated radiolabelled pantothenate in mitochondria seemed bound to free subunits rather than to the enzyme complexes. A similar explanation as was put forward to account for the observed transient modification of the matrix enzymes acetyl-CoA acetyltransferase and 3-oxoacyl-CoA thiolase with coenzyme A (Huth et al., 1991). Fully active enzymes were not found to be modified. These modifications described here may be distinguished from those factors required for the assembly of cytochrome *c* oxidase (determined by the *COX11* and *COX12* genes) that are responsible for the synthesis of a porphyrin group that ultimately resides in the final assembly.

Other post-translational modifications may play a role in assembly. The glycosylation state of proteins may be an important influence on the course of folding events and their interaction with chaperones. The chaperone-dependent refolding of chemically denatured glycosylated forms of invertase from *S. cerevisiae* is favored over nonglycosylated forms (Kern et al., 1992). Chaperonin 60 is unable to arrest the refolding of glycosylated invertase in the absence of ATP, in contrast to the nonglycosylated form, thus indicating that glycosylation modulates the interaction of chaperones with proteins.

C. Disassembly

In addition to the assembly of new mitochondrial complexes coordinated with import of precursors and protein synthesis, we must also take into consideration, as suggested by Ellis (1990), those cases where the normal function of oligomeric protein complexes involves rearrangements of protein-protein interactions. Regions previously involved in subunit interactions are transiently exposed to the environment and would need stabilization or protection by chaperones. Any complex controlled by reversible assembly of its structure would fall into this category, for example, the replisome catalysing mtDNA replication. Chaperone function (DnaJ, DnaK, and GrpE) has been implicated in specific disassembly reactions of proteins involved in the replication of DNA in prokaryotes (Zylicz, 1993). It is presumed that proteins with similar functions will be involved with the replication of mitochondrial DNA.

By extension of this notion, such proteins may also be involved in the managed disassembly of other mitochondrial enzyme complexes. This may occur at a basal rate, generating a 'routine' turnover of mitochondrial components, or may be subject to some regulation, for example in physiological responses or particular cell differentiation events. Such modulation of subunit levels would take place in the context of the general flow of imported proteins into mitochondria. Proteolysis is expected to play an important role in the management of assembled, missassembled, or disassembled subunits. Several reports have presented evidence that the turnover of mitochondrial translation products is mediated by novel proteases that are dependent on both ATP and metal ions for activity and which show association with the inner-membrane of the mitochondrion (Pajic et al., 1994; Yashuhara et al., 1994; Nakai et al, 1994, 1995; Pearce and Sherman, 1995). Misfolded proteins in the matrix of the mitochondrion are degraded by a matrix localized ATP-dependent protease, PIM1, in cooperation with the mitochondrial chaperone hsp70 (Wagner et al., 1994).

Understanding the cellular maintenance of appropriate subunit levels for each individual mitochondrial enzyme complex represents one of the major challenges for the future.

ACKNOWLEDGMENTS

Work of the authors cited in this review was supported by grants from the Australian Research Council to Phillip Nagley and Rodney J. Devenish.

REFERENCES

Abrahams, J. P., Leslie, A. G. W., Lutter, R., & Walker, J. E. (1994). Structure at 2.8 Å resolution of F_1-ATPase from bovine heart mitochondria. Nature 370, 621–628.

Ackerman, S. H., & Tzagoloff, A. (1990a). Identification of two nuclear genes (*ATP11, ATP12*) required for assembly of the yeast F_1-ATPase. Proc. Natl. Acad. Sci. USA 87, 4986–4990.

Ackerman, S. H., & Tzagoloff, A. (1990b). *ATP10*, a yeast nuclear gene required for the assembly of the mitochondrial F_1-F_0 complex. J. Biol. Chem. 265, 9952–9959.

Ackrell, B. A. C., Johnson, M. K., Gunsalus, R. P., & Cecchini, G. (1992). In: *Chemistry and Biochemistry of Flavoenzymes*, Vol. 3 (Mueller, F., Ed.). CRC Press, Boca Raton, FL, pp. 229–297.

Aggeler, R., & Capaldi, R. A. (1990). Yeast cytochrome *c* oxidase subunit VII is essential for assembly of an active enzyme. Cloning, sequencing and characterisation of the nuclear-encoded gene. J. Biol. Chem. 265, 16389–16393.

Altamura, N., Capitano, N., Bonnefoy, N., Papa, S., & Dujardin, G. (1996). The *Saccharomyces cerevisiae OXA1* gene is required for the correct assembly of cytochrome *c* oxidase and oligomycin-sensitive ATP synthase. FEBS Lett. 382, 111–115.

Alves, P. C., & Videira, A. (1994). Disruption of the gene coding for the 21.3 kDa subunit of the peripheral arm of complex I from *Neurospora crassa*. J. Biol. Chem. 269, 7777–7784.

Asson-Batres, M. A., & Hare, J. F. (1991). Effect of oxygen on the synthesis and assembly of mitochondrial encoded subunits of cytochrome oxidase and cytochrome bc_1 in mouse embryo fibroblasts. J. Biol. Chem. 266, 9932–9938.

Attardi, G., & Schatz, G. (1988). Biogenesis of mitochondria. Ann. Rev. Cell. Biol. 4, 289–333.

Barrera, C. R., Narihara, G., Hamilton, L., Munk, P., Eley, M. H., Linn, T. C., & Reed, L. J. (1972). α-keto acid dehydrogenase complexes. Studies on the subunit structure of the pyruvate dehydrogenase complexes from bovine kidney and heart. Arch. Biochem. Biophys. 148, 343–358.

Bateson, M., Devenish, R. J., Nagley, P., & Prescott, M. (1996). Entrapment by immobilised metal ion affinity chromatography of assembled yeast mitochondrial ATP synthase containing individual subunits tagged with hexahistidine. Anal. Biochem. 238, 14–18.

Bonnefoy, N., Chalvet, F., Hamel, P., Slonimski, P. P., & Dujardin, G. (1994). *OXA1*, a *Saccharomyces cerevisiae* nuclear gene whose sequence is conserved from prokaryotes to eukaryotes controls cytochrome oxidase biogenesis. J. Mol. Biol. 239, 201–212.

Bowman, S., Ackerman, S. H., Griffiths, D. E., & Tzagoloff, A. (1991). Characterisation of *ATP12*, a yeast nuclear gene required for the assembly of the mitochondrial F_1-ATPase. J. Biol. Chem. 266, 7517–7523.

Brambl, R., & Plesofsky-Vig, N. (1986). Pantothenate is required in Neurospora crassa for assembly of subunit peptides of cytochrome *c* oxidase and ATPase/ATP synthase. Proc. Natl. Acad. Sci. USA 83, 3644–3648.

Brandt, V., Yu, L., Yu, C-A., & Trumpower, B. L. (1993). The mitochondrial targeting presequence of the Rieske iron sulphur protein is processed in a single step after insertion into the cytochrome bc_1 complex in mammals and retained as a subunit of the complex. J. Biol. Chem. 268, 8387–8390.

Brandt, U., Uribe, S., Schagger, H., & Trumpower, B. L. (1994). Isolation and characterisation of *QCR10*, the nuclear gene encoding the 8.5 kDa subunit 10 of the *Saccharomyces cerevisiae* cytochrome bc_1 complex. J. Biol. Chem. 269, 12947–12953.

Brunori, M. G., Antonini, F., Malatesta, F., Sart, P., & Wilson, M. T. (1987). Cytochrome *c* oxidase. Subunit structure and proton pumping. Eur. J. Biochem. 169, 1–8.

Buchner, J., Schmidt, M., Fuchs, M., Jaenicke, R., Rudolph, R., Schmid, F. X., & Kiefhaber, T. (1991). *GroE* facilitates refolding of citrate synthase by suppressing aggregation. Biochemistry 30, 1586–1591.

Burns, D. J., & Lewin, A. S. (1986). The rate of import and assembly of F_1-ATPase in *Saccharomyces cerevisiae*. J. Biol. Chem. 268, 12066–12073.

Cabral, F., & Schatz, G. (1978). Identification of cytochrome *c* oxidase subunits in nuclear yeast mutants lacking the functional enzyme. J. Biol. Chem. 253, 4396–4401.

Calder, K. M., & McEwen, J. E. (1991). Deletion of the *COX7* gene in *Saccharomyces cerevisiae* reveals a role for the cytochrome *c* oxidase subunit VII in assembly of remaining subunits. Mol. Microbiol. 51, 1769–1777.

Capaldi, R. A., Gonzalez-Halphen, D., & Takamiya, S. (1986). Sequence homologies and structural similarities between the polypeptides of yeast and beef heart cytochrome *c* oxidase. FEBS Lett. 207, 11–17.

Capaldi, R. A. (1990a). Structure and function of cytochrome *c* oxidase. Ann. Rev. Biochem. 59, 569–596.

Capaldi, R. A. (1990b). Structure and assembly of cytochrome *c* oxidase. Arch. Biochem. Biophys. 280, 252–262.

Choo, W. M., Hadikusumo, R. G., & Marzuki, S. (1985). Mitochondrial adenosine triphosphatase in mit^- mutants of Saccharomyces cerevisiae with defective subunit 6 of the enzyme complex. Biochim. Biophys. Acta 806, 290–304.

Collinson, I. R., van Raaij, M. J., Runswick, M. J., Fearnley, I. M., Skehel, J. M., Orriss, G. L., Miroux, B., & Walker, J. E. (1994). ATP synthase from bovine heart mitochondria. *In vitro* assembly of a stalk complex in the presence of F_1-ATPase and in its absence. J. Mol. Biol. 242, 408–421.

Colson, A. M. (1993). Random mutant generation and its utility in uncovering structural and functional features of cytochrome *b* in Saccharomyces cerevisiae. J. Bioenerg. Biomembr. 25, 211–220.

Constanzo, M. C., & Fox, T. D. (1990). Control of mitochondrial gene expression in *Saccharomyces cerevisiae*. Ann. Rev. Genet. 24, 91–113.

Cooper, C. E., Nicholls, P., & Freedman, J. A. (1991). Cytochrome *c* oxidase: Structure, function and membrane topology of the polypeptide subunits. Biochem. Cell Biol. 69, 586–607.

Cox, G. B., Devenish, R. J., Gibson, F., Howitt, S. M., & Nagley, P. (1992). The structure and assembly of ATP synthase. In: *Molecular Mechanisms in Bioenergetics* (Ernster, L., Ed.). Elsevier Science Publishers B.V. pp. 283–316.

Crivellone, M. D., Wu, M., & Tzagoloff, A. (1988). Assembly of the mitochondrial membrane system. Analysis of structural mutants of the yeast coenzyme QH_2-cytochrome *c* reductase complex. J. Biol. Chem. 263, 14323–14333.

Crivellone, M. D. (1994). Characterization of *CBP4*, a new gene essential for the expression of ubiquinol-cytochrome *c* reductase in *Saccharomyces cerevisiae*. J. Biol. Chem. 269, 21284–21292.

Cumsky, M. G., Ko, C., Trueblood, C. E., & Poyton, R. O. (1985). Two non-identical forms of subunit V are functional in yeast cytochrome *c* oxidase. Proc. Natl. Acad. Sci. USA 82, 2235–2239.

Cumsky, M. G., McEwen, J. E., Ko, C., & Poyton, R. O. (1983). Nuclear genes for mitochondrial proteins. Identification and isolation of a structural gene for subunit V of yeast cytochrome *c* oxidase. J. Biol. Chem. 258, 13418–13421.

De Marcucci, O. L., & Lindsay, J. G. (1985). Component X. An immunologically distinct polypeptide associated with mammalian pyruvate dehydrogenase multi-enzyme complex. Eur. J. Biochem. 149, 641–648.

De Marcucci, O. L., DeBuysere, M. S., & Olson, M. S. (1995). Dissociation and reassembly of the dihydrolipoyl transacetylase component of the bovine heart pyruvate dehydrogenase complex. Arch. Biochem. Biophys. 323, 169–176.

De Vries, S., & Grivell, L. A. (1988). Purification and characterisation of a rotenone-insensitive NADH: Q6 oxidoreductase from mitochondria of *Saccharomyces cerevisiae*. Eur. J. Biochem. 176, 377–384.

Devenish, R. J., Galanis, M., Papakonstantinou, T., Law, R. H. P., Grasso, D. G., Helfenbaum, L., & Nagley, P. (1992). Molecular biology and assembly of yeast mitochondrial ATP synthase. In: *Adenine Nucleotides in Cellular Energy Transfer and Signal Transduction* (Papa, S., Azzi, A., & Tager, J. M., Eds.). Birkhauser, Basel, pp. 1–12.

Dowhan, W., Bibus, C. R., & Schatz, G. (1985). The cytoplasmically-made subunit IV is necessary for assembly of cytochrome *c* oxidase in yeast. EMBO J. 4, 179–184.

Duarte, M., Sousa, R., & Videira, A. (1995). Inactivation of genes encoding subunits of the peripheral and membrane arms of Neurospora mitochondrial complex I and effects on enzyme assembly. Genetics 139, 1211–1221.

Ellis, R. J. (1990). The molecular chaperone concept. Seminars in Cell Biology 1, 1–9.

Fecke, W., Sled, V. D., Ohnishi, T., & Weiss, H. (1994). Disruption of the gene encoding the NADH-binding subunit of NADH: Ubiquinone oxidoreductase in *Neurospora crassa*. Formation of a partially assembled enzyme without FMN and the iron-sulphur cluster N-3. Eur. J. Biochem. 220, 551–558.

Finnegan, P. M., Payne, M. J., Keramidaris, E., & Lukins, H. B. (1991). Characterization of a yeast nuclear gene, *AEP2*, required for accumulation of mitochondrial mRNA encoding subunit 9 of the ATP synthase. Curr. Genet. 20, 53–61.

Friedrich, T., Hofhaus, G., Ise, W., Nehls, U., Schmitz, B., & Weiss, H. (1989). A small isoform of NADH: Ubiquinone oxidoreductase (complex I) without mitochondrially encoded subunits is made in chloramphenicol-treated *Neurospora crassa*. Eur. J. Biochem. 180, 173–180.

Galanis, M., Devenish, R. J., & Nagley, P. N. (1991). Duplication of leader sequence for protein targeting to mitochondria leads to increased import efficiency. FEBS Letts. 282, 425–430.

Girdlestone, J., Bisson, R., & Capaldi, R. A. (1981). Interaction of succinate-ubuiquinone reductase (complex II) with (arylaido) phospho-lipids. Biochemistry 20, 152–156.

Glerum, D. M.. Koerner, T. J., & Tzagoloff, A. (1995). Cloning and characterisation of *COX14*, whose product is required for assembly of yeast cytochrome oxidase. J. Biol. Chem. 270, 15585–90.

Gonzalez-Halphen, D., Lindorfer, M. A., & Capaldi, R. A. (1988). Subunit arrangement in beef heart complex III. Biochemistry 27, 7021–7031.

Graham, L. A., Brandt, V., Sargeant, J., & Trumpower, B. L. (1993). Mutational analysis of assembly and function of the iron-sulphur protein of the cytochrome bc_1 complex in *Saccharomyces cerevisiae*. J. Bioenerg. Biomembr. 25, 245–257.

Graham, L. A., & Trumpower, B. L. (1991). Mutational analysis of the mitochondrial Rieske iron-sulfur protein of *Saccharomyces cerevisiae*. J. Biol. Chem. 266 (33), 22485–22492.

Grasso, D. G., Nero, D., Law, R. H. P., Devenish, R. J., & Nagley, P. (1991). The C-terminal positively charged region of subunit 8 of yeast mitochondrial ATP synthase is required for efficient assembly of this subunit with the membrane F_0 sector. Eur. J. Biochem. 199, 203–209.

Grivell, L. A. (1989). Nucleo-mitochondrial interactions in yeast mitochondrial biogenesis. Eur. J. Biochem. 182, 477–493.

Hadikusumo, R. G., Meltzer, S., Choo, W. M., Jean-Francois, M. J. B., Linnane, A. W., & Marzuki, S. (1988). The definition of mitochondrial H^+ ATPase assembly defects in mit^- mutants of *Saccharomyces cerevisiae* with a monoclonal antibody to the enzyme complex as an assembly probe. Biochim. Biophys. Acta 933, 212–222.

Hadikusumo, R. G., Hertzog, P. J., & Marzuki, S. (1984). Monoclonal antibodies against subunits of yeast mitochondrial H^+-ATPase. Biochim. Biophys. Acta. 765, 258–267.

Hall, R. E., & Hare, J. F. (1990). Respiratory chain-linked NADH dehydrogenase. Mechanisms of assembly. J. Biol. Chem. 265, 16984–16490.

Hartman, D. J., Hoogenraad, N. J., Condron, R., & Hoj, P. B. (1992). Identification of a mammalian 10-kDa heat shock protein, a mitochondrial chaperonin 10 homologue essential for assisted folding of trimeric ornithine transcarbamoylase *in vitro*. Proc. Natl. Acad. Sci. USA 89, 3394–3398.

Hederstedt, L., & Ohnishi, T. (1992). Progress in succinate: Quinone oxidoreductase research. In: *Molecular Mechanisms in Bioenergetics* (Ernster, L., Ed.). Elsevier Science Publishers, Amsterdam. pp. 163–198.

Hofhaus, G., Weiss, H., & Leonard. K. (1991). Electron microscopic analysis of the peripheral and membrane parts of mitochondrial NADH dehydrogenase (complex I). J. Mol. Biol. 221, 1027–1043.

Hofhaus, G., & Attardi, G. (1993). Lack of assembly of mitochondrial DNA-encoded subunits of respiratory NADH dehydrogenase and loss of enzyme activity in a human cell mutant lacking the mitochondrial ND4 gene product. EMBO J. 12, 3043–3048.

Hofhaus, G., & Attardi, G. (1995). Efficient selection and characterization of mutants of a human cell line which are defective in mitochondrial DNA-encoded subunits of respiratory NADH dehydrogenase. Mol. Cell Biol. 15, 964–974.

Huth, W., Worm-Breitgoff, C., Muller, V., & Wunderlich, I. (1991). Evidence for an *in vivo* modification of mitochondrial proteins by coenzyme A. Biochim. Biophys. Acta 1077, 1–10.

Japa, S., & Beattie, D. S. (1989). The iron-sulphur protein is necessary for the complete assembly of the low molecular weight subunits into the cytochrome b-c_1 complex of yeast mitochondria. Arch. Biochem. Biophys. 268, 716–720.

Jean-Francois, M. J. B., Lukins, H. B., & Marzuki, S. (1986). Post-transcriptional defects in the synthesis of the mitochondrial H^+-ATPase subunit 6 in yeast mutants with lesions in the subunit 9 structural gene. Biochim. Biophys. Acta 868, 178–182.

Jilka, J. M., Rahmatullah, M., Kazemi, M., & Roche, T. E. (1986). Properties of a newly characterized protein of the bovine kidney pyruvate dehydrogenase complex. J. Biol. Chem. 261, 1858–1867.

Kadenbach, B., & Merle, P. (1981). On the function of multiple subunits of cytochrome c oxidase from higher eukaryotes. FEBS Lett. 135, 1–11.

Kadenbach, B., Stroh, A., Ungibauer, M., Kuhn-Nentwig. L., Buge, U., & Jarausch, J. (1986). Isozymes of cytochrome-*c* oxidase: Characterisation and isolation from different tissues.

Kern, G., Schmidt, M., Buchner, J., & Jaenicke, R. (1992). Glycosylation inhibits the interaction of invertase with the chaperone *GroEL*. FEBS Lett. 305, 203–205.

Koerner, T. J., Homison, G., & Tzagoloff, A. (1985). Nuclear mutants of *Saccharomyces cerevisiae* with altered subunits 4,5 and 6 of cytochrome oxidase. J. Biol. Chem. 260, 5871–5874.

Krummeck, G., & Rödel, G. (1990). Yeast *SCO1* protein is required for a posttranslational step in the accumulation of mitochondrial cytochrome *c* oxidase subunits I and II. Curr. Genet. 18, 13–15.

La Marche, A. E. P., Abate, M. I., Chan, S. H. P., & Trumpower, B. L. (1992). Isolation and characterisation of *COX12*, the nuclear gene for a previously unrecognized subunit of *Saccharomyces cerevisiae* cytochrome c oxidase. J. Biol. Chem. 267, 22743–22480.

Lakin-Thomas, P. L., & Brody, S. (1985). A pantothenate derivative is covalently bound to mitochondrial proteins in *Neurospora crassa*. Eur. J. Biochem. 146, 141–147.

Law, R. H. P., Devenish, R. J., & Nagley, P. (1990). Assembly of imported subunit 8 into the ATP synthase complex of isolated yeast mitochondria. Eur. J. Biochem. 188, 421–429.

Lawson, J. E., Behal, R. H., & Reed, L. J. (1991). Disruption and mutagenesis of the *Saccharomyces cerevisiae PDX1* gene encoding protein X component of the pyruvate dehydrogenase complex. Biochemistry 30, 2834–2839.

Leonard, K., Wingfield, P., Arad, T., & Weiss, H. (1981). Three-dimensional structure of ubiquinol: Cytochrome *c* reductase from *Neurospora* mitochondria determined by electron microscopy of membrane crystals. J. Mol. Biol. 149, 259–274.

Lightowlers, R., Chrzanowska-Lightowlers, Z., Marvsich, M., & Capaldi, R. A. (1991). Subunit function in eukaryote cytochrome *c* oxidase. A mutation in the nuclear-coded subunit IV allows assembly but alters the function and stability of yeast cytochrome *c* oxidase. J. Biol. Chem. 266, 7688–7693.

Linnane, A. W., Lukins, H. B., Nagley, P., Marzuki, S., Hadikusumo, R. G., Jean-Francois, M. J. B., John, U. P., Ooi, B. G., Watkins, L., Willson, T. A., Wright, J., & Metzer, S. (1985). Assembly of the yeast mitochondrial H^+-ATPase: Correlation studies involving gene sequencing and immunochemical probes of assembly. In: *Achievements and Perspectives of Mitochondrial Research* (Quagliariello, E., Slater, E. C., Palmieri, F., Saccone, C., & Kroon, A. M., Eds.). Elsevier, Amsterdam, pp. 211–222.

Ljungdahl, P. O., Beckmann, J. D., & Trumpower, B. L. (1989). Mutational analysis of the mitochondrial Rieske iron-sulphur protein of *Saccharomyces cerevisiae*. Biochemical characterisation of temperature-sensitive *RIP1* mutations. J. Biol. Chem. 264, 3723–3731.

Lomax, M., & Grossman, L. I. (1989). Tissue-specific genes for respiratory proteins. Trends Biochem. Sci. 14, 501–503.

Lombardo, A., & Scheffler, I. E., (1989). Isolation and characterisation of a *Saccharomyces cerevisiae* mutant with a disrupted gene for the IP subunit of succinate dehydrogenase. J. Biol. Chem. 264, 18874–18877.

Lombardo, A., Carine, K., & Scheffler, I. E. (1990). Cloning and characterisation of the iron-sulfur subunit gene of structure succinate dehydrogenase from *Saccharomyces cerevisiae*. J. Biol. Chem. 265, 10419–10423.

McEwen, J. E., Hong, K. H., Park, S., & Preciado, G. T. (1993). Sequence and chromosomal localization of two *pet* genes required for cytochrome *c* oxidase assembly in *Saccharomyces cerevisiae*. Curr. Genet. 23, 9–14.

McEwen, J. E., Ko, C., Kleockner-Gruissem, B., & Poyton. R. O. (1986). Nuclear functions required for cytochrome *c* oxidase biogenesis in *Saccharomyces cerevisiae*. Characterization of mutants in 34 complementation groups. J. Biol. Chem. 261, 11872–11879.

Maarse, A. C., De Haaen, M., Schoppinck, P. J., Berden, J. A., & Grivell, L. A. (1988). Inactivation of the gene encoding the 11-kDa subunit VIII of the ubiquinol-cytochrome *c* oxidoreductase in *Saccharomyces cerevisiae*. Eur. J. Biochem. 172, 179–184.

Manon, S. T., & Guerin, M. G. (1989). Modifications of the relative proteolipid composition in the ATP synthase of a respiratory competent mutant of *Saccharomyces cerevisiae*. Biochim. Biophys. Acta 985, 127–132.

Marzuki, S., Watkins, L. C., & Choo, W. M. (1989). Mitochondrial H^+-ATPase in mutants of *Saccharomyces cerevisiae* with defective subunit 8 of the enzyme complex. Biochim. Biophys. Acta 975, 222–230.

Michon, T., Galante, M., & Velours, J. (1988). NH_2-terminal sequence of the isolated yeast ATP synthase subunit 6 reveals post-translational cleavage. Eur. J. Biochem. 172, 621–625.

Nagley, P. (1988). Eukaryote membrane genetics: The F_0 sector of mitochondrial ATP synthase. Trends Genet. 4, 46–52.

Nagley, P. (1991). Coordination of gene expression in the formation of mammalian mitochondria. Trends Genet. 7, 1–4.

Nagley, P., Farrell, L. B., Gearing, D. P., Nero, D. Meltzer, S., & Devenish, R. J. (1988). Assembly of functional proton-translocating ATPase complex in yeast mitochondria with cytoplasmically synthesised subunit 8, a polypeptide normally encoded within the organelle. Proc. Natl Acad. Sci. USA 85, 2091–2095.

Nagley, P., & Devenish, R. J. (1989). Leading organellar proteins along new pathways: The relocation of mitochondrial and chloroplast genes to the nucleus. Trends Biochem. Sci. 14, 31–35.

Nakai, T., Mera, Y., Yashuhara, T., & Ohashi, A. (1994). Divalent metal ion-dependent mitochondrial degradation of unassembled subunits 2 and 3 of cytochrome *c* oxidase. J. Biochem. 116, 752–758.

Nakai, T., Yasuhara, T., Fujiki, Y., & Ohashi, A. (1995). Multiple genes, including a member of the AAA family, are essential for degradation of the unassembled subunit 2 of cytochrome *c* oxidase in yeast mitochondria. Mol. Cell Biol. 15, 4441–4452.

Nalecz, M. J., Bolli, R., & Azzi, A. (1985). Molecular conversion between monomeric and chimeric states of the mitochondrial cytochrome bc_1 complex: Isolation of active monomers. Arch. Biochem. Biophys. 236, 619–628.

Nehls, U., Friedrich, T., Schmiede, A., Ohnishi, T., & Weiss, H. (1992). Characteristics of assembly intermediates of NADH: Ubiqinone oxidoreductase (complex I) accumulated in Neurospora mitochondria by gene disruption. J. Mol. Biol. 227, 1032–1042.

Nero, D., Ekkel, S. M., Wang, L-F., Grasso, D. G., & Nagley, P. (1990). Site directed mutagenesis of subunit 8 of yeast mitochondrial ATP synthase. Functional and import properties of a series of C-terminally truncated forms. FEBS Letts. 270, 62–66.

Nijtmans, L. G. J., Klement, P., Houstek, J., & van den Boogert, C. (1995). Assembly of mitochondrial ATP synthase in cultured human cells: Implications for mitochondrial diseases. Biochim. Biophys. Acta 1272, 190–198.

Nobrega, M. P., Nobrega, F. G., & Tzagoloff, A. (1990). *COX10* codes for a protein homologous to the *ORF1* product of *Paraccocus denitrificans* and is required for the synthesis of yeast cytochrome oxidase. J. Biol. Chem. 265, 14220–14226.

Nobrega, F. G., Nobrega, M. P., & Tzagoloff, A. (1992). *BCS1*, a novel gene required for the expression of functional Rieske iron-sulfur protein in *Saccharomyces cerevisiae*. EMBO J. 11, 3821–3829.

Norais, N., Prome, D., & Velours, J. (1991). ATP synthase of yeast mitochondria. Characterization of subunit d and sequence analysis of the structural gene *ATP7*. J. Biol. Chem. 266, 16541–16549.

Ohnishi, T. (1987). Structure of the succinate-ubiquinone oxidoreductase (complex II). Curr. Top. Bioenergetics 15, 37–65.

Orian, J. M., Hadikusumo, R. G., Marzuki, S., & Linnane, A. W. (1984). Biogenesis of mitochondria: Defective yeast H^+-ATPase assembled in the absence of mitochondrial protein synthesis is membrane associated. J. Bioenerg. Biomembranes 16, 561–581.

Oudshoorn, P., Van steeg, H., Swinkels, B. W., Schoppinck, P. J., & Grivell, L. A. (1987). Subunit II of yeast QH_2: Cytochrome *c* oxidoreductase nucleotide sequence of the gene and features of the protein. Eur. J. Biochem. 163, 97–103.

Pajic, A., Tauer, R., Feldman, H., Neupert, W., & Langer, T. (1994). Yta10p is required for the ATP-dependent degradation of polypeptides in the inner membrane of mitochondria. FEBS Lett. 353, 201–206.

Papakonstantinou, T., Galanis, M., Nagley, P., & Devenish, R. J. (1993). Each of the three positively-charged amino acids in the C-terminal region of yeast mitochondrial ATP synthase subunit 8 is required for assembly. Biochim. Biophys. Acta 1144, 22–32.

Patterson, T. E., & Poyton, R. O. (1986). *COX8*, the structural gene for yeast cytochrome *c* oxidase subunit VIII. J. Biol. Chem. 261, 17192–17197.

Paul, M. F., Velours, J., Arselin de Chateaubodeau, G., Aigle, M., & Guerin, B. (1989). The role of subunit 4, a nuclear-encoded protein of the F_0 sector of yeast mitochondrial ATP synthase, in the assembly of the whole complex. Eur. J. Biochem. 185, 163–171.

Paul, M. F., Guerin, B., & Velours, J. (1992). The C-terminal region of subunit 4 (subunit b) is essential for assembly of the F_0 portion of yeast mitochondrial ATP synthase. Eur. J. Biochem. 205, 163–172.

Paul, M-F., & Tzagoloff, A. (1995). Mutations in *RCA1* and *AFG3* inhibit F_1-ATPase assembly in *Saccharomyces cerevisiae*. FEBS Lett. 373, 66–70.

Payne, M. J., Schweizer, E., & Lukins, H. B. (1991). Properties of two nuclear *pet* mutants affecting expression of the mitochondrial *oli1* gene of *Saccharomyces cerevisiae*. Curr. Genet. 19, 343–351.

Pearce, D. A., & Sherman, F. (1995). Degradation of cytochrome oxidase subunits in mutants of yeast lacking cytochrome *c* and suppression of the degradation by mutation of *YME1*. J. Biol. Chem. 270, 20879–20882.

Pel, H. J., Tsagaloff, A., & Grivell, L. A. (1992). The identification of 18 nuclear genes required for the expression of the yeast mitochondrial gene encoding cytochrome *c* oxidase subunit I. Curr. Genet. 21, 139–149.

Pelissier, P. D., Camougrand, N. M., Manon, S. T., Velours, G. M., & Guerin, M. G. (1992). Regulation of nuclear genes of the mitochondrial synthesis of subunits 6 and 8 of ATP synthase of *Saccharomyces cerevisiae*. J. Biol. Chem. 267, 2467–2473.

Pelissier, P., Camougrand, N., Velours, G., & Guerin, M. (1995). *NCA3*, a nuclear gene involved in the mitochondrial expression of subunits 6 and 8 of the F_0F_1 ATP synthase of *S. cerevisiae*. Curr. Genet. 27, 409–416.

Pfanner, N., Craig, E. A., & Meijer, M. (1994). The protein import machinery of the mitochondrial inner membrane. Trends Biochem. Sci. 19, 368–372.

Phillips, J. D., Graham, L. A., & Trumpower, B. L. (1993). Subunit 9 of the *Saccharomyces cerevisiae* cytochrome bc_1 complex is required for insertion of EPR-detectable iron-sulfur cluster into the Rieske iron-sulphur protein. J. Biol. Chem. 268, 11727–11736.

Planques, Y., Capitanio. N., Capitanio, G., De Nitto, E., Villani, G., & Papa, S. (1989). Role of supernumary subunits in mitochondrial cytochrome *c* oxidase. FEBS Lett. 258, 285–288.

Plesofsky-Vig, N., & Brambl, R. (1984). Three subunit proteins of membrane enzymes in mitochondria of *Neurospora crassa* contain pantothenate derivative. J. Biol. Chem. 259, 10660–10663.

Power, S. D., Lochrie, M. A., Sevarino, K. A., Paterson, T. E., & Poyton, R. O. (1984). The nuclear-coded subunits of yeast cytochrome *c* oxidase. Fractionation of the holoenzyme into chemically pure polypeptides and the identification of two new subunits using solvent extraction and reversed phase high performance chromatography. J. Biol. Chem. 259, 6564–6570.

Poyton, R. O. (1980). Cooperative interaction between mitochondrial and nuclear genome: Cytochrome *c* oxidase assembly as a model. Curr. Top. Cell. Regulation. 17, 231–295.

Poyton, R. O., Trueblood, C. E., Wright, R. M., & Farrell, L. E. (1988). Expression and function of cytochrome *c* oxidase subunit isologues. Modulators of cellular energy production. Annal. N.Y. Acad. Sci. 550, 289–307.

Pratje, E., Mannhaupt, G., Michaelis, G., & Beyrenther, K. (1983). A nuclear mutation prevents processing of a mitochondrially encoded membrane protein in *Saccharomyces cerevisiae*. EMBO J. 2 1049–1054.

Prescott, M., Bush, N., Nagley, P., & Devenish, R. J. (1994). Properties of yeast cells depleted of the OSCP subunit of mitochondrial ATP synthase by regulated expression of the *ATP 5* gene. Biochem. Mol. Biol. Int. 34, 789–799.

Prescott, M., Higuti, T., Nagley, P., & Devenish, R. J. (1995). The functional expression of a rat cDNA encoding OSCP in the yeast *Saccharomyces cerevisisae*. Biochem. Biophys. Res. Commun. 207, 943–949.

Reed, L. J., & Hackert, M. L. (1990). Structure-function relationships in dihydrolipoamide acyltransferases. J. Biol. Chem. 265, 8971–8974.

Repetto, B., & Tzagoloff, A. (1991). *In vivo* assembly of yeast mitochondrial α-ketoglutavate dehydrogenase complex. Mol. Cell. Biol. 11, 3931–3939.

Robinson, K. M., & Lemire, B. D. (1992). Isolation and nucleotide sequence of the *Saccharomyces cerevisiae* gene for the succinate dehydrogenase flavoprotein subunit. J. Biol. Chem. 267, 10101–10107.

Saghbini, M., Broomfield, P. L. E., & Scheffler, I. E. (1994). Studies on the assembly of complex II in yeast mitochondria using chimeric human/yeast genes for the iron-sulfur protein subunit. Biochemistry 33, 159–165.

Saltzbgaber-Müller, J., & Schatz, G. (1978). Heme is necessary for the accumulation and assembly of cytochrome *c* oxidase subunit in *Saccharomyces cerevisiae*. J. Biol. Chem. 253, 305–310.

Schatz, G. (1968). Impaired binding of mitochondrial adenosine triphosphatase in the cytoplasmic "petite" mutant of *Saccharomyces cerevisiae*. J. Biol. Chem. 243, 2192–2199.

Schmidt, B., Henrig, B., Zimmerman, R., & Neupert, W. (1983). The biosynthetic pathway of mitochondrial ATPase subunit 9 in *Neurospora crassa*. J. Cell Biol. 96, 248–255.

Schmidt, D. M., Saghbini, M., & Scheffler, I. E. (1992a). The C-terminus of the succinate dehydrogenase IP peptide of *Saccharomyces cerevisiae* is significant for assembly of complex II. Biochemistry 31, 8442–8448.

Schmidt, M., Friedrick, T., Wallrath, J., Ohnishi, T., & Weiss, H. (1992b). Accumulation of the preassembled membrane arm of NADH: Ubiquinone oxidoreductase in mitochondria in mitochondria of manganese-limited grown *Neurospora crassa*. FEBS Lett. 313, 1–11.

Schmitt, M. E., Phillips, J. D., & Trumpower, B. L. (1990). Genetic controls of mitochondrial development and function. Genetic analysis of subunits 6 and 9 of yeast ubiquinol-cytochrome *c* oxidoreductase complex. Biochim. Biophys. Acta 1018, 119–123.

Schoppinck, P. J., Hemrika, W., Reynen, J. M., Grivell, L. A., & Berden, J. A. (1988). Yeast ubiquinol: Cytochrome *c* oxidorectase is still active after inactivation of the gene encoding the 17-kDa subunit VI. Eur. J. Biochem. 173, 115–122.

Schoppinck, P. J., Berden, J. A., & Grivell, L. A. (1989a). Inactivation of the gene encoding the 14 kDa subunit VII of yeast ubiquinol cytochrome *c* oxidoreductase and analysis of the resulting mutant. Eur. J. Biochem. 181, 475–483.

Schoppinck, P. J., de Jong, M. A., Berden, J. A., & Grivell, L. A. (1989b). The C-terminal half of the 11 kDa subunit VIII is not necessary for the enzymic activity of yeast cytochrome *c* oxidoreductase. Eur. J. Biochem. 181, 681–687.

Schoppinck, P. J., Hemrika, W., & Berden, J. A. (1989c). The effect of deletion of the gene encoding the 40 kDa subunit II or the 17 kDa subunit VI on the steady state kinetics of yeast ubiquinol cytochrome *c* oxidoreductase. Biochim. Biophys. Acta 974, 192–201.

Schulze, M., & Rodel, G. (1989). Accumulation of cytochrome *c* oxidase subunits I and II in yeast requires a mitochondrial membrane-associated protein, encoded by the nuclear *SCO1* gene. Mol. Gen. Genet. 216, 37–43.

Taanman, J-W., & Capaldi, R. A. (1992). Purification of yeast cytochrome *c* oxidase with a subunit composition resembling the mammalian enzyme. J. Biol. Chem. 267, 22481–22485.

Takeda, M., Vassarotti, A., & Douglas, M. G. (1985). Nuclear gene coding the yeast mitochondrial adenosine triphosphatase complex. Primary sequence analysis of *ATP2* encoding the F_1-ATPase β-subunit precursor. J. Biol. Chem. 260, 15458–15465.

Takeda, M., Chen, W.-J., Salzgaber, J., & Douglas, M. G. (1986). Nuclear genes encoding the yeast mitochondrial ATPase complex. Analysis of *ATP1* coding the F_1-ATPase β-subunit and its assembly. J. Biol. Chem. 261, 15126–15133.

Tanaka, N., Koike, K., Hamada, M., Otsuka, K., Suematsu, T., & Koike, M. (1972). Mammalian α-keto acid dehydrogenase complexes. Resolution and reconstitution of the pig heart 2-oxoglutarate dehydrogenase complex. J. Biol. Chem. 247, 4043–4049.

Tauer, R., Mannhaupt, G., Schnall, R., Pajic, A., Langer, T., & Feldmann, H. (1994). Ytal0p, a member of a novel ATPase family in yeast, is essential for mitochondrial function. FEBS Lett. 353, 197–200.

Trumpower, B. L. (1990). Cytochrome bc_1 complexes of microorganisms. Microbiol. Rev. 54, 102–129.

Tuschen, G., Sackmann, U., Nehls, Haiker, H., Buse, G., & Weiss, H. (1990). Assembly of NADH: Ubiquinone reductase (complex I) in *Neurospora* mitochondria. Independent pathways of nuclear-encoded and mitochondrially encoded subunits. J. Mol. Biol. 213, 845–857.

Tzagaloff, A., & Diekmann, C. L. (1990). *Pet* genes of *Saccharomyces cerevisiae*. Microbiol. Rev. 54, 211–225.

Tzagoloff, A., Capitanio, N., Nobrega, M. P., & Gatti, D. (1990). Cytochrome oxidase assembly in yeast requires the product of *COX11*, a homolog of the *P. denitrificans* protein encoded by ORF3. EMBO J. 9, 2759–2764.

Tzagoloff, A., Crivellone, M. D., Gampel, A., Muroff, I., Nishikimi, M., & Wu, M. (1988). Mutational analysis of the yeast coenzyme QH_2-cytochrome c reductase complex. Phil. Trans. R. Soc. Lond. 319, 107–120.

Tzagoloff, A., Nobrega, M., Gorman, N., & Sinclair, P. (1993). On the functions of the yeast *COX10* and *COX11* gene products. Biochem. Mol. Biol. Int. 31, 593–598.

Uh, M., Jones, D., & Mueller, D. M. (1990). The gene coding for the yeast oligomycin sensitivity-conferring protein. J. Biol. Chem. 265, 19047–19052.

Van Belzen, R., de Jong, A. M. Ph., & Albracht, S. J. P. (1992). On the stoichiometry of the iron-sulphur clusters in mitochondrial NADH: Ubiquinone oxidoreductase. Eur. J. Biochem. 209, 1019–1022.

Velours, J., Durrens, P., Aigle, M., & Guerin, B. (1988). *ATP4*, the structural gene for yeast F_oF_1 ATPase subunit 4. Eur. J. Biochem. 170, 637–642.

Videira, A., & Werner, J. F. (1989). Assembly kinetics and identification of precursor proteins of complex I from *Neurospora crassa*. Eur. J. Biochem. 181, 493–502.

Wagner, I., Arlt, H., van Dyck, L., Langer, T., & Neupert, W. (1994). Molecular chaperones cooperate with PIM1 protease in the degradation of misfolded proteins in mitochondria. EMBO J. 13, 5135–5145.

Walker, J. E. (1992). The NADH: Ubiquinone oxidoreductase (complex I) of respiratory chains. Q. Rev. Biophys. 25, 253–324.

Walker, J. E., & Collinson, I. R. (1994). The role of the stalk in the coupling mechanism of F_1F_0-ATPases. FEBS Lett. 346, 39–43.

Weiss, H. (1987). Structure of mitochondrial ubiquinol-cytochrome-*c* reductase (Complex III). Curr. Top. Bioenergetics 15, 67–90.

Weiss, H., Friedrich, T., Hofhaus, G., & Preis, D. (1991). The respiratory-chain NADH dehydrogenase complex 1) of mitochondria. Eur. J. Biochem. 197, 563–576.

Wielburski, A., & Nelson, B. D. (1983). Evidence for the sequential assembly of cytochrome *c* oxidase in rat liver mitochondria. Biochem. J. 212, 829–834.

Wielburski, A., Kuzeza, S., & Nelson, B. D. (1982). Studies on the assembly of cytochrome *c* oxidase in isolated rat hepatocytes. Biochem. J. 204, 239–245.

Woodrow, G., & Schatz, G. (1990). The role of oxygen in the biosynthesis of cytochrome *c* oxidase of yeast mitochondria. J. Biol. Chem. 254, 6088–6093.

Wright, R. M., Dircks, L. K., & Poyton, R. O. (1986). Characterisation of *COX 9*, the nuclear gene encoding the yeast mitochondrial protein cytochrome *c* oxidase, subunit VIIa. J. Biol. Chem. 261, 17183–17191.

Wu, M., & Tzagoloff, A. (1989). Identification and characterisation of a new gene (*CBP3*) required for the expression of yeast coenzyme QH_2-cytochrome *c* reductase. J. Biol. Chem. 264, 11122–11130.

Yang, X., & Trumpower, B. L. (1986). Purification of a three-subunit ubiquinol cytochrome *c* oxidoreductase complex from *Paracoccus denitrificans*. J. Biol. Chem. 261, 12282–12289.

Yang, M., & Trumpower, B. L. (1994). Deletion of *QCR6*, the gene encoding subunit six of the mitochondrial cytochrome bc_1 complex, blocks maturation of cytochrome c_1 and causes temperature-sensitive petite growth in *Saccharomyces cerevisiae*. J. Biol. Chem. 259, 1–6.

Yashuhara, T., Mera, Y., Nakai, T., & Ohashi, A. (1994). ATP-dependent proteolysis in yeast mitochondria. J. Biochem. 115, 1166–1171.

Yu, L., & Yu, C-A. (1981). The participation of primary amino groups of succinate dehydrogenase in the formation of succinate-Q reductase. Biochim. Biophys. Acta 637, 383–388.

Yu, L., Haley, P. E., Xu, J-X., & Yu, C-A. (1987). Properties of bovine heart mitochondrial b_{650}. J. Biol. Chem. 262, 1137–1143.

Zylicz, M. (1993). The *Escherichia coli* chaperones involved in DNA replication. Phil. Trans. R. Soc. Lond. B. 339, 271–278.

THE DIVISION AND INHERITANCE OF MITOCHONDRIA

Michael P. Yaffe

I. Introduction 341
II. Mitochondrial Form and Behavior in the Cell 342
III. Mitochondrial Movement and the Cytoskeleton 343
IV. Mutants of Mitochondrial Distribution and Morphology 345
V. Models of Mitochondrial Division 348
References 348

I. INTRODUCTION

The division and inheritance of mitochondria are essential elements of cell proliferation. During every cell cycle, mitochondrial mass is duplicated and the mitochondria are distributed among daughter cells. This mitochondrial division and inheritance ensures the continuation of mitochondrial populations and provides daughter cells with adequate mitochondria to perform a myriad of metabolic reactions throughout the cell cycle. Key components of mitochondrial growth, the replication and expression of the mitochondrial genetic system and the import of nuclear-encoded proteins into mitochondria, have been the focus of intense study,

Advances in Molecular and Cell Biology,
Volume 17, pages 341–350.

ISBN: 0-7623-0144-9

yet little is known about the mechanisms responsible for faithful transmission of mitochondria to daughter cells.

II. MITOCHONDRIAL FORM AND BEHAVIOR IN THE CELL

The consideration of mitochondrial division and inheritance benefits from an understanding of the shape and distribution of mitochondria within the eukaryotic cell. Electron microscopic studies combining analysis of serial sections with three-dimensional reconstructions (Stevens, 1977; Blank and Arnold, 1981) and fluorescence microscopy employing mitochondrial-specific dyes (Chen, 1988; Bereiter-Hahn, 1976) have revealed that many types of cells contain an elaborate, reticulated, mitochondrial network. Rather than the spherical or cigar-shaped structures suggested by electron micrographs of single cellular cross-sections, mitochondria often exist in cells as branched, tubular networks with connected, snake-like domains reaching numerous cytoplasmic regions. These mitochondrial tubules may be entirely interconnected so that the cells contain, in essence, a single giant mitochondrion. Such a situation has been documented for unicellular algae (Blank and Arnold, 1981) and yeast cells under certain growth conditions (Stevens, 1977). Extensive mitochondrial networks also have been described in a variety of mammalian cells (Johnson et al., 1980; Schnedl, 1974; Bakeeva et al., 1986).

Mitochondria, or the cellular mitochondrial reticulum, also displays a remarkable plasticity in form and positioning within cells. Microscopic studies of isolated plant cells and of animal cells in culture have revealed that mitochondria undergo frequent fusions and fissions throughout the cell cycle (Bereiter-Hahn, 1990). Morphological changes also can occur in response to altered nutritional status or growth state of cells (Stevens, 1977) during the differentiation of certain tissues (Munn, 1974) and in response to certain pathological conditions (Munn, 1974).

In the yeast *Saccharomyces cerevisiae*, mitochondrial number and morphology alter with changes in carbon source in the media and with the growth phase (Stevens, 1981). During exponential growth on glucose, yeast cells often contain one giant mitochondrion in the form of an extended reticulum, comprising the bulk of the mitochondrial mass. Several smaller mitochondria can also be present. When such cells are changed to a nonfermentable carbon-source, the total mass of mitochondria increases two to fourfold and is distributed as a giant mitochondrial reticulum along with 6–10 additional mitochondria. During stationary phase, the mitochondrial mass redistributes into 20–40 discrete organelles.

A number of intracellular mitochondrial movements that appear to be specifically controlled or programmed have been described in several diverse types of cells. In the fission yeast, *Schizosaccharomyces pombe*, the mitochondria aggregate and move as two masses to either end of the cell during mitosis (Hirano et al., 1988). Changes in mitochondrial positioning have been observed in newly plated fibroblasts in which mitochondria are initially found in the cell periphery and subsequently migrate to the perinuclear region (Bereiter-Hahn et al., 1990). Perinuclear aggregation of mitochondria was increased by stimulation of quiescent BALBc/3T3 cells with growth factors (Chen, 1988). As chick fibroblasts cease migration, their mitochondria redistribute from a tight perinuclear localization to a broader distribution throughout the cytoplasm (Couchman and Rees, 1982). During the development of *Xenopus oocytes*, many of the mitochondria aggregate into a prominent cytoplasmic mass, "the mitochondrial cloud," as the cells become stage I oocytes (Heasman et al., 1984). This mitochondrial mass divides into two portions as oogenesis proceeds: a group of organelles moves toward the plasma membrane and forms a cortical layer in the vegetal hemisphere, while the majority of mitochondria remain near the nucleus until late in oogenesis when they disperse throughout the cytoplasm of the animal hemisphere (Mignotte et al., 1989). Evidence of abnormal mitochondrial movement was detected in a study of epithelial cells from a nematode mutant (Hedgecock and Thomson, 1982). Mitochondria are normally dispersed through the cytoplasm in this cell-type but were found abnormally clustered in the centers of the mutant cells. This mutation did not appear to affect the segregation of mitochondria during cell division.

One of the earliest events in the cell division cycle of the yeast *Saccharomyces cerevisiae* is the movement of mitochondria into the growing bud (Stevens, 1981). Almost as soon as a bud is apparent, a portion of a mitochondrion is found in this region of the cell. As the bud grows, it is filled with more mitochondria (and other organelles) until, at cytokinesis, the daughter cell is provisioned with a mitochondrial content slightly exceeding that of the mother cell (Stevens, 1977). This pattern of mitochondrial movement into the yeast bud occurs without regard to the carbon source or environmental conditions (i.e., even in cells growing anaerobically on glucose) and appears to be independent of microtubule function (Huffaker et al., 1988) and nuclear division (Thomas and Botstein, 1986). The identification of recessive mutations that block mitochondrial movement into buds (McConnell et al., 1990) suggests that mitochondrial inheritance is a specific and active process. Many of the molecular details and control of this mitochondrial movement have yet to be described.

III. MITOCHONDRIAL MOVEMENT AND THE CYTOSKELETON

Studies with cells from a number of diverse organisms have implicated the cytoskeleton as playing a major role in the positioning and distribution of mitochondria. Mitochondria exhibit saltatory motion (Aufderheide, 1977; Adams, 1982) characteristic of transport along cytoskeletal components. Specific intracellular positions of mitochondria have been correlated with microtubules in some cell types (Heggeness et al., 1978; Ball and Singer, 1982; Couchman and Rees, 1982). In other types of cells, mitochondrial position correlates with the distribution of intermediate filaments (David-Ferreira and David-Ferreira, 1980; Mose-Larsen et al., 1982; Chen, 1988). A study of mitochondrial distribution in living cells implicated both microtubules and intermediate filament networks as contributing to the positioning of mitochondria (Summerhayes et al., 1983). Studies with mammalian neuronal cells have documented transport along microtubules as a key component of the axonal transport of mitochondria and other organelles (Vale, 1987), and movement of organelles along microtubules has been described for certain non-neuronal animal cells (Vale, 1987). Video and fluorescence microscopy of *Neurospora crassa* cells have indicated that mitochondria migrate along cytoplasmic microtubules (Steinberg and Schliwa, 1993). Electron microscopy of rapidly frozen, frog neurons revealed short cross-bridges of thin filaments between mitochondria and microtubules and between mitochondria and neurofilaments (a type of intermediate filament; Hirokawa, 1982). *In vitro*, purified tubulin binds directly to isolated mitochondria (Bernier-Valentin and Rousset, 1982), although the physiological relevance of this interaction is unknown.

Reconstitution studies using permeabilized cells or isolated cellular components have identified proteins involved in the movement of particles along microtubules (Vale, 1987). These investigations have implicated two key proteins, kinesin (Vale, 1987) and cytoplasmic dynein (Lye et al., 1987; Schroer et al., 1989), as microtubule-based motors found ubiquitously in eukaryotic cells. Bound kinesin will drive the movement of vesicles or even latex beads toward the plus-end of microtubules while dynein drives movement toward the minus-end. The *in vitro* studies have suggested also that other, as yet unidentified, proteins are required for the similar movements of purified organelles (Schroer et al., 1988). Although the force-generating proteins, kinesin and cytoplasmic dynein, have been implicated in the positioning of organelles within nondividing cells and in the axonal transport of cellular particles, little is known of their role in organellar distribution during mitosis.

Cytoplasmic microtubules largely disassemble prior to mitosis (Saxton et al., 1984), so the role of microtubules in the distribution of organelles during the mitotic phase of the cell cycle is obscure. Studies with the filamentous fungus *Aspergillus nidulans* demonstrated that the inhibition of microtubule function with the agent benomyl or by mutations in β-tubulin blocked the movement and division of the nucleus but had no effect on the movement of mitochondria (Oakley and Reinhart, 1985). Additionally, mitochondria and other organelles migrate into the growing buds of the yeast *Saccharomyces cerevisiae* even in the presence of inhibitors or mutations that disrupt microtubule function (Huffaker et al., 1988; Jacobs et al., 1988).

Actin microfilaments may also play a role in mitochondrial distribution. In some mammalian cells, regions with mitochondrial and organellar streaming movement correlate with areas of change in the structure of actin-like microfilament bundles (Wang and Goldman, 1978). Actin-mediated organellar transport has been characterized in characean algal cells (Kachar and Reese, 1988) where endoplasmic reticulum and other organelles contact actin filament bundles during cytoplasmic streaming. Using an *in vitro* reconstituted system, Adams and Pollard (1986) demonstrated the function of myosin-I in the active translocation of organelles from *Acanthamoeba* along actin microfilaments. In *Saccharomyces cerevisiae*, mutations in the single actin gene cause aberrant transport of materials into a new bud (Novick and Botstein, 1985), but it is difficult to assess the specific role of actin in mitochondrial movement in these mutants since a number of cellular processes are disrupted and buds do not form. Drubin and colleagues (1993) have described clumped mitochondria in yeast cells harboring certain mutant actin alleles, and these observations suggest a role for the actin-myosin cytoskeleton in the distribution or organization of mitochondria. The specific binding of actin to the mitochondrial outer membrane has been described (Pardo et al., 1983), but the functional significance of this interaction remains to be determined.

Mitochondrial behavior may depend on several different cytoskeletal systems. For example, microtubules may mediate the positioning and division of mitochondria during interphase, while interactions with intermediate filaments could determine mitochondrial distribution during mitosis.

An alternative mechanism of mitochondrial movement (discussed by Bereiter-Hahn, 1990) is that mitochondria move by changing shape and "creeping" through the cytoplasm. Such creeping would employ components and processes internal to the organelle and might be directed by gradients of ADP or other small molecules in the cytoplasm. This model envisions cytoskeletal components as playing a passive role by providing a structure upon which the mitochondria can maneuver.

IV. MUTANTS OF MITOCHONDRIAL DISTRIBUTION AND MORPHOLOGY

One approach to the study of mitochondrial division and inheritance has been the isolation and analysis of *Saccharomyces cerevisiae* mutants possessing conditional defects in mitochondrial morphology and distribution (McConnell et al., 1990). These *mdm* mutants were isolated by screening a collection of temperature-sensitive yeast strains by fluorescence microscopy to identify cells that failed to transfer mitochondria into developing buds during incubation at 37 °C. A number of such mutant strains have been identified (McConnell et al., 1990), and many of these also display aberrant mitochondrial morphologies. Genetic analysis has revealed that the morphological, distribution, and temperature-sensitive growth phenotypes in *mdm* mutants result from single, recessive, nuclear mutations. The mutations define at least 20 complementation groups.

Microscopic studies employing indirect immunofluorescence and electron microscopy have confirmed the absence of mitochondria from buds that develop at the nonpermissive temperature in the *mdm* strains (McConnell et al., 1990). These studies also allowed an examination of the effect of the mutations on other cellular structures. Two classes of *mdm* mutants have been identified: those in which only the mitochondria are affected by the mutant lesions and those in which the transfer of both mitochondria and nuclei into daughter buds are defective at the nonpermissive temperature. None of the mutants have been found to affect the secretory process, the development of vacuoles or the transfer of vacuolar material into buds, the assembly and function of microtubules, or the cellular distribution of the actin cytoskeleton.

The analysis of one mutant, *mdm1*, has led to the identification of a new cytoskeletal component that mediates mitochondrial inheritance in yeast (McConnell and Yaffe, 1992). The *mdm1* mutation affects transfer of both mitochondria and nuclei into daughter buds. Isolation and analysis of the wild-type *MDM1* gene revealed its product to be a 52 kDa protein with modest sequence similarity to intermediate filament proteins of animal cells. The protein, Mdm1p, is localized to a unique series of punctate structures distributed throughout the yeast cytoplasm. These structures disappear or disassemble in the *mdm1-1* mutant during incubation at the nonpermissive temperature (37°C), although Mdm1p protein levels remain constant in the cells.

Mdm1p is an intermediate filament-forming protein. Mdm1p expressed in *E. coli* and purified from bacterial inclusion bodies readily formed into filaments of 10 nm diameter *in vitro* (McConnell and Yaffe, 1993). Conditions of filament self-assembly were essentially those under which the animal proteins, vimentin and desmin, formed intermediate-sized filaments. Mutant Mdm1p, purified from bacteria expressing the *mdm1-1* gene, was temperature-sensitive for intermediate filament assembly *in vitro*: 10 nm filaments formed at 4°C or 23°C but failed to form at 37°C. However, filaments formed of the mutant protein at 4°C were stable when subsequently incubated at 37°C, indicating that the *mdm1-1* mutation prevents filament assembly. The relationship between the filaments formed of Mdm1p *in vitro* and the punctate Mdm1p structures in the cell is unclear. Mdm1p may assemble into a different type of structure in the cellular environment or, alternatively, the punctate structures may be intersections of a number of filaments, with individual filaments being undetectable by currently available methods. For mitochondrial inheritance, the Mdm1p structures may constitute a network upon which mitochondria are transported.

A second *mdm* mutant has revealed a role for unsaturated fatty acids in mitochondrial movement. The *mdm2* mutation appeared to affect specifically the inheritance of mitochondria at the nonpermissive temperature (McConnell et al., 1990). This lesion also blocked a second type of mitochondrial movement: distribution of mitochondria throughout cytoplasmic projections induced by exposure of yeast cells to mating pheromones. Mitochondria in *mdm2* cells appeared to aggregate following a shift of cells to the nonpermissive temperature. Cloning and analysis of *MDM2* (Stewart and Yaffe, 1991) revealed its identity with a previously identified gene, *OLE1*, encoding fatty acid desaturase. Consistent with this identification, cellular levels of unsaturated fatty acids decreased following a shift of *mdm2* cells to the nonpermissive temperature. Furthermore, the addition of oleic acid to culture media cured both the temperature-sensitive growth and mitochondrial distribution defects of *mdm2* cells. The function of unsaturated fatty acids in mitochondrial movement might be related to some gross physical properties of the mitochondrial membranes (e.g., membrane fluidity) or could reflect a more specific requirement for unsaturated fatty acid in sustaining the activity or structure of a protein that mediates interaction of mitochondria with other cellular components.

Analysis of a third *mdm* mutant has highlighted the relationship between mitochondrial morphology and mitochondrial inheritance. Yeast cells with the *mdm10* lesion possess giant, spherical mitochondria, as well as a defect in mitochondrial inheritance (Sogo and Yaffe, 1994). The giant mitochondria contain classical structural features and are competent for respiration. *MDM10* encodes a 57 kDa polypeptide that is an integral protein of the mitochondrial outer membrane. A portion of this protein is exposed to the cytoplasm and may mediate the interaction of mitochondria with cytoskeletal components.

Function of the Mdm10p was further revealed by analyzing the cellular distribution and morphology of mitochondria in cells following a controlled cessation of *MDM10* expression. The depletion of Mdm10p led to a progressive condensation of snake-like mitochondria into thicker structures and, eventually, into one or a few giant, mitochondrial balls. Empty daughter buds, reflecting defective mitochondrial inheritance, also appeared as Mdm10p levels dropped. Re-expression of *MDM10* in cells with giant mitochondria led to a rapid return of normal mitochondrial morphology. This return to snake-like mitochondria seemed to occur by a stretching-out and fragmentation of the giant organelles.

A second approach further illustrated the reversible nature of the giant mitochondrial phenotype. Haploid *mdm10* cells (containing giant mitochondria) were mated to wild-type cells of the opposite mating type. The mitochondrial morphologies in the resulting diploid zygotes were analyzed by fluorescence microscopy. Initially, zygotes contained normal mitochondria on one side and a giant mitochondrion on the other side of the cell. Over the subsequent hour, the giant mitochondrion fragmented into numerous snake-like organelles until both sides of the zygote displayed wild-type mitochondrial morphologies.

The giant mitochondria in the *mdm10* mutant appeared to be defective for both movement and division. One model to explain how all of these defects could result from a loss of Mdm10p would be that interactions between mitochondria and cytoskeleton mediate the control of mitochondrial morphology, mitochondrial division, and mitochondrial movement. Mdm10p might serve as an anchor-point on the mitochondrial surface for attachment of the organelle to cytoskeletal networks. Another explanation for all of the phenotypes of the *mdm10* mutant is that mitochondrial movement and division require an extended, snake-like morphology, and their failure to occur in *mdm10* is a secondary consequence of the dramatic morphological changes.

A second, distinct mutant with altered mitochondrial morphology, *mmm1*, was recently described by Burgess and colleagues (1994). This mutant displays phenotypes very similar to those of *mdm10*, and *MMM1* also encodes an integral protein of the mitochondrial outer membrane. A third new mutant, *mdm12*, possesses giant mitochondria and defective mitochondrial inheritance (Burger and Yaffe, 1994). These various mutants may define subunits of a protein complex localized to the surface of mitochondria.

V. MODELS OF MITOCHONDRIAL DIVISION

The continuous increase in mitochondrial mass throughout the cell cycle is accompanied by a parallel increase in mitochondrial number (Schnedl, 1974; Posakony et al., 1977). This increase in number occurs by the division of pre-existing mitochondria. Additionally, mitochondria divide in response to various physiological changes (as described previously). Mitochondrial division does not appear to be tightly coordinated with the cell division cycle. Although the amount (i.e., mass) of mitochondria is likely to be regulated by metabolic needs of the cell, factors governing the number of mitochondria are unknown. Additionally, as with mitochondrial inheritance, the mechanism of mitochondrial division has yet to be described.

Electron microscopic images of mitochondria in a variety of cells have suggested several possible models for division. One possibility is that extended mitochondrial structures are pulled apart by molecular motors attached to opposite sides of the organelle and moving in opposite directions along a cytoskeletal track. Such a mechanism might involve kinesin, dynein, or myosin-like motor proteins attached to the outer membrane and could proceed by an initial stretching of the organelle followed by the development of a division furrow. A second possible mechanism is that an encircling band of cytoskeletal proteins constricts the organelle and eventually pinches it in two. This model resembles cytokinesis and might employ an actinomyosin system. A third model is that a division septum or similar structure is assembled within the mitochondrion, facilitating a separation into two distinct organelles. This mechanism might recall the prokaryotic origins of mitochondria and rely on activities internal to the organelle. A further understanding of mitochondrial division may emerge from genetic studies of mutant cells defective in mitochondrial fission and biochemical analysis using systems reconstituted from purified components.

REFERENCES

Adams, R. J. (1982). Organelle movement in axons depends on ATP. Nature 297, 327–329.

Adams, R. J., & Pollard, T. D. (1986). Propulsion of organelles isolated from Acanthamoeba along actin filaments by myosin-I. Nature 322, 754–756.

Aufderheide, K. J. (1977). Saltatory motility of uninserted trichocysts and mitochondria in *Paramecium tetraurelia*. Science 198, 299–300.

Bakeeva, L. E., Chentsov, U. S., & Skulachev, V. P. (1986). Mitochondrial framework (reticulum mitochondriale) in rat diaphragm muscle. Biochim. Biophys. Acta 501, 349–369.

Ball, E. H., & Singer, S. J. (1982). Mitochondria are associated with microtubules and not with intermediate filaments in cultured fibroblasts. Proc. Natl. Acad. Sci. USA 79, 123–126.

Bereiter-Hahn, J. (1976). Dimethylaminostyrylmethylpyridiniumiodine (DASPMI), a fluorescent probe for mitochondria in situ. Biochim. Biophys. Acta 423, 1–14.

Bereiter-Hahn, J. (1990). Behavior of mitochondria in the living cell. Int. Rev. Cytol. 122, 1–64.

Bereiter-Hahn, J., Luck, M., Miebach, T., Stelzer, K. E., & Voth, M. (1990). Spreading of trypsinized cells: Cytoskeletal dynamics and energy requirements. J. Cell Sci. 96, 171–188.

Bernier-Valentin, F., & Rousset, B. (1982). Interaction of tubulin with rat liver mitochondria. J. Biol. Chem. 257, 7092–7099.

Blank, R., & Arnold, C.-G. (1981). Structural changes of mitochondria in Chlamydomonas reinhardii after chloramphenicol treatment. Eur. J. Cell Biol. 24, 244–251.

Burger, K. H., & Yaffe, M. P. (1994). Unpublished results.

Burgess, S. M., Delannoy, M., & Jensen, R. E. (1994). *MMM1* encodes a mitochondrial outer membrane protein essential for establishing and maintaining the structure of yeast mitochondria. J. Cell Biol. 126, 1375–1391.

Chen, L. B. (1988). Mitochondrial membrane potential in living cells. Ann. Rev. Cell Biol. 4, 155–181.

Couchman, J. R., & Rees, D. A. (1982). Organelle-cytoskeleton relationships in fibroblasts: Mitochondria, Golgi apparatus, and endoplasmic reticulum in phases of movement and growth. Eur. J. Cell Biol. 27, 47–54.

David-Ferreira, K. L., & David-Ferreira, J. F. (1980). Association between intermediate-sized filaments and mitochondria in rat Leidyg cells. Int. Cell Biol. Rep. 4, 655–662.

Drubin, D. G., Jones, H. D., & Wertman, K. F. (1993). Actin structure and function: roles in mitochondrial organization and morphogenesis in budding yeast and identification of the phalloidin-binding site. Mol. Biol. Cell 4, 1277–1294.

Heasman, J., Quarmby, J., & Wylie, C. C. (1984). The mitochondrial cloud of *Xenopus* oocytes: The source of germinal granule material. Dev. Biol. 105, 458–469.

Hedgecock, E. M., & Thomson, J. N. (1982). A gene required for nuclear and mitochondrial attachment in the nematode Caenorhabditis elegans. Cell 30, 321–330.

Heggeness, M. H., Simon, M., & Singer, S. J. (1978). Association of mitochondria with microtubules in cultured cells. Proc. Natl. Acad. Sci. USA 75, 3863–3866.

Hirano, T., Hiraoka, Y., & Yanagida, M. (1988). A temperature-sensitive mutation of the *Schizosaccharomyces pombe* gene nuc2+ that encodes a nuclear scaffold-like protein blocks spindle elongation in mitotic anaphase. J. Cell Biol. 106, 1171–1183.

Hirokawa, N. (1982). Cross-linker system between neurofilaments, microtubules, and membranous organelles in frog axons revealed by the quick-freeze, deep-etch method. J. Cell Biol. 94, 129–142.

Huffaker, T. C., Thomas, J. H., & Botstein, D. (1988). Diverse effects of β-tubulin mutations on microtubule formation and function. J. Cell Biol. 106, 1997–2010.

Jacobs, C. W., Adams, A. E. M., Szaniszlo, P. J., & Pringle, J. R. (1988). Functions of microtubules in the *Saccharomyces cerevisiae* cell cycle. J. Cell Biol. 107, 1409–1426.

Johnson, L. V., Walsh, M. L., & Chen, L. B. (1980). Localization of mitochondria in living cells with rhodamine 123. Proc. Natl. Acad. Sci. USA 77, 990.

Kachar, B., & Reese, T. S. (1988). The mechanism of cytoplasmic streaming in Characean algal cells: Sliding of endoplasmic reticulum along actin filaments. J. Cell Biol. 106, 1545–1552.

Lye, R. J., Porter, M. E., Scholey, J. M., & McIntosh, J. R. (1987). Identification of a microtubule-based cytoplasmic motor in the nematode C. elegans. Cell 51, 309–318.

McConnell, S. J., Stewart, L. C., Talin, A., & Yaffe, M. P. (1990). Temperature-sensitive yeast mutants defective in mitochondrial inheritance. J. Cell Biol. 111, 967–976.

McConnell, S. J., & Yaffe, M. P. (1992). Nuclear and mitochondrial inheritance in yeast depends on novel cytoplasmic structures defined by the MDM1 protein. J. Cell Biol. 118, 385–395.

McConnell, S. J., & Yaffe, M. P. (1993). Intermediate filament formation by a yeast protein essential for organelle inheritance. Science 260, 687–689.

Mignotte, F., Tourte, M., & Mounolou, J-C. (1989). Segregation of mitochondria in the cytoplasm of *Xenopus vitellogenic* oocytes. Biol. Cell 60, 97–102.

Mose-Larsen, P., Bravo, R., Fey, S. J., Small, J. V., & Celis, J. E. (1982). Putative association of mitochondria with a subpopulation of intermediate-sized filaments in cultured human skin fibroblasts. Cell 31, 681–692.

Munn, E. A. (1974). The *Structure of Mitochondria*. Academic Press, London.

Novick, P., & Botstein, D. (1985). Phenotypic analysis of temperature-sensitive yeast actin mutants. Cell 40, 405–416.

Oakley, B. R., & Reinhart, J. E. (1985). Mitochondria and nuclei move by different mechanisms in *Aspergillus nidulans*. J. Cell Biol. 101, 2392–2397.

Pardo, J. V., Pittenger, M. F., & Craig, S. W. (1983). Subcellular sorting of isoactins: Selective association of Θ-actin with skeletal muscle mitochondria. Cell 32, 1093–1103.

Posakony, J. W., England, J. M., & Attardi, G. (1977). Mitochondrial growth and division during the cell cycle in HeLa cells. J. Cell. Biol. 74, 468–491.

Saxton, W. M., Stemple, D. L., Leslie, R. J., Salmon, E. D., Zavortink, M., & McIntosh, J. R. (1984). Tubulin dynamics in cultured mammalian cells. J. Cell Biol. 99, 2175–2186.

Schnedl, W. (1974). Changes in mitochondria during the cell cycle. Cytobiol. 8, 403–411.

Schroer, T. A., Schnapp, B. J., Reese, T. S., & Sheetz, M. P. (1988). The role of kinesin and other soluble factors in organelle movement along microtubules. J. Cell Biol. 107, 1785–1792.

Schroer, T. A., Steuer, E. R., & Sheetz, M. P. (1989). Cytoplasmic dynein is a minus end-directed motor for membranous organelles. Cell 56, 937–946.

Sogo, L. F., & Yaffe, M. P. (1994). Regulation of mitochondrial morphology and inheritance by Mdm10p, a protein of the mitochondrial outer membrane. J. Cell Biol. 126, 1361–1373.

Steinberg, G., & Schliwa, M. (1993). Organelle movements in the wild type and wall-less fz; sg; os-1 mutants of *Neurospora crassa* are mediated by cytoplasmic microtubules. J. Cell Sci. 106, 555–564.

Stevens, B. (1981). Mitochondrial structure. In: *The Molecular Biology of the Yeast Saccharomyces cerevisiae*, Vol. 1 (Strathern, J. N., Jones, E. W., & Broach, J. R., Eds.). Cold Spring Harbor Laboratory, Cold Spring Harbor, New York, pp. 471–504.

Stevens, B. J. (1977). Variation in number and volume of the mitochondria in yeast according to growth conditions. A study based on serial sectioning and computer graphics reconstruction. Biol. Cell 28, 37–56.

Stewart, L. C., & Yaffe, M. P. (1991). A role for unsaturated fatty acids in mitochondrial movement and inheritance. J. Cell Biol. 115, 1249–1257.

Summerhayes, I. C., Wong, D., & Chen, L. B. (1983). Effect of microtubules and intermediate filaments on mitochondrial distribution. J. Cell Sci. 61, 87–105.

Thomas, J. H., & Botstein, D. (1986). A gene required for the separation of chromosomes on the spindle apparatus. Cell 44, 65–76.

Vale, R. D. (1987). Intracellular transport using microtubule-based motors. Ann. Rev. Cell Biol. 3, 347–378.

Wang, E., & Goldman, R. D. (1978). Functions of cytoplasmic fibers in intracellular movements in BHK-21 cells. J. Cell Biol. 79, 708–726.

INDEX

ADP/ATP carrier (AAC), 7, 53, 89, 91-92, 129, 150
Adrenodoxin precursor (pAd), 36-44, 46, 166-167
Amphiphilic helix, 4-5, 7, 44, 86, 164-165, 167, 220
Apocytochrome c, 90, 103-104, 106, 108-109, 194-195
Arg-at-2 rule, 164-165, 179
Aspartate aminotransferase, 36, 264
ATP (Adenosine triphosphate) (*see also* "Intermembrane space; Protein export; Protein import..."), 38, 130
  depletion of, 56-58, 57, 149-150
  functions of matrix ATP in protein import, 232-237
  and protein import, 224-228

Biochemical studies, 18-21, 36-46, 84-94, 153-155, 162-183, 268-269, 314-316
Bipartite presequences (*see also* "Intermembrane space; Sorting signals...")

CCHL (*see also* "Cytochrome c heme lyase...")
CCPO (*see also* "Cytochrome c peroxidase...")
Chaperones (*see also* "Molecular chaperones...")
Chaperonins (*see also* "Hsp60...")
Chloroplast transit peptide (cTP), 3-4, 86-87
Cleavage sites (*see also* "Intermembrane space; Mitochondrial outer membrane machinery; Mitochondrial targeting peptides...")
Conservative sorting (*see also* "Sorting signals...")
Contact sites (*see also* "Intermembrane Space; Translocation contact sites...")
Cotranslational export of precursors, 21-22, 248-260
cTP (*see also* "Chloroplast transit peptide...")
Cytochrome b2, 116-118, 151-152, 202-208, 314
Cytochrome c
  cytochrome c1 import pathways, 116-118, 151
  evolutionary relationships of targeting proteins, 118-120
  export, 258-264

import, 106, 109-113, 194-195
targeting signals, 113-116
Cytochrome c heme lyase (CCHL), 60-63, 103-109, 150-152, 233
and import into mitochondrial membrane, 61, 194-195
Cytochrome c oxidase, 257, 259-260, 321
Cytochrome c peroxidase (CCPO), 151-152, 194, 201-202, 264
Cytosolic chaperones (*see also* "Molecular chaperones...")
Cytosolic factors (*see also* "Cytochrome c; Molecular chaperones; Protein import..."), 226-230, 289-290

Dynamic channel model, 147-152

Endoplasmic reticulum (ER), 16-17, 74, 76, 129, 201
Enzyme assembly (*see also* "Mitochondrial assembly of subunit complexes...")
ER (*see also* "Endoplasmic reticulum...")
Escherichia coli (*E. coli*), 2, 17-19, 24, 35, 39, 105, 107, 118, 171, 174, 182, 201, 210, 214, 227, 231

Fusion proteins (*see also* "Genetic studies..."), 23-24, 28, 55, 114, 199, 204-205, 209

General insertion pore (GIP) (*see also* "Mitochondrial receptor complex; Protein export; Protein import..."), 54, 60-64
Genetic screens, 24-26, 154-156, 202
Genetic studies (*see also* "*Saccharomyces cerevisiae*; *Neurospora crassa*..."), 5-7, 117-119, 130-140, 152-157, 165-172, 180-183, 230-231, 248-268, 299-327
GIP (*see also* "General insertion pore...")

Heat shock proteins, 16-19, 109-110, 130, 137, 140, 155-157
Hsp60 (*see also* "Heat shock proteins; Molecular chaperones; Protein folding..."), 284-289
Hsp70 (*see also* "Heat shock proteins; Mitochondrial inner membrane machinery; Molecular chaperones..."), 8, 14, 34, 52, 60, 155-157
cytosolic Hsp70 chaperone system, 226-228
model for mhsp70 mechanism, 156
role in import of IMS proteins, 138, 205-208, 281-284
role in preprotein folding in matrix, 137-140, 281-289

IM (Inner membrane) (*see also* "Mitochondrial inner membrane machinery...")
Immunoprecipitation, 87, 91, 93, 136, 154-155, 289
IMP (*see also* "Inner membrane peptidase...")
IMS (*see also* "Intermembrane space...")
Inner membrane peptidase (IMP) (*see also* "Intermembrane space..."), 182-183, 196, 201-202, 251-252, 262, 266

Inner membrane protease (*see also* "Inner membrane peptidase...")
Inner membrane proteins (*see also* "Mitochondrial inner membrane machinery; Protein export..."), 55-58
Intermediate processing peptidase (*see also* "Mitochondrial intermediate peptidase...")
Intermembrane space (IMS) (*see also* "Mitochondrial inner membrane machinery; Protein export..."), 5-7, 60, 102, 117, 192-195, 212-214
  cytochrome b2 sorting intermediate, 202-205
  import of cytochrome c, 194-195, 233-237
  and inner membrane peptidases (IMP), 201-202
  and membrane potential, 109, 205-207
  molecular chaperones and import of IMS proteins, 205-210
  protein import and ATP requirements, 230-237
  role of the Rieske FeS protein, 210-212
  sorting of precursors with presequences, 195-197
  and sorting signals in bipartite presequences, 197-201, 234

Lipid vesicles, 108-109, 115

Malate dehydrogenase, 16, 36, 264
Matrix processing peptidase (*see also* "Mitochondrial processing peptidase..."), 5, 58, 63, 168, 196, 286, 289
Matrix proteins (*see also* "Intermembrane space; Protein export; Precursor proteins..."), 55-58
Membrane potential (*see also* "Intermembrane space; Mitochondrial inner membrane machinery; Protein export...")
Membrane vesicles, 62-66, 149
Methotrexate, 65, 202-203
  and sorting models for intermembrane space, 204-205
MIM (*see also* "Mitochondrial inner membrane machinery...")
MIP (*see also* "Mitochondrial intermediate peptidase...")
Mitochondrial assembly of subunit complexes, 298-301, 324
  ATP synthase, 309-313
  encoded by the nucleus and mitochondria, 305-323
  encoded exclusively by the nucleus, 301-305
  factors controlling assembly, 323-327
  nuclear genes and post-translational assembly, 300
  of the respiratory chain, 304-309, 313-323
Mitochondrial division and inheritance, 338-341, 344
Mitochondrial import stimulation factor (MSF), 21, 36-43, 45, 229-230
  molecular properties of MSF subunits, 44, 46
  precursor protein-induced ATPase activity of, 42-43
Mitochondrial inner membrane machinery (MIM) (*see also* "Inner membrane proteins; Protein export..."), 65, 126-127, 147-151
  dynamic channel model and protein sorting, 151-152

and membrane potential, 52, 127-130, 149-150
MIM proteins and their functions, 55, 64, 130-137, 152-157
role of HSP70 in protein folding and translocation, 137-141
Mitochondrial intermediate peptidase (MIP), 5, 175-182, 196
Mitochondrial outer membrane machinery (MOM) (*see also* "Mitochondrial receptor complex; Precursor proteins; Protein import...")
general mechanism of, 52-57
import pathway of heme lyases, 60-62
import reactions with outer membrane vesicles, 62-66
insertion and assembly of outer membrane proteins, 58-60
MOM protein complexes, 42, 53-60, 62, 65, 72-73, 76, 117, 151
Mitochondrial processing peptidase (MPP) (*see also* "Matrix processing peptidase..."), 82-89, 163-175
Mitochondrial receptor complex (*see also* "General insertion pore; Mitochondrial outer membrane machinery; Protein import; Precursor proteins..."), 52-53, 84-96
Mitochondrial ribosomes (*see also* "Protein export..."), 247-248, 258
Mitochondrial targeting peptides (mTPs), 3-8, 52, 55
cleavage sites, 5, 164-168
cytochrome c, 4, 23-25, 36, 38, 44, 108-109, 113-120, 166-169, 194-195
pOMD29, 73-75
Mitoplasts, 55-56, 149, 154
Molecular chaperones (*see also* "Heat shock proteins; Hsp70; Intermembrane space..."), 16-21, 26, 46-47, 208-210
MOM (*see also* "Mitochondrial outer membrane machinery...")
MPP (*see also* "Mitochondrial processing peptidase...")
mRNA (messenger RNA) and protein translocation, 15, 23, 26-29
MSF (*see also* "Mitochondrial import stimulation factor...")
mTPs (*see also* "Mitochondrial targeting peptides...")
Mutagenesis (*see also* "Genetic studies; *Saccharomyces cerevisiae*; *Neurospora crassa*..."), 3-4, 7, 73-74, 165-168

N-ethylmaleimide (NEM), 34, 38-39, 45, 114-115
Neurospora crassa (*N. crassa*) (*see also* "Biochemical studies; Genetic studies..."), 52-55, 72-76, 82-97, 103-104, 107, 110-111, 248-268, 299-327

OMM (*see also* "Signal anchor sequences...")
OM (Outer membrane) (*see also* "Mitochondrial outer membrane machinery...")
Ornithine carbamoyltransferase, 35-36
Ornithine carbamyl transferase (OCT), 7, 76-77
Ornithine transcarbamylase (OTC), 165-168, 176-179, 285-286

pAd (*see also* "Adrenodoxin precursor...")
PBF (*see also* "Presequence binding factor...")
pOMD29, 73-75, 77
pO-OMD, 75, 77-78
Porin, 24, 42, 58, 60, 62, 71-72, 89, 94, 111
Precursor proteins, 14-15, 22, 34, 58, 268-269
  and the import process, overview, 162-163
  and inner membrane peptidase, 182-183
  and MIP, 175-182
  and MPP, 163-175
  nomenclature of, 163
  and other peptidases, 183-184
  and sorting to the intermembrane space, 192-214
Precytochrome P450 (SCC), 37-38, 43-44
Preproteins (*see also* "Precursor proteins...")
Presequence binding factor (PBF), 20-21, 34-36, 46, 229-230
Presubunit II, 250-257, 259-260
Protein export
  to the cytosol, 263-264
  energy for insertion process, 257-260
  to the inner membrane, 260-261
  to the intermembrane space, 261-263
  and leader peptides, 252-257
  mechanism, 264-270
  and mitochondrial gene products, 247-248, 265
  pathways, an overview, 246-248
  and protein precursors, 248-252
  relationship between import and export, 270-271
Protein folding, 18-19, 58, 102, 137-140, 280-291
Protein import (*see also* "Intermembrane space; Mitochondrial outer membrane machinery; Molecular chaperones..."), 14-24
  ATP requirements for, 15-16, 224-226
  cytosolic import stimulation factors, 21-26, 34-46, 226-230
  and Hsp70 and ATP interaction, 228, 230-238
  and membrane potential, 221-224
  and presequence-binding factors in the cytosol, 46-47
Protein transport (*see also* "Mitochondrial inner membrane machinery; Mitochondrial outer membrane machinery; Protein export; Protein import...")

Rieske Fe/S protein, 116-118, 208, 210-212, 314-316, 321

Saccharomyces cerevisiae (*see also* "Biochemical studies; Genetic studies..."), 16-18, 23-26, 36-37, 52-55, 72-76, 82-97, 102, 111-113, 130-140, 201-202, 248-268, 299-327, 338-344
Secretory signal peptides (*see also* "Protein export; Precursor proteins; Sorting signals..."), 2-6
Signal anchor sequences, 71-78
Sorting signals, 2-8, 14-15, 192-202
  bipartite presequences and, 3, 5-7, 197-201, 234
  conservative model for, 196, 200-201, 210-212, 234, 262-263

in cytochrome c, 113-114
and IMS, 5-7, 195-201
in the outer membrane, 71-78
requirements of the IMS and, 197-199
stop-transfer model and, 196-197, 199-200, 233-237, 262-263
SSC (*see also* "Precytochrome P450...")
Stop-transfer model (*see also* "Sorting signals...")

Targeting factor, 35, 46
Targeting signals (*see also* "Sorting signals...")
TOM (Translocase of the mitochondrial outer membrane) (*see also* "Mitochondrial receptor complex...")
Translocation channels (*see also* "General insertion pore; Dynamic channel model; Mitochondrial receptor complex; Intermembrane space..."), 269-270
Translocation contact sites, 52, 127, 135-137, 148-149

Uncoupler protein (UCP), 22-23

www.ingramcontent.com/pod-product-compliance
Ingram Content Group UK Ltd.
Pitfield, Milton Keynes, MK11 3LW, UK
UKHW020233290926
15470UKWH00002B/520

9 780762 301447